Philosophy and the Evolution of Virtue and Vice

This volume explores what evolutionary theory can contribute to our philosophical understanding of virtue and vice.

The last twenty years have seen an uptick in philosophical discussions of the biological and cultural evolution of morality. Yet there has been surprisingly little discussion of virtue ethics and the evolution of human character, even though human virtues are almost certainly one of the markers of our species' distinctive sociality. This volume investigates the evolution of human character traits and its implications for moral philosophy and value theory. The chapters address central questions such as: To what extent can evolutionary theories explain virtue, vice, or specific virtues/vices? Are philosophical accounts of virtue compatible with evolutionary explanations? How might tensions between them be resolved? What implications might the evolution of virtue have concerning debates in value theory more broadly? What implications does the evolution of virtue have for living well in the digital age? And how can we re-engineer the conceptual machinery of virtue/vice to our advantage? The volume serves as a first step toward evaluating the influence of evolution on human conceptions of what kind of people we ought to be.

Philosophy and the Evolution of Virtue and Vice will appeal to researchers and graduate students working in normative ethics, virtue ethics, metaethics, moral psychology, philosophy of social science, philosophy of biology, and evolutionary theory.

Michael T. Dale is an Assistant Professor of Philosophy at Hampden-Sydney College. His research focuses on moral psychology and the evolution of morality.

Isaac Wiegman is an Associate Professor of Instruction at Texas State University. His work explores the evolution of morally interesting traits like revenge, disgust, spite, and reciprocity.

Routledge Studies in Ethics and Moral Theory

Taking Moral Theorising Seriously
A Theory-First Approach to Moral Philosophy
Andrew Sneddon

Public Health Virtue Ethics
Institutions, Structures, and Political Virtue for the Good Society
Kathryn L. MacKay

Goodness and Tradition
Irene Liu

Living Ethics in Theory and Practice
Edited by Eric Racine, Bénédicte D'Anjou, Caroline Favron-Godbout, Giulia Inguaggiato, Suzanne Metselaar, and Gabriel Saso-Baudaux

Virtuous Responses to Suffering, Tragedy, and Evil
Edited by Eric J. Silverman

Because You Promised
A Non-Reductive Account of the Normativity of Promises
Daniele Bruno

Moral Impossibility
Ethical, Political, and Psychological Perspectives
Silvia Caprioglio Panizza

Philosophy and the Evolution of Virtue and Vice
Michael T. Dale and Isaac Wiegman

For more information about this series, please visit: www.routledge.com/Routledge-Studies-in-Ethics-and-Moral-Theory/book-series/SE0423

Philosophy and the Evolution of Virtue and Vice

Edited by Michael T. Dale
and Isaac Wiegman

Routledge
Taylor & Francis Group
NEW YORK AND LONDON

First published 2027
by Routledge
605 Third Avenue, New York, NY 10158

and by Routledge
4 Park Square, Milton Park, Abingdon, Oxon, OX14 4RN

Routledge is an imprint of the Taylor & Francis Group, an informa business

ISBN: 978-1-041-05839-7 (hbk)
ISBN: 978-1-041-05838-0 (pbk)
ISBN: 978-1-003-63264-1 (ebk)

DOI: 10.4324/9781003632641

Typeset in Sabon
By KnowledgeWorks Global Ltd.

This book is dedicated to our students and teachers, our cultural and biological forebears and descendants, and to the many communities of thinkers who sustain us. It is in them that our cognitive technologies take root, and our shared intentions find their shape.

Michael dedicates this volume especially to Crystal and Ted.

Isaac dedicates this volume especially to Ron and John, Karen and Ralph, Sofia, Anabel, and Sam.

Contents

Figures and Tables

Figures

Tables

List of Contributors

Justin P. Bruner is an Associate Professor in the Department of Philosophy at the University at Buffalo. He works in the philosophy of science and philosophy, politics, and economics (PPE).

Howard J. Curzer is a President's Excellence in Research Professor at Texas Tech University. His publications include *Aristotle and the Virtues* (Oxford, 2012), *Virtue Ethics for the Real World* (Routledge, 2023), *Difficult Virtues* (Routledge, 2024), and various articles on ancient philosophy, contemporary virtue ethics, and applied ethics.

Jill B. Delston is an Assistant Professor of Philosophy at the University of Missouri-St. Louis. She is the author of *Medical Sexism: Contraception Access, Reproductive Medicine, and Health Care* (2019) and the co-editor of *Applied Ethics: A Multicultural Approach* (Editions 5 and 6).

Iskra Fileva is an Associate Professor of Philosophy at the University of Colorado in Boulder. She works primarily on issues in moral psychology. She also does some work in aesthetics and epistemology.

Patrick Forber is an Associate Professor of Philosophy at Tufts University. Forber's research interests include philosophy of biology, philosophy of science, and evolutionary game theory. His current work focuses on games in dynamics networks and the evolution of spite, and he is working on a joint monograph with Rory Smead.

Daniel Kelly is a Professor of Philosophy at Purdue University. His research focuses on issues at the intersection of philosophy of mind and cognitive science, moral theory, and evolution.

Charlie Kurth is a Professor of Philosophy at Western Michigan University and a Fellow at the Helsinki Collegium for Advanced Studies. His research focuses on questions at the intersection of ethical theory, moral psychology, and emotion science.

Mark LeBar is a Professor of Philosophy at Florida State University and the author of *Just People* (Oxford University Press, 2025).

John Lemos is the McCabe Professor of Philosophy at Coe College. He is the author of four books, including *Commonsense Darwinism: Evolution, Morality, and the Human Condition* (Open Court, 2008) and, most recently, *Free Will's Value: Criminal Justice, Pride, and Love* (Routledge, 2023), as well as 40 published articles/book chapters.

Qiannan Li is an Assistant Professor of Philosophy at the University of California, Davis. Her research focuses on normative ethics, particularly how early Chinese philosophy and feminist theories can deepen our understanding of well-being. She also specializes in moral psychology, exploring the value and nature of emotions such as respect, shame, hope, and loneliness.

Ron Mallon is a Professor of Philosophy at Washington University in Saint Louis. He is the author of *The Construction of Human Kinds* (OUP).

Joshua May is a Professor of Philosophy and Psychology at the University of Alabama at Birmingham and author of *Neuroethics* (OUP 2023) and *Regard for Reason in the Moral Mind* (OUP 2018). He writes about moral controversies and social change.

Brian Robinson is an Associate Professor of Philosophy at Texas A&M University-Kingsville, whose research has focused on moral psychology, philosophy of language, and philosophy of science. His work is funded by the National Science Foundation and has appeared in *Philosophy of Science*, *AI & Society*, *PLOS One*, *The Monist*, and *Philosophical Studies*.

Armin Schulz is an interdisciplinary researcher investigating what we can learn from linking evolutionary biology and the cognitive and social sciences. His particular focus is on the evolutionary pressures on high-level decision-making, both with a view toward cognitive psychology and economics.

Michael Slote is UST Professor of Ethics at the University of Miami and a member of the Royal Irish Academy. He has defended an overall sentimentalist approach to ethics and has most recently attempted to integrate Chinese and Western philosophy by showing that our understanding of ethics, epistemology, and the philosophy of mind can benefit from the use of the Chinese categories of yin and yang.

Rory Smead is an Associate Professor of Philosophy and the Ronald L. and Linda A. Rossetti Professor for the Humanities at Northeastern

University. Smead's research interests include philosophy of science, philosophy of biology, logic, and game theory. His current research is focused on the evolution of spite, costly behavior that harms others, as well as the emergence of social conventions and ethical behavior.

Joel D. Velasco is an Associate Professor and Department Chair of Philosophy at Texas Tech University. He specializes in the philosophy of biology as well as more general philosophy of science.

Maria Waggoner is an Assistant Professor of Philosophy at Purdue University. Her research lies within the domain of moral psychology, focusing on developing empirically informed accounts of moral improvement and virtue.

Acknowledgments

For their advice, feedback, and support, the editors would like to thank Mark Alfano, Ted Dale, Steve Dale, John Doris, Monica Hand, Marc Hight, Dan Kelly, Raina Mansfield, Ron Mallon, Christian Miller, Rosaleah Stammler, Crystal Thurston, Valerie Tiberius, Andrew Weckenmann, and Patrick Wilson.

Introduction

Evolving Characters

*Michael T. Dale and Isaac Wiegman**

According to some views of evolution, organisms are nothing more than survival machines, programmed by blind biological processes to serve the interests of selfish replicators (Dawkins, 1976). Whatever the accuracy of this picture, it applies to humans no less than any other creature—except that we are also capable of reflection. We can think about what is reasonable, good, right, and true, and act accordingly, even when such action runs counter to the interests of the replicators that shaped us. How are these images to be reconciled? How can we integrate what Sellars (1963) called the *manifest image*—our self-conception as agents who reflect on motives, virtues, and norms—with the *scientific image* presented by evolutionary biology (among other sciences)? This volume is a small step toward merging these perspectives into the stereoscopic view that philosophy so often seeks.

These two images—one normative, the other explanatory—have developed largely in isolation. The ancient *manifest image* depicts humans as rational and social animals, possessed of virtues integral to flourishing, a picture developed independently in both Eastern and Western traditions (Aristotle, 2000; Confucius, 2003; Mencius, 2008). In the last century, this tradition has expanded into fields as diverse as environmental philosophy, business ethics, feminist ethics, virtue epistemology, and even empirical moral psychology. The *scientific image*, by contrast, portrays humans as products of blind selection pressures acting on selfish genes. Recent decades have enriched this image with sophisticated evolutionary accounts of cooperation, altruism, cultural learning, and social

* Author order is alphabetical by last name and does not reflect level of contribution. IW had primary responsibility for the conceptual framing of the introduction, the articulation of central tensions and harmonies among included articles, and the development of questions for future research. MD drafted the initial version of the introduction and wrote section 3 of the final draft.

DOI: 10.4324/9781003632641-1

cognition, drawing on evolutionary biology, anthropology, cultural psychology, and archaeology.

All of this has resulted in an uptick in philosophical discussions of the biological and cultural evolution of morality, but there has been surprisingly little discussion of virtue ethics and the evolution of human character, even though human virtues are almost certainly one of the markers of our species' distinctive sociality. Most of this work falls into three domains (FitzPatrick, 2008): *descriptive ethics*, *normative ethics*, and *metaethics*. Metaethical discussions have been especially dominant. Street (2006) and Joyce (2006), for example, argue that if natural selection shaped our moral beliefs, and if natural selection tracks fitness rather than moral truth, then many of our moral beliefs lack justification—arguments now known as evolutionary debunking arguments. In normative ethics, philosophers attempt to debunk specific moral doctrines—such as retributivist and deontological principles—without compromising others—such as utilitarian commitments (see de Lazari-Radek & Singer, 2012, 2014; Greene, 2008, 2014; Singer, 2005; Wiegman, 2017). Still others explore the possibility of moral progress through gene-culture co-evolution (Buchanan & Powell, 2018; Henrich, 2017; Richerson & Boyd, 2005). Yet engagement with virtue theory remains largely absent across these literatures.

Of course, there are other prominent debates that bring virtue ethics into conversation with science. In recent years, discussions about *situationism* in social and personality psychology have raised questions about whether behavior is more strongly influenced by personality traits—including moral virtues—or situations. This has led philosophers to ask whether (or to what extent) people have reliable character traits (Doris, 2002, 2022; Kamtekar, 2004; Miller, 2003; Sabini & Silver, 2005; Sreenivasan, 2002; Waggoner et al., 2022). *Virtue skeptics* argue that virtues either do not exist or exert far less influence on behavior than Aristotle and others have supposed. However, evolutionary theory has for the most part been left out of these discussions.

As a result, attempts to reconcile the *manifest* and *scientific images* of human morality have seldom integrated virtue ethics with evolutionary explanations (though cf. Casebeer, 2003; Lemos, 2008; and Mizzoni, 2017). This volume aims to put them into direct conversation: how do we reconcile the images of humanity rendered to us by evolutionary sciences with normative theories of virtue?

Here is one way these images might collide: According to many evolutionary models, virtues exist as cooperation-stabilizing adaptations. Consider models of *indirect reciprocity*, which ask whether cooperation can evolve among individuals without family ties when repeated interactions are scarce. These models show that cooperation will reliably emerge if reputation can be tracked publicly—often via gossip—and their specific psychological predictions have been reliably confirmed (Nowak & Sigmund, 1998a; 1998b). On this view, the language of virtue functions as a system for tracking reputation, helping individuals identify trustworthy partners and avoid exploiters. Thus, the organizing principle of the virtues would be cooperation (an idea that Robinson, this volume, also explores).

But on an Aristotelian conception of virtues—the most influential version of virtue ethics—the organizing principle of the virtues is living a *Good Life*, not merely stabilizing cooperation. Indeed, cooperation and mutual benefit form only one element of the *Good Life*. After all, mafiosos and humanitarian workers alike track reputation, reward cooperation, and punish defection, yet only the latter come close to virtue. If an evolutionary explanation of virtue is correct, we cannot assume that cooperation *per se* has moral significance, or that the traits that stabilize it will typically constitute a *Good Life*. On the *scientific image*, evolution equips us to track virtues and vices because it supports cooperation; on the *manifest image*, virtue helps us to reflectively shape our lives toward flourishing.

Although no single contribution to this volume resolves this exact tension, each shows how the Darwinian/*scientific image* of human evolution intersects with the *manifest image* of human beings as bearers of virtue and vice.

Chapter Summaries

The volume brings these images into a common frame through four interconnected parts. The first examines theoretical frameworks and developments in evolutionary thought that explicate the origins and structure of virtue and vice—not only their influence on behavior but also the human tendency to think in these evaluative terms, what Schulz (this volume) calls *virtue-based thinking*. The second concerns the sentimental foundations of character, tracing how evolution has shaped our passions and emotional capacities. The third considers questions about the nature of justice and the emergence of spontaneous social order. The fourth section shifts the focus to the future, assessing how virtue and vice continue to evolve in the digital age.

Each chapter shows how evolutionary theories and explanations interact with normative questions about character. Because one purpose of this volume is to stimulate further research and discussion, we also highlight questions for future research that arise from these contributions and their many potential interactions—some best approached philosophically, others requiring empirical study. The fruitfulness of these discussions is evident not only in the abundance of questions they prompt us to ask but also in the potential to enrich both the *manifest* and *scientific images* of humankind as bearers of virtue and vice.

Part I: Theoretical Frameworks—The Mechanics of Evolved Character

Part I places the Darwinian image at the center. It examines the evolutionary forces—both specific mechanisms and generalized pressures—that shaped the phenomena studied within virtue ethics and moral psychology. These chapters show how virtues can be understood as *cognitive technologies* that enable cooperation, stabilize social expectations, and structure character, even as those same forces sustain spiteful tendencies and situational frailties.

In Chapter 1, Armin Schulz begins with a foundational question: how can we account for the uniquely human tendency to think in terms of virtue and vice while still recognizing both continuities with non-human cognition and variation within our own species? Schulz's answer appeals to a dynamic, interactionist process in which evolved psychological capacities, cultural learning, and cognitive technology mutually reinforce one another: stable character traits help bring about patterns of character-guided action, which in turn strengthen those evaluative practices.

Schulz's interactionist model introduces a theme that reappears throughout the volume and represents one of its distinctive contributions: Genes and culture constitute two separate modes of inheritance that can reciprocally interact, often producing feedback loops and ratchet effects that can lead to rapid construction of new niches and changes in the adaptedness of human groups to their environment. Cultural learning allows the acquisition of behaviors, myths, techniques, and even the perpetuation of institutions. These cultural technologies then shape which genes will be adaptive in the resulting niches and ecologies, a theme that recurs in work by Smead and Forber, Waggoner and Kelly, and—more implicitly—Li, Delston, May, and Curzer and Velasco.

Schulz concludes by drawing out some consequences of this feedback-based view, arguing that character-centered deliberation should not be set in opposition to situation-based decision-making. Instead, both emerge from a shared psychological endowment, with some cultures emphasizing situational demands over character-based consistency. Similarly, Schulz also suggests that a culture's cognitive technologies can substantially reshape how individuals make decisions and how those decisions are morally appraised. In short, for humans, cultural technologies shape character, and character-based behavior makes virtue—by making virtue-based concepts more predictive.

In the second chapter, Brian Robinson raises the question of which traits count as virtues and why. He tackles Schulz's challenge of explaining cross-cultural variation in virtue concepts while also explaining shifts over time. Robinson's proposal draws on the framework of Morality as Cooperation (MAC) (e.g., Curry, 2016), which distinguishes seven fundamental forms of cooperative behavior observable across cultures. He argues that these cooperative behaviors also correspond to seven core virtues. On this account, virtues are enduring dispositions that incline agents toward cooperative responses to recurring social coordination problems. Because the dominant forms of cooperation differ across societies and historical contexts, so too do the virtues they emphasize. Robinson closes by using humility as a case study to illustrate and assess this evolutionary account of virtue.

Readers will notice some strong similarities in approach between these first two contributions: both claim that virtue is unique to humans, both aim to explain cultural variation, and both appeal to foundational evolutionary principles.

However, there are also some important differences—which underlie divergences in the content of the respective accounts. Schulz focuses on cognitive architecture and proximate stabilizing mechanisms for virtue, whereas Robinson takes stability of character for granted within a given ecology and highlights the survival value and evolutionary functions of virtue-based thinking and character-based action selection. Robinson takes certain virtues—or families of virtue—to be universal, while Schulz does not. Schulz emphasizes cognitive technologies that shape the cultural evolution and individual development of virtues; Robinson focuses on ecological variation—particularly in material and social resources—that influences how societies solve cooperative dilemmas and which virtues are most effective in doing so. Robinson treats character-based action as central to solving such dilemmas and guiding cooperative partner choice, whereas Schulz remains neutral about its adaptive value and focuses instead on the proximate psychological role of virtue-based thinking in evaluating actions and stabilizing character–based dispositions.

These contrasts suggest that important aspects of the *scientific image* of virtue remain unresolved, and they indicate that the task of integrating the *scientific* and *manifest images* may be correspondingly more complex.

In Chapter 3, Rory Smead and Patrick Forber begin with a puzzle about the evolution of cooperation: if helping others is costly and creates opportunities for free riders, why doesn't cooperation unravel under selective pressure? From the standpoint of short-term individual advantage, cooperative behavior appears unstable. Nevertheless, humans cooperate extensively and have evolved sophisticated capacities for acquiring, sharing, and exploiting cultural information—capacities that support rapid cultural change, enable complex tool use, foster locally adapted subsistence strategies, and underwrite the emergence of highly structured social systems.

But the very evolutionary mechanisms that explain the persistence of cooperation and cumulative culture also produce morally troubling patterns of behavior. In particular, they show how these mechanisms can give rise to spite: actions that impose a cost on an actor while inflicting greater harm on a rival. Building on and extending existing evolutionary models, Smead and Forber demonstrate that spite can evolve reliably in environments where relative advantage over competitors matters more than absolute payoff. The authors distinguish several learning dynamics that stabilize spiteful strategies and suggest that human learning approximates these dynamics. Their analysis highlights a sobering conclusion: the same evolutionary pressures that sustain cooperation and cumulative culture can also foster some of the darker tendencies of human character.

Smead and Forber's contribution introduces several key themes and explanatory paradigms. Most centrally, their work addresses models in evolutionary game theory that capture the dynamics of *frequency-dependent selection*. As Mallon later explains, this is one species of *balancing selection*, which is especially useful in identifying equilibria among competing traits in a population. These equilibria are a form of *spontaneous order*, and equilibria can be *monomorphic*—where a population consists entirely of one trait—or *polymorphic*—consisting of several traits. Although the concept of spontaneous order first arose in economics, evolutionary game theory has greatly extended the theoretical resources for explaining it. Part III emphasizes this, especially in the contributions of LeBar, Bruner, and Mallon. While Smead and Forber focus on emerging order in the form of polymorphisms between spiteful and cooperative strategies, Mallon and Bruner discuss the importance of stable polymorphisms for divisions of labor that result in benefits at higher levels of organization (groups, populations, etc.). These discussions of polymorphisms—and their implications for concepts of virtue and vice within the *manifest image*—are a distinctive contribution of this volume.

In the fourth chapter, we (the editors) pose a challenge to Aristotelian virtue ethics—which views virtues as robust and consistent traits—by extending the situationist critique from psychology into evolutionary biology. Work in evolutionary game theory, anthropology, and cognitive science suggests that spiteful behaviors become advantageous under conditions of scarcity, small population size, and local competition—all cases in which relative gains become more important for reproduction than absolute gains (a key point of resonance with Smead and Forber). Historical climate variability during the Pleistocene likely favored flexible behavioral responses over fixed moral dispositions, with heritable mechanisms involving dopamine and serotonin modulating social preferences. These findings undermine the assumption that cooperation is always best explained by stable virtues, supporting a situationist account of virtue-relevant behavior. This account also suggests the possibility of universal situational sensitivity, a tension for accounts that emphasize the stability of virtues, even if stability is culturally variable (i.e., Robinson and Schulz).

Questions for Future Research

Part I raises several questions for future inquiry. One set of questions concerns how virtue-based thinking fits within broader evolutionary models of cooperation. How should Schulz's account incorporate adaptive explanations of cooperative behavior, such as models of indirect reciprocity and reputation tracking? And how might cultural technologies of self-worth—as found in cultures of honor, face, and dignity (Cohen & Leung, 2011; Nisbett & Cohen, 1996)—systematically influence virtue-based thinking?

Another set of questions arises from the MAC framework. Aside from humility, to what extent can MAC explain additional virtues—or patterns of vice—across different ecologies? How might ecological and cultural variation in cooperative challenges shape the range and stability of virtue and vice concepts across human societies, and are specific *failures* of cooperation also consistently regarded as vices across cultures?

Finally, Part I raises deep questions about stability and variability in character. Which variations in learning rules best explain the prevalence and transmission of spite in human groups? Can the situational variability of spite be reconciled with the stability of virtue? Or might the stability of virtue-based thinking itself be better explained by the stability of vicious character in some individuals? Together, these questions highlight the importance of clarifying how the *scientific image* illuminates and constrains virtue and vice in the *manifest image*.

Part II: Sentimental Foundations

If virtue exists, there are strong reasons to suppose that it would be anchored in our emotional architecture. Part II explores how evolution has shaped our passions and sympathies, illuminating what moves us to care for one another, protect ourselves, and tend our networks of relationships. These chapters reveal both the power and the limits of our emotional architecture—showing how trust is scaffolded, how it sometimes overextends, and where its boundaries lie.

In Chapter 5, Michael Slote expands on his widely discussed sentimentalist virtue ethics, which anchors the meaning of moral language to human sentiments rather than the rational capacities emphasized in Kantian approaches to morality and metaethics. Slote's sentimentalism bases morality on empathy, characterizing moral judgment in empathic terms.

This version of sentimentalism, he maintains, is immune to the evolutionary debunking arguments that Sharon Street (2006, among others) has directed at ethical rationalism. By contrast with rationalism, Slote's sentimentalism offers a Kripkean reference-fixing account of moral judgment. This identification of the relevant properties (i.e., "agential warm-heartedness") inoculates it against evolutionary arguments like Street's. Slote also maintains that the nature of empathy can fully account for the motivational force of moral considerations and moral judgments. On this view, humans evolved with distinctive capacities for empathy and also evolved to deploy this capacity in moral appraisal.

Charlie Kurth follows this with a chapter on the cultivation of virtue. Cultivating virtue requires adjusting our emotional responses in various ways, but virtue theories are divided over how emotions can and should be shaped. Whereas Aristotle held that emotions themselves could be cultivated so that the virtuous person feels the right things at the right time and to the right degree (etc.), Kant held that some emotions could at best be controlled or managed. Kurth surveys empirical work on specific emotions, revealing that disgust is far less plastic than shame or anxiety. He then considers the evolutionary origins of these differences to uncover the deep architectural distinctions that produce them. On these empirical and evolutionary bases, Kurth argues that the Kantian picture of moral cultivation is the most plausible, since it makes room for multiple strategies for managing emotion in the cultivation of virtue.

Qiannan Li expands the discussion of shame in Chapter 7. Whereas Kurth and others view shame as a co-option of evolutionarily ancient appeasement responses (submission to dominant others), Li offers a Confucian alternative. On her view, shame is an "affective alarm" triggered by threats to the relational self. Indeed, the appeasement model fails to account for the full range of shame-related phenomena. When ashamed, people do not always withdraw and appease, nor are they always concerned with their own inability to meet some standard. In some cases, people become aggressive in response to shame, and in others, they feel vicarious, or second-hand shame, when others fail to meet social standards. Drawing on the *Analects* and *Mencius*, Li argues that shame motivates agents to tend to the network of relationships that defines one's identity and self-worth. Moreover, she suggests that this concern for relational integrity is a product of cultural evolution. In Schulz's terms, shame finds its distinctive shape in Eastern cultures through various cognitive technologies, such as ritual, reflection, and explicit instruction. Li suggests that the resulting cultural ecology for shame transforms it into "a resource for virtue, relational harmony, and moral agency."

In Chapter 8, Iskra Fileva draws together several different but related senses in which emotional responses can be irrational. In some sense, we hold it to be irrational to feel regret or guilt for something completely out of our control. Self-blaming reactions make more sense when one acted intentionally or negligently to bring about some harm. Fileva suggests an underlying continuity between this kind of irrational regret and the gratitude or resentment we sometimes feel toward inanimate objects. While irrational in a sense, Fileva notes that Adam Smith viewed them as deeply tied to a virtuous character, whereas the more "rational" response in which these sentiments are absent can seem inhuman or unnatural in many cases. Fileva considers existing attempts to resolve this apparent inconsistency—including Smith's—arguing that they are incomplete. She expands on the sense that these responses are natural, arguing further that an evolutionary perspective on emotion is necessary to understand just how. When we understand the ways that evolution favors caution (especially in favoring false positives over false negatives) even in our moral emotions, we can see a clear sense in which this unnaturalness can reveal moral deficiency, or "unnatural vice." Fileva cautions against scientific attempts to engineer these emotions and eliminate these "irrational virtues," but she leaves room for argument on this question.

Questions for Future Research

Part II raises a particularly rich set of questions for future inquiry. First, sentimentalist metaethics must address the tensions resulting from the tribalistic patterns in our empathic responses: how might a sentimentalist framework justify the sense that our empathic responses should be extended to humankind as a whole, or even beyond it? From a naturalistic approach, why not think that these extensions of empathy are aberrant or that they are merely one point on a range of appropriate responses? In addition, questions about emotional plasticity remain: Is there a range of emotional responses or intensities at which neither control nor emotional change is possible, and how might the Kantian picture apply to the attribution and cultivation of virtue in individuals—such as those with PTSD—whose emotional repertoires have been profoundly shaped by trauma?

A further set of questions concerns shame and cultural logics of selfworth. How do honor, face, and dignity cultures shape the development of emotions like pride and shame and their associated virtues and vices? Is it possible to articulate an evolutionary account of shame's precultural potential or its norms of reaction across cultural ecologies, and to what extent would such an account resemble the appeasement model or Li's relational threat-detection model?

Finally, there are challenging questions about so-called irrational emotions. What is the architecture of emotions such as regret and gratitude that makes these cautious, overshooting responses so central to virtues, and their absence a mark of moral deformity? Could such emotions ever be made more "rational" without changing the shape of morality itself? And can explanations of irrational and arational emotional responses co-exist with cognitivist accounts of emotion, or must they rely on non-cognitivist motivational architectures (e.g., Wiegman, 2020)? Taken together, these questions demonstrate that the sentimental foundations of virtue are a fertile ground for integrating evolutionary explanations with normative theories of virtue.

Part III: Justice and the Communal Architecture of Virtue

Part III turns to justice and other forms of spontaneous order that emerge from evolved social dynamics. These chapters reveal how much of our moral and epistemic scaffolding arises outside our awareness, how it provides the conditions for joint knowledge and shared intention, and how it grounds the modes of accountability required for treating one another with dignity.

In Chapter 9, Mark LeBar explores interpersonal justice, advancing an account of justice as a personal virtue shaped by cultural evolution. Building from an Aristotelian conception of virtue, he argues that justice is best understood as grounded in mutual recognition of equal authority to obligate one another. Because just conduct requires reciprocal coordination, its norms cannot be reduced to abstract principles, but must be seen as emergent from concrete human interactions. Drawing on the work of Tomasello and Hayek, LeBar traces how such norms could have evolved through processes of intragroup learning, perspective-taking, and arbitration, as well as through intergroup competition and cultural transmission. Justice, in this sense, resembles the common law: a spontaneous order generated through bottom-up "experiments in living," continuously refined through social learning rather than intentionally designed.

A central theme that emerges from LeBar's chapter—and that resonates throughout this part of the volume—is that just communities are not composed of morally identical actors. Instead, they depend on forms of specialization and differentiation that allow norms to function at scale. This theme is taken up and sharpened in Justin Bruner's contribution, which turns from moral coordination to epistemic coordination. Bruner asks how the burdens of inquiry are distributed in populations lacking formal epistemic institutions, and how epistemic virtues might evolve under such conditions. Standard models in cultural evolution imply that widespread social learning cannot, by itself, improve collective epistemic performance—a phenomenon often referred to as Rogers' paradox. Bruner revises this picture by introducing heterogeneity in agents' capacities for successful individual learning. When individuals vary in reliability, evolutionary dynamics favor stable outcomes in which those best suited to inquiry bear its costs, while others appropriately depend on socially transmitted information. This competence-sensitive division of epistemic labor resolves Rogers' paradox and clarifies when reliance on others amounts to epistemic free-riding and when it reflects an efficient allocation of effort.

Bruner's model highlights how epistemic communities depend on a stable mix of complementary virtues and vices. Conscientiousness, diligence, and competence on the "discovery side" must be matched by humility and trust on the "testimonial side" if collective inquiry is to succeed. This heterogeneity is not a defect but a functional feature of epistemic communities. Whereas this division of labor might seem similar to a form of epistemic free-riding, in Bruner's view the reliance of knowledge consumers on more competent knowledge producers is a kind of humility—properly calibrating one's confidence to one's competence—which governs one's epistemic decisions, whatever one's role in the epistemic order. In this way, Bruner's chapter reinforces a broader theme of polymorphism: the idea that the "beauty of the symphony" depends on players performing distinct parts.

In the eleventh chapter, Ron Mallon develops this theme directly by challenging the assumption, common in moral philosophy and psychology, that dispositions underlying virtue and vice are fundamentally uniform and that variation in judgment or character across individuals is a deviation from a central norm. Mallon instead explores the possibility that polymorphic moral dispositions can be sustained by lawlike processes within a population. On this view, the persistence of certain traits may depend on the continued presence of other, even conflicting, traits. Such diversity can be understood as a division of moral labor, in which individuals come to occupy and sustain different moral niches that they both respond to and actively shape. The result is a resilient moral ecology whose stability cannot be explained solely at the level of individual psychology; instead, it requires attention to population-level dynamics—an insight with important implications for both moral psychology and moral theory.

This account of moral polymorphism introduces productive tension with more monomorphic approaches to virtue, such as Robinson's account of core virtues, while also creating bridges to other contributions in the volume—notably Bruner's epistemic models and the work of Forber and Smead on polymorphic equilibria involving traits like spite and fairness (Forber & Smead, 2014). Bruner's view, by contrast, reveals the possibility of a division of labor based on competence alongside a univocal—rather than polymorphic—account of the epistemic virtue of humility (i.e., calibrating confidence to competence). Together, these chapters suggest that justice may depend less on universal convergence in moral character than on the stable coordination of diverse moral roles.

The section concludes with John Lemos's chapter, which shifts focus to the problem of moral responsibility within such heterogeneous and evolutionarily shaped systems. Lemos asks whether it is right to blame and punish individuals for actions that flow from bad character, and argues that such practices are justified only if agents possess desert-grounding control over their actions. He contends that compatibilist accounts of free will fail to secure this form of control and that libertarian free will is both coherent and compatible with evolutionary naturalism. On Lemos's view, libertarian agency may itself be an adaptive trait as well as—to borrow from Schulz—a cognitive technology necessary for sustaining human dignity and the normative force of desert within a naturalistic framework.

Questions for Future Research

Part III raises several lines of inquiry at the population level. One set of questions has to do with the empirical grounding of polymorphism. Can we find evidence for the kind of population-level polymorphisms that Mallon describes? For example, could Forber and Smead's (2014) work on spite and justice explain the substantial percentage of spiteful tendencies in human populations (McCarthy-Jones, 2021, chs. 2–4)?

A second cluster concerns free will and moral responsibility. Can Lemos's pragmatic defense of libertarianism be justified in evolutionary terms? Is it perhaps better, from the standpoint of cooperation and punishment dynamics, to act *as if* people have libertarian free will because of its functional benefits in deterring free riders?

Finally, Part III raises questions about how theories of virtue, justice, and responsibility should adapt once we take population-level dynamics seriously. How can individual moral responsibility be reconciled with divisions of moral and epistemic labor that are partly shaped by frequency-dependent selection? If agents come to occupy particular moral roles because those roles are stabilized at the population level—as Mallon and Bruner suggest—then in what sense are individuals responsible for the "parts" they play?

Taken together, the chapters in Part III portray justice—within both the *scientific* and the *manifest images*—as a fragile but powerful achievement of social evolution; one that depends on the stable coordination of diverse—rather than homogeneous—dispositions in a population.

Part IV: Applied Ethics—Evolving Characters in the Digital Age

Part IV looks ahead, asking how contemporary forces are carrying us into new moral terrain shaped by digital and cultural change. These chapters identify the cultural attractors that steer us away from feared selves, the structures that isolate us, the institutions that sustain cooperation, the polarizing dynamics that threaten it, and the emerging forms of community we may yet create.

Maria Waggoner and Daniel Kelly begin Part IV with a chapter on social learning and moral education, which concerns a point of strong overlap in *manifest* and *scientific images* of virtue. Imitation, social learning, and explicit teaching in moral education take central importance in both. As we develop a deeper scientific understanding of how ideas and stories (etc.) spread and influence moral judgment and action, we can tackle larger questions about which ones are worth spreading. Waggoner and Kelly focus on contemporary antihero stories in television and film, considering what we might learn from them. Like many ideas that spread through cultural evolution, these stories might spread even if they are not beneficial to the individuals who consume or internalize them. However, they point to the possibility that these ideas spread because of the benefits they confer, and they consider one particular benefit of antihero stories, which they call *moral inoculation*. Waggoner and Kelly go on to argue that antihero stories allow us to identify with bad actors in antihero narratives in ways we would not with bad actors that we see or hear about in real life. As such, these stories provide us with a sustained encounter with morally important models—not of ideals we wish to embody but of feared selves that we are then motivated to avoid. Further elaborating themes around cultural evolution and social learning, Waggoner and Kelly identify antiheroes as cultural attractors that can help us avoid vicious behavior that we might otherwise fall into unawares.

In the fourteenth chapter, Jill Delston sustains the focus on moral education and the social cultivation of virtue and avoidance of vice, arguing that community and social connections are vital to human flourishing in a number of critical domains. Either by accident or design, modern society has shaped our social lives to be more isolated and segregated. Given our evolutionary history as social creatures, Delston argues that this isolation is "a significant and avoidable danger to human health, happiness, and wellbeing." She defends a "sociality solution" to many problems in public policy, pointing to evidence that a certain kind of social connection—involving intimacy, vulnerability, and authenticity—mitigates the risks of vice and mental illness. On her view, healthy aging, addiction, and incarceration demonstrate the importance of social connection and the harms of isolation. This is unsurprising given the evolution of human sociality. We evolved to be the sort of creatures that thrive in community.

The book concludes with two chapters that bring many of the themes and tensions we've discussed to a climax. May's chapter focuses on the problem of polarization, its causes, and a set of virtue ethical solutions.

He wades through a vast interdisciplinary literature on polarization, especially along political lines. Despite evidence that some degree of polarization can be rational and even beneficial, he argues that it is having increasingly corrosive effects on modern democracies. Moreover, the causes of polarization are legion and largely structural, including group-level threats, divisive ideologies, misinformation, social media, and political institutions.

Nevertheless, May argues that individuals have an important responsibility to be part of the solution to polarization by controlling the media we consume and by cultivating virtues of intellectual humility and civility. Just as healthy eating requires us to restrain ourselves from the binge-eating that was almost certainly adaptive for our ancestors who lived in an unpredictable famine/feast climate, so healthy media consumption requires us to restrain ourselves from consuming media that rewards our tribalistic sentiments. However, this is not enough to heal the scars that polarization inflames. Individuals must cultivate virtues, too. May argues that we need intellectual humility and a *scout mindset* to prioritize truth and curiosity over the soldier-like defense of identity-defining beliefs; we must cultivate civility over our tendency to divide the world into "us and them." On May's view, the necessary structural changes can only persist when individuals begin changing their preferences and habits in these ways.

These prescriptions resonate with a number of the contributions. Consider Robinson's claim that humility has been adapted to solve modern cooperative dilemmas, especially those related to living in Western, Educated, Industrialized, Rich, and Democratic (WEIRD) societies, where a diminished focus on the self allows one to fit into a crowded and interdependent social milieu. This illuminates how humility can come to play the important role that May gives it. His emphasis on controlling one's media environment may be a way of recognizing the sort of situational dependencies governing moral behavior emphasized by Wiegman and Dale. This kind of ecological control is also exactly the kind of external scaffolding that is required, on Schulz's view, to sustain character-based action. Waggoner and Kelly's ideas about moral inoculation suggest that one element of ecological control could be antihero stories, such as *Joker* (2019), that take on the risks of political radicalization. May's appeal to Julia Galef's work on scout mindset also provides a psychological foundation for the discovery side virtues modeled by Bruner, which protect the circulation of knowledge from failures of epistemic coordination. Finally, May's focus on the restoration of trust through interpersonal interactions and dialogue reinforces Delston's sociality solution. Perhaps polarization is another disease of isolation, and restoring social connection across political barriers can help us to prevent it.

May's account also brings out many illuminating and productive tensions. First, if the language of moral disgust and contempt has become common currency in political discourse and if Kurth is right that disgust is far less malleable than other emotions, there may be limits on the cultivation of virtues like civility. Or perhaps this only emphasizes the importance we must place on controlling our informational diets in terms of how political groups and nationalities (etc.) are represented. Finally, whereas May appears to give a "monomorphic" account of the importance of intellectual humility and civility, Mallon's contribution suggests the possibility that more militant virtues (*soldier mindset*) may be a stable part of our moral ecosystem, perhaps playing an important role alongside the virtues of humility and civility. Perhaps the human superorganism needs the analog of soldier ants alongside the worker ants. If so, then the question becomes not how to fix the problem of polarization but instead how to find an adaptive fit with the new technological and social realities that may be reshaping our species.

This shift from mismatch with our environment toward metamorphosis to a new kind of species captures the essence of Curzer and Velasco's contribution. They argue that humanity is in the midst of a major evolutionary transition toward a kind of hive species, whose flourishing consists in a new set of hive virtues. Taking on a broadly Aristotelian framework—whereby virtues are the best bet for flourishing as the kind of social creatures we are—they argue that human nature is in a process of radical change in which our rational capacities (theoretical, practical, and productive reason) are themselves being shaped by the addition of what they call an *e-self*, the part of one's life that exists on the internet and other computer-based technologies. This addition changes what it means to flourish by necessitating a new set of strategies for living in human communities. Instead of relying on virtues like generosity, self-reliance, and open-mindedness, the realities of digital life and commerce require new virtues that involve selective trust (*e-collaborative-open-mindedness*) and reliance upon experts in one's e-community (*e-crowdsourcery*), along with selective generosity and empathy extended toward a limited set of individuals and social ills (*e-generosity*).

Moreover, Curzer and Velasco argue for the emergence of a new species of human hives that function like superorganisms. Importantly, and by contrast with Aristotle's view of the state, hives do not exist to promote the flourishing of the individuals that comprise them. Rather, hives exist to promote their own flourishing and require their own set of virtues to flourish. Moreover, virtues of societies—virtues that allow for the society itself to flourish—might not promote, or even co-exist easily with, traditional virtues. Among these virtues are greed, inspirational mythmaking, and limitless ambition. Notably absent from them are the virtues of honesty and justice, which may not in the end be necessary for human hives to flourish.

This provocative contribution brings to the fore many immediate tensions. Rather than viewing polarization as a problem, Curzer and Velasco might view it as a symptom of the emergence of distinct superorganisms of a hive species, some liberal and others conservative. Rather than resisting polarization with the cultivation of virtues like intellectual humility and civility, they might suggest that we should accept it along with newer virtues of e-crowdsourcery and perhaps even participation in the inspirational mythmaking that benefits the superorganism of which we are a part. Whereas May places emphasis on the individual's responsibility to cultivate virtues that resist polarization, perhaps the individual had better accept its emerging functional role within the superorganism—the new locus of selection, bearer of adaptation, and subject of flourishing—and pull its weight in the survival war of superorganisms. Perhaps the culture wars of modernity will and should resemble the global war between the Argentine ant (*Linepithema humile*) and the red imported fire ant (*Solenopsis invicta*).

Whereas the tension with May is perhaps the most (shall we say?) polarized, Curzer and Velasco's contribution interacts productively with a number of other contributions as well. For example, like Mallon, LeBar, and Bruner, they also touch on themes of division of labor and spontaneous order in their discussions of superorganisms. Curzer and Velasco are likely to acknowledge that superorganisms must develop in ways that are constrained by their evolutionary past. But if, as LeBar argues, the virtue of justice has been emerging for eons, structuring human interactions in both the interpersonal and legal domain, it may be as generatively entrenched as the genetic code or the early embryonic stages of development (Wimsatt, 1986) and so also as difficult to modify or jettison. Or perhaps the virtue of justice can only continue to benefit superorganisms just as it may have benefited human groups in our ancestral past—as LeBar suggests.

Curzer and Velasco's presentation of e-collaborative-openmindedness and e-crowdsourcery as novel epistemic virtues represents a surface tension with Bruner's analysis of knowledge in the state of nature. Bruner's models seem to already require reliance on experts to assess novel claims (or "high-types" who have higher competence in individual learning, as Bruner puts it) and restricting self-reliance to the assessment of claims within one's own area of expertise. His account thus suggests that little fundamentally new enters the scene with the *e-self*; that the hazards of epistemic vice (poorly calibrated learning strategies and misplaced trust) have always threatened epistemic communities and are thus obstacles to knowledge for human and post-human cyborg alike. The deeper tension with Bruner's contribution might be in the superorganism's lack of concern for knowledge or truth according to Curzer and Velasco.

Questions for Future Research

Part IV raises emerging questions about how cultural and technological forces constrain virtue, vice, and moral education. One set of questions concerns moral education and the cultural transmission of virtue and vice. What are the mechanisms by which antihero stories become widespread, and are these mechanisms consistent with the moral inoculation hypothesis? Is their spread best explained by real benefits to the individuals and groups that consume them? And in what ways might such stories actually make people or groups more cooperative or successful, in ways appropriately connected with their propagation?

A second set of questions concerns the evolution of sociality and the kinds of connections essential to human flourishing. How do honesty, authenticity, and vulnerability figure into evolutionary accounts of sociality? Are there particular reasons why our social connections exhibit these qualities in abundance? For example, how might evolutionary models of partner choice (e.g., Baumard et al., 2013) help explain the significance of vulnerability in human social connections?

Finally, Part IV raises difficult questions about humility, civility, the epistemic division of labor, and the emergence of human superorganisms. How much overlap might there be between May's virtue of civility and the social virtues in Delston's "sociality solution?" How might the importance of virtues of humility and civility shift across epistemic roles—should civility have a different shape or priority on the discovery side than on the testimony side? Which kinds of stories and ideas might best propagate humility and civility via social learning? And at the level of the collective, how much inaccuracy can a successful superorganism tolerate in conjunction with its inspirational mythmaking? Might knowledge function for the superorganism as oxygen does for multicellular creatures—raising the question of whether a superorganism can thrive without supplying knowledge to its parts?

Conclusion: Re-engineering the Soul

Perhaps the relationship of individuals to wholes is the perfect place to wind down these reflections. Sellars's own resolution of the discrepancies between the *scientific* and *manifest images* aims to place the community of persons within the *scientific image*. This begins by recognizing that persons are irreducible to a purely scientific picture: to view another "featherless biped" as part of one's community is already to see oneself as bound up with the other in a normative web of obligations and rights; to see oneself and others as part of a community, an "us," with joint intentions by which we are mutually bound. To reconcile the images, then, we must place "us," the community, into the scientific picture—using the language of science as a way to articulate our shared intentions. On this rendering, we are both products of blind evolutionary forces—with all their accompanying aptitudes, dispositions, biases, sympathies, and limitations—and also members of a community committed to viewing each other as participants: participants whose virtue we may trust, whose vice we must beware, and by whom we may be vindicated or betrayed. This volume continues the process of placing our shared intentions within the bounds of our scientific frames.

If we come to see clearly how evolution has shaped our aspirations and our situational frailties, we are empowered. Such clarity offers us a tool kit with which to re-engineer our conceptual repertoire of virtue, and the environments we inhabit, so that we may better conform to our shared intentions.

Likewise, if we come to see how evolution has shaped our passions and sympathies, we gain a fresh perspective on what moves us to care for one another, protect the self, and tend our networks of relationships—along with our natural tendencies to overextend these dispositions. This gives us a clearer vision of how to be trustworthy, how to trust one another, and how to understand the scaffold and limits of that trust.

So too with justice. We become aware of the external forces that constituted us as a *we* in the first place, forces that set the stage for joint knowledge and joint intention. In becoming aware of these forces, we also deepen our understanding of the modes of accountability that enable our intention to treat each other with dignity.

But if evolutionary forces have accelerated us onto a trajectory that outstrips our current knowledge and intentions, it becomes critical to gain a clearer picture of where they are taking us—to cultivate the foresight needed to navigate terrain that is at once treacherous and promising for our joint intentions. We must attend to the cultural attractors that steer us away from our feared selves; the structures that incline us toward isolation and vice; the systems of conflict and polarization that threaten our coordination and the institutions that sustain it; and the new forms of community we might yet form. Insights like these are critical to our species' emerging community and ongoing existence.

As a whole, this volume presents new ways of understanding what we are and what we might become. Through the *scientific image*, we can see with increasing clarity our frailties, polymorphisms, learning biases, patterns of thought, and cognitive technologies, along with their effects on the communities we inhabit. And we can use this picture to chart our own course into the future. Let us form our intentions with knowledge and care.

References

Aristotle (2000). *Nicomachean Ethics* (R. Crisp, Trans.). Cambridge University Press.

Baumard, N., André, J.-B., & Sperber, D. (2013). A mutualistic approach to morality: The evolution of fairness by partner choice. *Behavioral and Brain Sciences*, *36*(1), 59–78.

Buchanan, A., & Powell, R. (2018). *The evolution of moral progress*. Oxford University Press.

Casebeer, W. D. (2003). *Natural ethical facts: Evolution, connectionism, and moral cognition*. Bradford Books.

Cohen, D., & Leung, A. K.-Y. (2011). Violence and character: A culture–personality–situation loop. In P. R. Shaver & M. Mikulincer (Eds.), *Human aggression and violence: Causes, manifestations, and consequences* (pp. 167–186). American Psychological Association.

Confucius (2003). *Analects: With selections from traditional commentaries*. Hackett Publishing Company.

Curry, O. S. (2016). Morality as cooperation: A problem-centred approach. In T. K. Shackelford & R. D. Hansen (Eds.), *The evolution of morality* (pp. 27–51). Springer International Publishing.

Dawkins, R. (1976). *The selfish gene*. Oxford University Press.

de Lazari-Radek, K., & Singer, P. (2012). The objectivity of ethics and the unity of practical reason. *Ethics*, *123*(1), 9–31.

de Lazari-Radek, K., & Singer, P. (2014). *The point of view of the universe: Sidgwick and contemporary ethics*. Oxford University Press.

Doris, J. (2002). *Lack of character*. Cambridge University Press.

Doris, J. (2022). *Character trouble*. Oxford University Press.

FitzPatrick, W. (2025). Morality and evolutionary biology. In Edward N. Zalta & Uri Nodelman (Eds.), *The Stanford encyclopedia of philosophy* (Fall 2025 Edition). https://plato.stanford.edu/archives/fall2025/entries/morality-biology

Forber, P., & Smead, R. (2014). The evolution of fairness through spite. *Proceedings of the Royal Society B: Biological Sciences*, *281*(1780), 20132439.

Greene, J. (2008). The secret joke of Kant's soul. In W. Sinnot-Armstrong (Ed.), *Moral psychology, vol. 3: The neuroscience of morality* (pp. 35–80). MIT Press.

Greene, J. (2014). Beyond point-and-shoot morality: Why cognitive (neuro)science matters for ethics. *Ethics*, *124*, 695–726.

Henrich, J. (2017). *The secret of our success: How culture is driving human evolution, domesticating our species, and making us smarter*. Princeton University Press.

Joyce, R. (2006). *The evolution of morality*. MIT Press.

Kamtekar, R. (2004). Situationism and virtue ethics on the content of our character. *Ethics*, *114*(3), 458–491.

Lemos, J. (2008). *Commonsense Darwinism: Evolution, morality, and the human condition*. Open Court.

McCarthy-Jones, S. (2021). *Spite: The upside of your dark side*. Basic Books.

Mencius (2008). *Mengzi: With selections from traditional commentaries*. Hackett Publishing.

Miller, C. (2003). Social psychology and virtue ethics. *The Journal of Ethics*, *7*(4), 365–392.
Mizzoni, J. (2017). *Evolution and the foundations of ethics: Evolutionary perspectives on contemporary normative and metaethical theories*. Lexington Books.
Nisbett, R. E., & Cohen, D. (1996). *Culture of honor: The psychology of violence in the South*. Westview Press.
Nowak, M., & Sigmund, K. (1998a). The dynamics of indirect reciprocity. *Journal of Theoretical Biology*, *194*, 561–574.
Nowak, M., & Sigmund, K. (1998b). Evolution of indirect reciprocity by image scoring. *Nature*, *393*, 573–577.
Richerson, P. J., & Boyd, R. (2005). *Not by genes alone: How culture transformed human evolution*. University of Chicago Press.
Sabini, J., & Silver, M. (2005). Lack of character? Situationism critiqued. *Ethics*, *115*(3), 535–562.
Sellars, W. (1963). Philosophy and the scientific image of man. In: *Science, perception and reality* (pp. 1–40). Routledge & Kegan Paul.
Singer, P. (2005). Ethics and intuitions. *The Journal of Ethics*, *9*, 331–352.
Sreenivasan, G. (2002). Errors about errors: Virtue theory and trait attribution. *Mind*, *111*(441), 47–68.
Street, S. (2006). A Darwinian dilemma for realist theories of value. *Philosophical Studies*, *127*, 109–166.
Waggoner, M., Doris, J., & Vargas, M. (2022). Situationism, moral improvement, and moral responsibility. In M. Vargas & J. Doris (Eds.), *The Oxford handbook of moral psychology* (629–658). Oxford University Press.
Wiegman, I. (2017). The evolution of retribution: Intuitions undermined. *Pacific Philosophical Quarterly*, *98*(2), 193–218.
Wiegman, I. (2020). Emotional actions without goals. *Erkenntnis*, *87*, 393–423.
Wimsatt, W. C. (1986). Developmental constraints, generative entrenchment, and the innate-acquired distinction. In W. Bechtel (Ed.), *Integrating scientific disciplines* (pp. 185–208). Martinus Nijhoff.

Part I

Theoretical Frameworks

The Mechanics of Evolved Character

1 Characterizing Humans

The Co-Evolution of Virtue

Armin Schulz

1.1 Introduction

Humans, at least sometimes, think in terms of virtue and vice: they assess character traits as being morally praiseworthy (or the opposite). This is interesting for two reasons. First, it differs from what is found in non-human animals: even though the latter may rely on some moral concepts in their decision-making, and even though non-human animals may show individual psychological variation, they do not appear to reason in terms of virtues and vices (though this is, of course, difficult to test with precision). Second, while virtue-based thinking is part of many cultures (including Ancient China and Greece), it is not clearly part of every (sub-)culture: for example, virtue ethical reasoning is controversial in contemporary Western philosophy. Indeed, it is not even the case that all humans, both across and within cultures, consistently act on particular, small sets of character traits—not to mention reason in terms of these. These points thus lead to a puzzle: how can we explain the distinctively human reliance on concepts of virtue and vice, all the while doing justice to the commonalities between human and non-human cognition and allowing for intra-specific variation in the latter?

To answer this question, this chapter argues that the evolution of virtue should be approached from two angles: (a) the evolution of a psychological apparatus that *sometimes* generates behaviors from stable character traits, as opposed to merely relying on social role expectations or momentary evaluations of the world, and (b) the evolution of the reliance on abstract and complex mental representations of "virtue" and "vice" that are used to assess these kinds of psychologies. The chapter goes on to show that (a) and (b) are best explained *conjointly*. That is to say, the goal of this chapter is to argue that the evolution of virtue-based thought (VBT) is driven by an interactionist feedback loop involving innate psychological structures and expectations, cultural learning, and cognitive technology (see also Schulz, 2020, 2025). (Implicit in this is thus the claim that thinking in terms of virtues of vices is not directly adaptive; it is merely a facet

DOI: 10.4324/9781003632641-3

of a more general psychology tailored toward navigating a mentalistically saturated social environment.)

The chapter is structured as follows. In Section 1.2, I make the puzzle of the evolution of virtue clearer: what is it, exactly, that needs to be explained here? In Section 1.3, I lay out a framework for understanding the evolution of virtue-based thinking. In Section 1.4, I draw out some novel implications of this framework and consider some possible responses to the account of the chapter. I conclude in Section 1.5.

1.2 The Puzzle of the Evolution of Virtue in Humans

What, exactly, do we need to explain when we seek to explain the evolution of "virtue"? What is the biological trait of which we are seeking an account here? Why is providing this explanation difficult? To answer these questions, it is best to begin by noting that I here treat the question of the explanation of the evolution of "virtue" as the question of the explanation of the human ability to *think* in terms of virtue. After all, since virtue is a normative concept, asking for the evolution of virtue would be like asking for the evolution of *justice* (the actual thing, not our concept of it). While it may be that there is something to say about the latter kind of case, for present purposes, the focus is on the more straightforward—but still quite complex—case of the evolution of a particular kind of psychology.

With this in mind, note that it is useful to break down VBT into two component elements:

a We see human thought and action as often and largely based on a relatively small number of relatively stable dispositions—i.e., ways of thinking or acting, such as courageousness or neuroticism.
b We assess some of these character traits as morally praiseworthy and others as blameworthy.

Component (a) is the *psychological* element of VBT. It is a claim about the psychological apparatus responsible for the *generation* or *production* of behavior. Instead of seeing human action as driven by a set of mental states that differ from situation to situation without clustering in particular narrow ways, VBT sees human action as often driven by a (relatively) small cluster of psychological dispositions. The exact nature of these states is a point of some controversy (see, e.g., Doris, 2002; Harman, 1999; Miller, 2014), but for present purposes, all that matters is that, for VBT to be at all a substantive thesis, the extent and reach of these character traits have to have some bite. So, we can admit only fairly few such stable dispositions (Aristotle's ten virtues, say) or allow a larger number (such as Mischel's "cognitive-affective personality system"—see, e.g., Mischel, 1968, 1973; see also Miller, 2014)—but we cannot admit so many that the appeal to stable character traits does little work. If every situation is characterized by a different such trait, the appeal to character traits loses its value: it would amount to the claim that whatever people do is what they tend to (Doris, 2002; Elster, 2015). Similarly and independently, we can allow these character traits to be more or less stable—i.e., affecting many or fewer decision situations the agent finds itself in—but there has to be a significant number of cases where these character traits are the key factors driving the agent's behavior. Otherwise, there would again be no need to appeal to the traits in the first place.

In short, there is no need to see VBT as implying that all human actions are driven by a very small set of character traits, but it should be seen to imply that people are at least prone to acting according to a specific, relatively narrow set of character traits in many different situations (Doris, 2002; Harman, 1999; Miller, 2014). Exactly how to make these terms precise can be left open here; for what follows, it is sufficient to note that it is at least not implausible that they can be spelled out in a sufficiently stringent manner to make it at all worthwhile to appeal to the notion of character.

Note that one might question to what extent component (a) of virtue-based thinking is *accurate*. That is, must virtue-based thinking imply that people *actually* act on a relatively small number of relatively stable dispositions, or is it possible that all VBT is based on a fundamental attribution error? Here, there are two points to be noted in response. On the one hand, and most importantly, the rest of the chapter provides an account of *why* component (a) of VBT may be accurate; it is thus a way to respond to error theoretic views of VBT. On the other hand, though, everything that follows is consistent with there being much error about how people think about how they and others make decisions. Just because, in some cultures at least, character-based decision-making is common, this does not mean that it operates in exactly the ways we ordinarily think. Put differently, the account in the rest of the chapter does not presume that people are 100% accurate in their everyday life assessments of how humans think and act. Still, the account is a way to make sense of the fact that character-based views of human action are at least sometimes reasonably close to the mark.

Part (b) of VBT concerns the psychological apparatus responsible for the *interpretation* or *evaluation* of behavior. Instead of merely seeing actions as stemming from more or less stable character traits, we also assess the moral worth of these character traits. Again, there are several different issues that can be distinguished here. On the one hand, the moral assessment of characters can be second- or third-personal (Cokelet, 2023; Darwall, 1996): we can ask whether a person acts in ways that a virtuous person would, or we can ask how a virtuous person would respond to the actions of the agent in question. On the other hand, we can ask whether these character assessments are an (or even *the*) independent basis of moral theorizing, or merely a part of the package of considerations that morality rests on—a package that also includes appeals to rights, duties, freedoms, or utilities (Cokelet, 2023). Again, though, deciding this is not necessary at this point (though I will come back to related issues in Section 1.4).[1] What matters here is just that the appeal to *virtue* is the appeal to a concept that is deeply embedded in our moral theorizing, and which functions to evaluate the character traits a person acts on.

It is furthermore important that (a) and (b) are *treated together*. It is not just that humans (sometimes) act on some character traits and that they sometimes morally evaluate some character traits. It is that the character traits (some) humans act on are the very same ones that they morally evaluate. Put the other way around, it is not just that VBT is based on morally evaluating some character traits or others—it is that they morally evaluate certain, specific character traits. While the details of which character traits these are differ across different virtue ethical accounts, what matters here is just that not *all* stable character traits are morally evaluated; only some are. So, in general, when explaining the evolution of VBT, we need not explain the evolution of the "Big Five" personality traits—these are generally not given a moral gloss—but the evolution of character traits like courage, humility, and patience.

With this in mind, the problem of the evolution of VBT can be stated more clearly. On a broad level, this problem can be stated as the fact that VBT seems to be *uniquely* human, but not *universally* human. That is, in the biological world, it is only humans that engage in significant VBT—but not all humans do so all the time. The task, then, is to make sense of these twin facts. Put differently, we are trying to account for both inter-specific and intra-specific differences in VBT: VBT needs to be seen to be embedded in human cognitive evolution more generally. This does not mean that nonhuman animals cannot have precursors or alternative versions of virtues—but the explanation in the chapter focuses specifically on human virtue.[2]

When it comes to this problem, it is furthermore important to note that these questions arise for *both* components of VBT. So, when it comes to component (a), it is a major finding of work in social and cross-cultural psychology that much of human behavior is situationally variable in ways not well captured by the character model built into (a)—with some people's actions seemingly driven rarely by their character (Doris, 2002; Harman, 1999; Henrich, 2020; Mischel, 1968, 1973; Smith & Apicella, 2020). While there are questions about numerous aspects of these studies, the overall upshot here is still widely accepted: it is not the case that all humans consistently act on a small set of stable character traits. Focusing on the cross-cultural studies, Henrich (2020, p. 33) summarizes these findings succinctly: "being consistent across relationships—'being yourself'—pays off more in America [than in Korea]." So, in Korea, whether you are a courageous or generous person is not thought to be a relevant factor when deciding how to interact with elders, say: the situational demands dominate the appropriate ways of acting. The opposite is true in the United States.

Ipso facto, of course, this also affects component (b): it is not the case that humans consistently evaluate each other's character traits. While moral evaluation in some cases clearly seems character-based, in other cases, it is not. Indeed, in some cultures, moral evaluation is largely non-character-based (Doris, 1998, 2002; Henrich, 2020).

Finally, even where virtues are prevalent, the kinds of virtues appealed to can differ drastically. As has been frequently pointed out, the Aristotelian virtue of "magnificence" (spending large sums of money for appropriate purposes—like being a major public donor to important cultural institutions like the New York City Ballet) is particular to the ancient Greek cultural milieu Aristotle was working in (but see also Athanassoulis, 2016), and parallel remarks can be made for some of the Confucian virtues like "li" (which includes engaging in the appropriate actions to honor and remember ancestors, such as the creation of household altars) (Angle & Slote, 2013).

Something similar holds for the inter-specific case: this also affects both components (a) and (b). Here, the situation with regard to component (b) is straightforward. While it is at least arguable that non-humans can have a proto-morality and that they, at least sometimes, respond to each other as distinct individuals (Andrews, 2015, 2020; Andrews et al., 2024; Brown, 2018; Monsó & Andrews, 2022; Westra & Andrews, 2022; Westra et al., 2024), it does not seem to be the case that they morally evaluate others' personality traits. Of course, establishing a negative claim like this empirically is very difficult (indeed, it is hard to think of clear ways of testing it); however, it is at least the case that researchers have not yet found any positive reasons for attributing virtue-focused thoughts to non-human animals.

However, there are also important inter-specific differences when it comes to the psychological component (a) of VBT. On the one hand, it is noteworthy that while some non-human animals are recognized to have personalities—i.e., character traits—not all of them are: for example, while many mammals and birds do seem to have personalities, many insects and crustaceans do not (Buss & Hawley, 2010).[3] In this sense, therefore, while humans are not *unique* in sometimes acting on stable character traits, this is something that still has to be explained, as not *all* animals do this. On the other hand, and more importantly, humans seem to be standouts even among the animals that have recognized personalities. For while non-human animals may have personalities, it is not clear that they have personalities *in the sense of VBT*. As noted earlier, VBT is not just about the existence of some character traits—it is about the existence of the kinds of character traits that are morally evaluated. As noted earlier, "being social," "being risk-averse," or "being curious," though, are not typically among the latter—though these are typically exactly the dimensions along which animal personalities differ. One exception may be the fact that some animals can track cooperators and non-cooperators (see, e.g., Carter & Wilkinson, 2013; but see also Hammerstein, 2003). However, even if this is granted, it is clear that humans are standouts in the detail with which they track personalities: we do not just assess each other as to whether we are cooperators or not, but also whether we are courageous when cooperating (or not cooperating), whether we are generous, etc.

On top of this, only in humans do character-based actions seem to vary *intra*-specifically. That is to say, in non-human animals—unlike in the human case—character-based action does not seem to differ within a group of animals, with some groups of dogs (say) showing personalities, and other groups do not. The way personalities interact with other psychological features thus seems different in the human case as opposed to the non-human case. I return to this point in Section 1.3. In short, part (a) of VBT is reasonably seen as uniquely human, too—even if it has homologs in non-human animals.

Therefore, the puzzle of the evolution of VBT in humans can be seen to consist in providing an account that can make sense of the following two sets of findings:

1 Humans sometimes, but not always, act largely on relatively few stable character traits, while non-human animals either do not seem to have stable personalities of the relevant kind at all or do not show intra-specific variation in these personalities in the way that humans do.
2 Humans sometimes, but not always, morally evaluate the relatively few relatively stable character traits they and others act on (or fail to do so), while non-human animals do not seem to do so—even though they may display a proto-morality in other regards, and even though they may still (sometimes) act based on stable character traits.

The next section provides an account that can explain these two sets of findings.

1.3 The Evolution of Distinctively Human Virtue-Based Thought

To explain the evolution of VBT, it is best to treat its two explanatory components—the psychological and the moral one—separately at first. However, as will become clearer below, ultimately, a compelling explanation of the evolution of VBT has to treat these together: the key to the explanation of the evolution of VBT is to relate the evolution of the abilities underlying its two components *to each other*. So, while this section will begin by considering the evolution of the capacities underlying the second, moral-evaluative element of VBT, it is important to emphasize that this is only done for expository reasons.

To understand the evolution of the capacities and disposition to employ the concepts of *virtue* and *vice* to assess character traits, begin by considering the evolution of complex and abstract moral concepts more generally. After all, the concepts of *virtue* and *vice* share some key features with other moral concepts, like *justice*, *duty*, or *honor*. In particular, these concepts are abstract: whether something is *just* or *virtuous* is, famously, not necessarily obvious—it may depend on historical or unobservable details of the situation (such as the past state of affairs of the agent's mental state).[4] Similarly, and as noted earlier, these concepts are *complex* in the sense that they are embedded in a network of other (often highly abstract) concepts. A *duty* is a *[demand] to [do or think something specific]*; a *virtue* is a *[character trait] [appropriate] [to the situation at hand]* (or the like). The details of these networks of concepts are not so important here; what matters is that attributing *virtue* or *duty*, etc. cannot be done unless a number of other abstract concepts are also attributed. In this way, these concepts differ from more peripheral, perceptual concepts, like *tastes sour*, which we can attribute without attributing as many other concepts—though of course the latter concepts are also not isolated. It is just that moral concepts like *virtue* and *vice* are particularly deeply embedded in our web of beliefs. Importantly, this is independent of the question of whether there is a specific domain of *morality* to which concepts like *virtue* and *vice* belong. This is a controversial issue (see, e.g., Campbell & Kumar, 2012, pp. 97–98; Haidt & Graham, 2007; Joyce, 2005; Machery & Mallon, 2010; Sarkissian et al., 2011; Stanford, 2018; Stich, 2018; Suhler & Churchland, 2011), made even more difficult by the fact that some concepts are "thick" (Geertz, 1973). However, nothing of significance hangs on this here: the point is just that these are concepts that are (a) deeply connected to other concepts and (b) directly or indirectly involve an evaluative component. The details of this evaluative component can differ from case to case and are not central to what follows.

Importantly also, humans rely on these concepts very frequently and in cases of major significance to their health, well-being, social standing, and family relations. Infamously, cultures of honor can lead to death and murder. More generally, though, decisions based on what is virtuous or just or our duty can be all-consuming, life-altering events for many humans. Sophie Scholl, Rosa Parks, Nelson Mandela, and many other humans organized their entire lives around what is just, right, or virtuous. These are thus not concepts of limited significance and rare usage in human living—they are at the very heart of what makes us, us.

What needs to be explained, therefore, is how humans came to rely on these kinds of concepts—and to do so as frequently as they do. Here, it is important to again note that it is increasingly clear that precursors of these kinds of concepts—e.g., in the form of basic concepts of fairness—can be found in non-human animals (Andrews, 2015, 2020; Andrews et al., 2024; Brown, 2018; Monsó & Andrews, 2022; Westra & Andrews, 2022; Westra et al., 2024). While details of the relevant studies can be questioned, it is becoming clear that the overall picture is one that supports the idea that humans are not alone in relying on broadly "moral" concepts. However, given the above, it is also clear that this is not sufficient to explain the *frequent* reliance on *particular abstract and complex* moral concepts in humans.

To do this, the first step is to see that the specifics of the ecological conditions humans evolved in favored the evolution of an enriched moral conceptual framework. Humans are social organisms of a particular sort: it is not just that we are dependent on others to survive and thrive, it is that, in order to do so, we need to respond to others as agents of their own—with their own volitions, opinions, and emotions (Henrich, 2015; Richerson & Boyd, 2005; Schulz, 2025; Sterelny, 2003; Whiten & Byrne, 1997). We cannot just respond to specific behaviors, because the same behaviors may have very different intentions behind them. Responding to and shaping these intentions is thus a key problem of human social living. One way of doing so is through the reliance on more abstract and complex versions of the basic moral concepts we are likely to have inherited from our common ancestor with chimps and bonobos. This expanded conceptual repertoire allows us to not just respond to being given unequal rewards for the same work, but also to the fact that getting the same rewards for the same work can be *unjust*—e.g., if the difficulties of accomplishing the task are unequal. The details of these moral concepts will differ on the specifics of the situation: they are about navigating a complex cultural environment (this is a point to which I return below). What matters here is just that, in the psychologically structured ecology of humans, where intentions and effort often matter more than outcomes, the ability to rely on abstract and complex moral concepts is crucial for biologically and culturally successful behavior.[5]

In turn, these conditions make the *innate priming* of these more complex and abstract moral concepts adaptive. This is due to the combined workings of two well-known phenomena: the Frame Problem and the Baldwin Effect. On the one hand, without having some innate expectations about how to evaluate others' (and our own) thoughts and actions—and their outcomes—there would simply be too many options to make this feasible in real time. If we had to come up, from scratch, with all the possible parameters to consider when assessing thoughts, actions, and their outcomes (day and time of the week? cloud cover? proximity to the equator?), it would be very difficult, if not impossible, to make these assessments in real time. Having richer concepts of *rights*, *duties*, *justice*, and so on allows for these inferences to get off the ground in the first place. Further—and this is where the Baldwin Effect comes in—since making these assessments is important for human living, even if abstract and complex moral concepts could be learned, humans with appropriate innate biases would have a leg up, as they would avoid the costs of learning, such as time and errors.

It is very important to be clear about the scope of the point here, though. The claim is not that all moral concepts are fully innate. This would be implausible, since many moral concepts are relatively new (*microaggression, systemic injustice, human rights*), and more generally, since many moral concepts are specific to certain cultural or other circumstances. Put the other way around: the point here is not to deny that many moral concepts are plausibly culturally (or sometimes maybe even individually) learned. Furthermore, and for similar reasons, the claim here is also not that humans are born with *very rich and specific* moral concepts of *justice*, *right*, *duty*, etc. We know that these concepts get enriched over time (and differently in different cultures), and it is again not plausible that highly specific versions of these concepts would be selected for, due to the fact that different enrichments are favored in different circumstances. On top of this—and this will be made clearer below—actually employing very complex and abstract (moral) concepts in decision-making is computationally very costly.

Rather, the point here is just that our—human—moral psychological starting point is plausibly innately *slightly* enriched from non-human animals. We have innate dispositions to form more abstract and complex moral concepts relative to what is true in non-human animals. Put differently, our moral *expectations*—what we look for when figuring out (by ourselves or with others' guidance) how to assess others' thoughts, actions, and dispositions—are innately more abstract and complex than what is found in non-human animals. Importantly, though, these differences remain ones of degree and may in fact often be quite small: there is no clear measure of the abstractness and complexity of human and non-human moral concepts. Moreover, even with innate predispositions, the reliance on such abstract and complex concepts as *justice*, *duty*, or *virtue* is difficult in real time. Hence, by themselves, innate differences between human and non-human animals cannot be used to make sense of claim 2 from the previous section—though they are part of that explanation (see also my Schulz, 2025).[6]

To see how human moral thinking can actually get off the ground, two further elements need to be introduced into the picture here: cultural learning and tool use. Both of these are widely recognized to be two key elements of distinctively human cognition—though there is dispute over exactly how to understand them or the precise centrality for explaining distinctively human cognition.[7] However, what is not in dispute is the fact that they are both well-supported empirically and theoretically. So, while there are differences in the form that cultural learning takes—explicit instruction, copying, apprentice learning, etc. (see, e.g., Boyd & Richerson, 1985, 2005; Henrich, 2015; Heyes, 2018; Sterelny, 2012; Tomasello, 1999)—there really is no dispute *that* it takes place and is key to explaining human thought and action (Tooby & Cosmides, 1992). Similarly, while there is dispute about whether human technological competence and ability for cultural learning have an innate basis and how different they are from what is found in non-human animals (see, e.g., Birch, 2021; Heyes, 2018; Shumaker et al., 2011; Sterelny, 2017), there is no question *that* human cultural learning and technological competence are particularly pronounced—whatever, exactly, their source is. (For more on this, see also my Schulz, 2025.) These enhanced human abilities for cultural learning and technological use and development are important here for two reasons.

On the one hand, the human ability for enhanced cultural learning makes it possible to enhance and expand our innate set of moral concepts. This point is relatively familiar, but a few remarks are still helpful to make here. By paying attention to the specifics of our culture, we can come to be attuned, say, to the importance (or lack thereof) of honor and how this is to be spelled out here. Similarly, cultural learning can tell us about the major duties to pay attention to in our society, exactly what these duties demand, and what the consequences are of non-conformance.[8]

On the other hand—and this is perhaps slightly less widely appreciated—cultural learning can also make available the development of cognitive tools like myths, stories, theories, and social institutions which can also support moral thinking—and cultural learning itself. The idea that cumulative cultural learning is central to human tool development is widely accepted (see, e.g., Tomasello, 1999). However, it is important to realize that these tools should not be seen to be restricted to physical tools (swords and ploughshares), but also cognitive tools that can support the use of complex and abstract concepts. This support can take many forms. In particular, humans create tools that support the memory demands that come with the reliance on complex and abstract concepts like those of *justice*, *desert*, and *duty*. For example, symbolic writing systems (like hieroglyphics) are useful as storage of who owns what when. In turn, this makes the determination of whether a given property arrangement is *just* much easier (indeed, it may not even be feasible without such records). Similarly, computational tools like calculators, sundials, and accounting conventions further support these kinds of determinations: for example, if we know that a property arrangement is (a) the result of well-functioning markets, and (b) from a just initial endowment, it becomes easier to determine whether it is *just*: we do not need to track every individual interaction, but can rely on the features of the market mechanism to evaluate the final outcomes.[9]

These points are especially helpful when it comes to the concepts of *virtue* and *vice*: reliance on these kinds of concepts famously comes with major cognitive demands (Hursthouse, 2010). Learning to recognize the virtues is generally thought to require much practice (often guided by others), and actually recognizing them in real-time adds further complexities and difficulties. Myths, writing, and music can help with exactly this. When assessing someone else's action, we can consult a written text (like the Analects) or a song (like Robyn's "Call Your Girlfriend") to tell us what the virtuous agent would do. This is especially important in the learning phase: it guides the shape of this learning and makes frequent practice easier. This is much like how written music can make it easier to learn to and actually play a piano piece: by seeing the notes in front of us, we can concentrate on controlling our hands, instead of also having to memorize what the next note is. In this way, the ability to build and refine cognitive tools makes it possible to overcome the costs associated with the use of complex and abstract concepts like *justice*, *virtue*, and *right*. As noted earlier, this thus solves one of the main problems associated with the use of these kinds of concepts.[10]

All in all, this thus explains component 2 of the evolution of distinctively human VBT: humans—but not non-human animals—are able to rely on abstract and complex moral concepts like *virtue*, due to the fact that they can rely on both cultural learning directly and cognitive technology (itself both a product and supporter of cultural learning) to enrich their innate moral expectations. Importantly also, since this enrichment is dependent on the specifics of the relevant culture, we do not find all humans relying on the same set of moral concepts in the same ways.

This then sets the stage for the explanation of component 1 of the evolution of distinctively human VBT: why humans sometimes, but not always, act on relatively few relatively stable character traits. The explanation of this component begins, like that of the previous component, by noting that much of our innate psychological endowment is shared with non-human animals. As noted in the previous section, there is no question that, at least when it comes to mammals, there is a lot of individual variation in the psychological machinery underlying decision-making (Barrows, 2011; Buss & Hawley, 2010).

Now, of course, variation is ubiquitous in the biological world, so as such this may not seem surprising. However, the point here is not that there is *some* variation in the cognitive machinery underlying animal decision-making, but that this variation is *sufficiently extensive* in mammalian psychology that it is often worth paying attention to when explaining their behaviors. The same, for example, does not seem to be true when it comes to many insects or fish. For example, cleaner ants of the same species act in ways that are largely indistinguishable from each other (Wilson, 1971). That said, as also noted earlier, even in most non-human mammals, there are limitations to the extent in which there is significant character variation: for example, dominance status in a social hierarchy generally trumps individual differences in many social animals—sub-dominant hyenas, wolfs, chimpanzees, etc. typically act in similar ways in many interactions with dominant individuals, making the appeal to any individual differences less helpful here.

All in all, what this means is that humans are likely to have inherited, from their common mammalian ancestors, a sufficiently complex psychological machinery that makes appealing to individual differences at least potentially very useful. However, they are *also* likely to have evolved from their common ancestor a lability in the importance of these individual differences: social circumstances—such as the existence of dominance hierarchies—may well swamp these individual differences. That is to say, humans inherited an action-generating machinery that is broad enough to encompass different ways of working in different bio-social environments: the evolution of human decision-making produced a pluralistic set of mechanisms. We sometimes rely on mappings between particular situations and particular behavioral responses, and sometimes infer what to do based on intermediate decision-principles—which may well be based on what the virtuous agent would do (Schulz, 2018).[11]

It is at this junction that the previous discussion concerning cultural learning and cognitive technology again becomes important. It is precisely these latter processes that, in humans, are key factors in our bio-social ecology that determine when we rely on situation-behavior mappings and when on (virtuous agent-based) rule-mediated behavioral inferences. That is to say, in some cultures, we are often encouraged to aim for behaviors that can be related to a relatively stable character, whereas in other cultures, we are often encouraged to act in line with (our assessment of) the situational demands (see also Zawidzki, 2018). In turn, this cultural learning is aided by cognitive technology like myths, writing, and music (which are themselves products of cultural learning) that can fine-tune which situations fall in either category. So, in some cultures, stories like the Mahabharata—where Arjuna follows his mother's commands, despite their being based on a misunderstanding, leading to his polyandrous marriage to Draupadi—might emphasize acting based on situational, social role expectations, at least when it comes to socially extremely important decisions like marriage. In other cultures, we might hear "I know that 5 years is a long time, and that times change; but I think that you'll find: people are basically the same" (Depeche Mode, *See You*), and are thus nudged toward acting based on stable character traits, even in similarly important social decisions.

This cultural learning and this cognitive technology are further important due to the fact that they are the fulcrum that connects the explanation of the two components of VBT that is at the heart of the present account. In particular, as just noted, it is through cultural learning and cognitive technology that we are able to rely on abstract and complex concepts like *virtue* and *vice* with which to assess and encourage character-based decision-making. This is important here, as, once enabled, the reliance on these kinds of concepts further alters the nature of cultural learning and cognitive technology toward character-based decision-making and thereby further entrenches this way of thinking and acting. In turn, this of course further makes the use of concepts like *virtue* and *vice* socially even more useful—and thus entrenches them even further, including leading to the construction of further cognitive technology underwriting both their use of and reliance on character traits in decision-making.[12]

In this manner, the two-sided puzzle at the heart of the chapter can be answered. The distinctively human reliance on concepts of *virtue* and *vice* can be explained by the fact that character-based decision-making is (1) evolutionary enabled and (2) encouraged in certain cultures, which (3) paves the ground for the usefulness of the deployment of concepts like virtue and vice—which, in turn, (4) entrench character-based decision-making. Put differently, in humans, advances in virtue thinking can advance character-psychology—which makes virtue thinking more useful, creating a positive feedback loop. Figure 1.1. summarizes this overall feedback loop, as well as (some of) the subsidiary feedback loops making it up.

Implicit in this picture is the fact that small, unplanned differences in the available cognitive technology—e.g., myths such as the Mahabharata, which, as a byproduct, encourage character-based decision-making—may

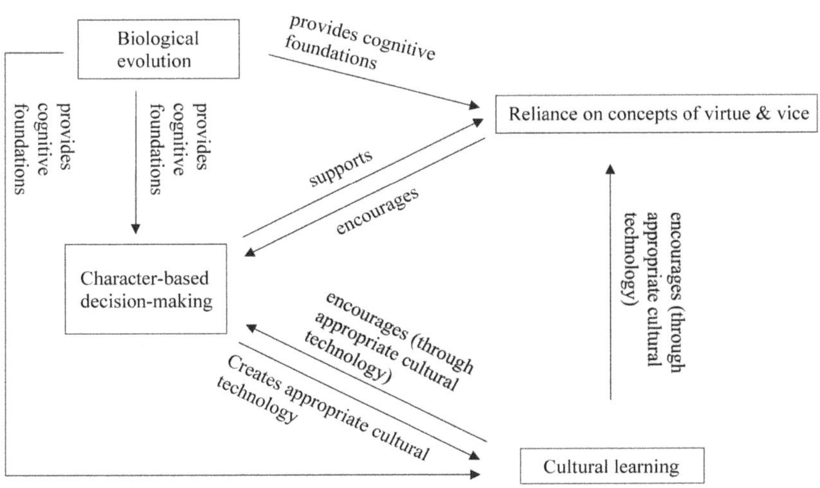

Figure 1.1 The feedback loop surrounding the evolution of VBT

set the feedback process in motion, leading to ever bigger changes. Put the other way around, the reason why VBT is more prevalent in some cultures than others is simply that, in some cultures, small initial cultural or technological biases set the feedback loop off in one direction, whereas in other cultures, the opposite happened.

Now, there are two important points to note about this feedback loop. First, this feedback loop is a natural outgrowth of an *independently well-supported* account of the evolution of distinctively human cognition more generally (Schulz, 2025). That is, the first key point of this chapter is really that the evolution of VBT is a paradigm case of how, in the human lineage, innate representational expectations, cultural learning, and (cognitive) technology interact in complex ways to bring about uniquely human, but still culturally specific ways of thinking. In turn, this congruence between common virtue ethical thinking and evolutionary reasoning about human cognition supports *both* of these further.[13]

Second, it is key to note that implicit in this feedback loop is the fact that it is not just the appeal to virtues that is variable and determined by culture and technology—it is also character-based action in and of itself. The latter is more general than what has been noted in the literature so far. So, while it has been noted before that virtue-based education can make virtue-based action more likely (Alfano, 2013; Burnyeat, 1980; Hursthouse, 2010; Miller, 2014), what is new here is that virtue-based education can make character-based action in general more likely. That is, it is not just the case that being told that you are courageous in fact makes you courageous (Alfano, 2013), but also that, in a culture where there are lots of references to *courage*, it is more likely that you act based on stable character traits in the first place—whether or not that is courage specifically. Moreover, this also goes the other way around: more references to character-based action make it more likely that people appeal to virtues to assess these character traits. So, when Depeche Mode emphasizes that people do not change much, and that this matters to how they act, this also makes it more likely that we wonder about how to *assess* people's actions *as* character-determined responses to the world.[14] To better understand some of the details of this account of the evolution of VBT, and to provide further support to it, it is now useful to consider and clarify some of its implications.

1.4 The Feedback Account of the Evolution of Virtue-Based Thinking: Some Applications and Clarifications

There are some intrinsic features of this feedback-based account of distinctively human VBT that it is worthwhile to make explicit here. First and most obviously, the account is designed to explain cultural variation in both character-based decision-making and the appeal to virtues in moral assessment. Whether character-based decision-making—and, thus, the feedback loop involving concepts of virtue and vice—is culturally favored depends on the details of the relevant cultures (including the technological possibilities the latter make available, as well as their particular histories). The present account thus avoids the twin pitfalls of either not being able to account for the existence of these two aspects of human cognition or overgeneralizing the pervasiveness of these ways of thinking. The avoidance of this dilemma is a key feature that any compelling explanation of the evolution of VBT needs to have.

Second, the account here further underwrites the contention that we should not assume that humans either have or do not have stable character traits (see also Elster, 2015; Harman, 1999; Miller, 2014; Mischel, 1968; for a related point, see also Doris, 2002). Rather, the right thing to think here is that humans can *develop* the disposition to frequently and systematically act on stable character traits—if this development is culturally encouraged (e.g., through appropriate cognitive technology). In this way, the chapter advances the interactionist view concerning character traits and situational features (Doris, 2002; Elster, 2015) by making the specifics of that interaction clearer: the account here laid out shows that how much weight is given to the situation or the character traits, how often, and which character traits exactly is a culturally mediated issue that depends crucially on the nature of the available cognitive technology (among other things).

Third and relatedly, the chapter makes clear that virtues and vices—and stable character traits more generally—are unlike some other mental states, such as emotions, motivations, and some representational expectations. The latter are pan-human and thus, in a sense, a "deeper" part of our cognitive nature. Virtues, vices, and stable character traits, by contrast, are shallower features of our thinking; while the disposition to develop these—and significant individual variation in decision-making—is pan-human, their actual development in their systematic and frequent use is not.

The chapter also makes some more specific predictions. In particular, it is a direct implication of the account here laid out that by changing the available cognitive technology in a given culture, it becomes possible to make VBT more prevalent. Of course, the idea of using books, say, to teach about virtues is not new and something that is widely accepted in the virtue ethical space (see, e.g., Hursthouse, 2010; Miller, 2014). However, what the present account implies is that this phenomenon in fact scales up to the entire cultural level. That is to say, to the extent that there are changes in the cognitive technology that is widely relied on in a given culture—also for other, non-directly VBT-related reasons, such as supporting economic or trading decisions—it is plausible that the culture as a whole shifts more toward (or, as the case may be, away from) character-based decision-making.

In this context, it is also useful to forestall two possible misunderstandings about this prediction and the account more generally. First, one might think that appeals to virtues have been common in Chinese and other Asian societies for a long time, but—at least arguably—these are societies that are less individualistic in outlook. This would be a problem if the account here predicted that there would be a correlation between VBT and individualism.

Here, however, it is important to recognize that the present account does not, in fact, predict this: on the present theory, the individualism/holism distinction is not predicted to correlate with the virtue/non-virtue distinction. Indeed, the former distinction concerns how much emphasis there is on individual vs. the larger society in decision-making and evaluation; the latter is about whether the person comes first or if the situation does. These are logically independent: we could reason about what to do by appealing to what the virtuous person would do—and note that the virtuous person would put society first. (This is, in fact, an aspect of Confucian thought: Angle & Slote, 2013.) We could also map situations to behavioral responses, but this mapping could be tailored towards favoring the individual. Rather, then, the prediction the account makes is that more talk of virtue will create more character-based action. This is a new prediction that has not been as clearly tested (though see also Smith & Apicella, 2020).

Second, one might be concerned that key aspects of the VBT/non-VBT distinction are sub-cultural rather than cultural. For example, while there are aspects of the situationism/character-psychology debate that concern cultural differences (Smith & Apicella, 2020), many seem to concern differences within a culture. For example, it seems that people *just in the United States* (say) do not clearly always act on stable character traits (Doris, 2002). Given this, one may wonder how the present account can explain—and how this squares with the above prediction concerning cultural differences.

In response, the key thing to note is that the account here does not just appeal to cultural learning, but also the available cognitive technology. This matters, as technology cuts more finely than culture—though it interacts with the latter. So, as noted earlier, cultural learning is a key driver that makes cognitive technology available, but *using* the technology is still up to a person and may thus differ sub-culturally. So, while some cultures make virtue and character-based cognitive technology (like appropriate songs and stories) more widely available than others, even among the former, the uptake of that technology can differ widely across people and situations. So, even if someone is likely to be disposed to think that a virtuous person would help someone in need, when pressed for time, they may not be able to access that belief—even if it is supported with appropriate stories (e.g., of the good Samaritan).

1.5 Conclusion

In all, what this chapter has tried to argue is that distinctively human VBT is not just a tool for *assessing* character traits—it is also for *creating* them in the first place.[15] This explains the two sets of puzzling features of distinctively human VBT: humans sometimes, but not always, act largely on relatively few relatively stable character traits, and they sometimes, but not always, morally evaluate the relatively few relatively stable character traits they and others act on (or fail to do so). At the heart of this explanation is a positive feedback loop comprising (1) innate endowments for abstract and complex moral thought and heavily character-based decision-making, (2) cultural learning, and (3) cognitive technology such as myths and appropriate social institutions. The upshot of this positive feedback loop is the idea that cultural encouragement toward virtue-based assessment of character traits partly underwrites the development of character-based acting in the first place—which further reinforces the cultural encouragement toward virtue-based assessment of character traits. The chapter concludes by noting some important implications of this feedback-based account, such as the fact that character-based decision-making should be seen as not deeply opposed to situational decision-making, but merely as different cultural developments of the same multifaceted psychological endowment. Similarly, the chapter predicts that changes to the cognitive technology prevalent in a given culture can lead to big changes in how people make decisions and morally assess these decisions. In short, in humans, virtue makes character, and character makes virtue.

Notes

1 As indicated earlier, in what follows, I will also not speak to any metaethical or normative ethical questions—e.g., whether character assessments in terms of virtue and vice are morally correct, or what, exactly, the nature is of virtue ethical discourse.

2 In the background here are the contentions that (a) not everything that is uniquely human is also universally human and (b) not everything that is uniquely human is different in kind from what is found among non-human animals. For an extended defense of these assumptions, see, e.g, Schulz (2025), but for present purposes, the remarks in the text are sufficient.

3 It is tempting to see this difference as resulting from general differences in the complexity of the psychologies of these different groups of animals. However, since assessing this complexity systematically is notoriously difficult, I will not pursue this further here. However, see also Section 1.3.

4 To be sure, on some conceptions, the truly virtuous agent would be able to recognize virtues directly; see, e.g., Burnyeat (1980); Hursthouse (2010). Importantly, though, a typical human might not. Most importantly, even a truly virtuous person need not be able to directly observe all the relevant features of the situation. It is just that they would immediately recognize a person as virtuous *once they would know about the latter's mental states*—which may be far from easy.

5 Clearly, reliance on "moral" concepts is not the only thing that is important. The reliance on psychological "mindreading" concepts (motivations, opinions, emotions, etc.), as well as concepts concerning the biological, physical, and (perhaps) non-moral social world (such as *cause*, *number*, and *property*), allows for the representation of and efficient reaction to a large number of different kinds of causal, social, and moral relationships (Schulz, 2018, 2025). However, since the focus here is on moral concepts, I will not discuss these other concepts further here (for more on these, see Spelke, 2022; and my Schulz, forthcoming). Distinctively human cognition is also partly characterized by some specific low-level features—i.e., cognitive traits that are non-representational, generally sub-personal, and generally automatic (Fodor, 1983; Kahneman, 2003; Leitgeb, 2004). Among these are the fact that humans have a large range of aspects of their environment that they can focus their attention on (from musical instruments to distant planets), their patience (they can engage in costly tasks for years for uncertain rewards years in the future), and their fine motor control (they can manipulate guitars and complex machinery). On top of this, there are cognitive suites that combine low- and high-level features. Since these issues are not directly relevant to the discussion of this chapter, though, I will not discuss them further here; for more on them, see my Schulz (2025).

6 Another way to put this is that, while it *is* plausible that we have a moral grammar (Mikhail, 2011), grammar needs to be seen as having lots of open variables. Indeed, like with language, the specifics of the actual moralities (as with actual languages) used are culturally and sub-culturally determined.

7 Cultural learning is sometimes seen to be the central source of human cognition (see, e.g., Tomasello, 1999; Henrich, 2015; Heyes, 2018)—but this does not do justice to the above facts or the data supporting some innate concepts; see also my Schulz (2025).

8 This point may raise the question of whether the difference between virtue- and character-based cultures and those that are not virtue- and character-based can be accommodated just in terms of normative differences (see, e.g., Henrich, 2020; Westra et al., 2024). As pointed out in the text, this is reasonable *in the sense* that normative differences may be part of the mechanism that drives this difference. However, as is made clearer below, the present account also goes beyond purely normative thought. On the one hand, the point here is not just that people just follow different norms; the idea is that different people might actually be more or less character-driven "underneath" their behavior (this may or may not be a major difference, depending on how human norm psychology is spelled out). On the other hand, the present account emphasizes technology more than most norm-focused views do.

9 Another relationship between cognitive technology and the application of moral concepts may come in the fact that the *crafting* of tools comes with internal standards, which reinforce or create the conditions for normative thinking in the first place. For more on this, see Witt (2023) and Birch (2021).

10 Of course, cognitive technology also supports cultural learning itself. This thus creates a further positive feedback loop here. See also below and my Schulz (2025).

11 This distinction among cultures—with some more character-driven and some less so—is related to, but different from, other such distinctions: e.g., concerning the "tightness" of the culture (Elster & Gelfand, 2021), its kinship intensity (Schulz et al., 2019), or its individualism (Nisbett et al., 2001). While the explanation of why some cultures are more character-driven than others may

also be transferred to these other cases, there may still be some specific differences that need to be taken into account. At any rate, this is the topic for a future investigation.

12 Merritt (2000) defends a Humean virtue ethics on somewhat related grounds (though the details of the argument there are quite different).

13 A quick remark concerning the account of Heyes (2018) is helpful here. The latter might suggest that some cultures have developed a "cognitive gadget"—a culturally evolved set of psychological dispositions for "character thinking." What makes the account defended in this chapter different is that it places greater emphasis on the above feedback loop, with significantly more appeal to innate representational components. However, a more detailed comparison does indeed have to be left to another occasion (see also Schulz, 2025).

14 In the same way, this is slightly different from, e.g., Carruthers's (2006) view of mindreading: it is not just that we "turn inwards" something we normally focus on others. It is rather that the availability of tools for the *assessment* of (others' or one's own) characters makes character-based action-generation more likely—and the reverse.

15 In this sense, therefore, the account of the chapter can also be expressed with the help of the idea of a self-fulfilling prophecy. Some cultures have a collective belief that individuals will and should behave in character-based ways, and they develop a range of evaluative tools and practices that bring those kinds of behaviors into existence and make the beliefs about them true.

Bibliography

Alfano, M. (2013). *Character as moral fiction* (new ed.). Cambridge University Press.

Andrews, K. (2015). *The animal mind: An introduction to the philosophy of animal cognition*. Routledge.

Andrews, K. (2020). Naïve normativity: The social foundation of moral cognition. *Journal of the American Philosophical Association*, 6(1), 36–56. https://doi.org/10.1017/apa.2019.30

Andrews, K., Fitzpatrick, S., & Westra, E. (2024). Human and nonhuman norms: A dimensional framework. *Philosophical Transactions of the Royal Society B: Biological Sciences*, *379*(1897), 20230026. https://doi.org/doi:10.1098/rstb.2023.0026 10.1098/rstb.2023.0026

Angle, S. C., & Slote, M. (Eds.). (2013). *Virtue ethics and Confucianism*. Routledge.

Athanassoulis, N. (2016). A defence of the Aristotelian virtue of magnificence. *Journal of Value Inquiry*, *50*, 781–795.

Barrows, E. M. (2011). *Animal behavior desk reference: A dictionary of animal behavior, ecology, and evolution* (3rd ed.). Taylor & Francis. https://books.google.com/books?id=2vf7ZbLENzEC

Birch, J. (2021). Toolmaking and the evolution of normative cognition. *Biology and Philosophy*, *36*(1), 1–26.

Boyd, R., & Richerson, P. (1985). *Culture and the evolutionary process*. University of Chicago Press.

Boyd, R., & Richerson, P. (2005). *The origin and evolution of cultures*. Oxford University Press.

Brown, R. L. (2018). Animal traditions: What they are, and why they matter. In K. Andrews & J. Beck (Eds.), *The Routledge handbook of philosophy of animal minds* (pp. 362–371). Routledge.

Burnyeat, M. (1980). Aristotle on learning to be good. In A. Rorty (Ed.), *Essays on Aristotle's ethics* (pp. 69–92). University of California Press.
Buss, D. M., & Hawley, P. H. (Eds.). (2010). *The evolution of personality and individual differences*. Oxford University Press.
Campbell, R., & Kumar, V. (2012). Moral reasoning on the ground. *Ethics*, *122*(2), 273–312.
Carruthers, P. (2006). *The architecture of the mind*. Oxford University Press.
Carter, G., & Wilkinson, G. (2013). Food sharing in vampire bats: Reciprocal help predicts donations more than relatedness or harassment. *Proceedings of the Royal Society B*, *280*, 20122573.
Cokelet, B. (2023). Competitive virtue ethics and narrow morality. *Philosophical Studies*, *180*(12), 3567–3591. https://doi.org/10.1007/s11098-023-02048-9
Darwall, S. (1996). *The second person standpoint: Morality, respect, and accountability*. Harvard University Press.
Doris, J. (1998). Persons, situations, and virtue ethics. *Noûs*, *32*(4), 504–530.
Doris, J. (2002). *Lack of character: Personality and moral behavior*. Cambridge University Press.
Elster, A., & Gelfand, M. J. (2021). When guiding principles do not guide: The moderating effects of cultural tightness on value-behavior links. *Journal of Personality*, *89*(2), 325–337. https://doi.org/10.1111/jopy.12584
Elster, J. (2015). *Explaining social behavior: More nuts and bolts for the social sciences*. Cambridge University Press.
Fodor, J. (1983). *The modularity of mind: An essay on faculty psychology*. MIT Press.
Geertz, C. (1973). Thick description: Toward an interpretive theory of culture. In *The interpretation of cultures: Selected essays* (pp. 3–30). Basic Books.
Haidt, J., & Graham, J. (2007). When morality opposes justice: Conservatives have moral intuitions that liberals may not recognize. *Social Justice Research*, *20*(1), 98–116. https://doi.org/10.1007/s11211-007-0034-z
Hammerstein, P. (2003). Why is reciprocity so rare in social animals? A protestant appeal. In P. Hammerstein (Ed.), *Genetic and cultural evolution of cooperation* (pp. 83–94). MIT Press.
Harman, G. (1999). Moral philosophy meets social psychology: Virtue ethics and the fundamental attribution error. *Proceedings of the Aristotelian Society*, *99*, 315–331.
Henrich, J. (2015). *The secret of our success: How culture is driving human evolution, domesticating our species, and making us smarter*. Princeton University Press.
Henrich, J. (2020). *The WEIRDest people in the world*. Farrar, Straus and Giroux.
Heyes, C. M. (2018). *Cognitive gadgets: The cultural evolution of thinking*. Harvard University Press.
Hursthouse, R. (2010). *On virtue ethics*. Oxford University Press.
Joyce, R. (2005). *The evolution of morality*. MIT Press.
Kahneman, D. (2003). Maps of bounded rationality: Psychology for behavioral economics. *American Economic Review*, *93*, 1449–1475.
Leitgeb, H. (2004). *Inference on the low level*. Kluwer.
Machery, E., & Mallon, R. (2010). Evolution of morality. In *The moral psychology handbook* (pp. 3–46). Oxford University Press. https://doi.org/10.1093/acprof:oso/9780199582143.003.0002
Merritt, M. (2000). Virtue ethics and situationist personality psychology. *Ethical Theory and Moral Practice*, *3*(4), 365–383.

Mikhail, J. (2011). *Elements of moral cognition: Rawls' linguistic analogy and the cognitive science of moral and legal judgment*. Cambridge University Press.
Miller, C. (2014). *Character and moral psychology*. Oxford University Press.
Mischel, W. (1968). *Personality and assessment*. John Wiley and Sons.
Mischel, W. (1973). Toward a cognitive social learning reconceptualization of personality. *Psychological Review*, *80*, 252–283.
Monsó, S., & Andrews, K. (2022). Animal moral psychologies. In M. Vargas & J. Doris (Eds.), *The Oxford handbook of moral psychology* (pp. 388–420). Oxford University Press.
Nisbett, R. E., Peng, K., Choi, I., & Norenzayan, A. (2001). Culture and systems of thought. *Psychological Review*, *108*(2), 291–310.
Richerson, P., & Boyd, R. (2005). *Not by genes alone*. University of Chicago Press.
Sarkissian, H., Park, J., Tien, D., Wright, J. C., & Knobe, J. (2011). Folk moral relativism. *Mind & Language*, *26*(4), 482–505. https://doi.org/10.1111/j.1468-0017.2011.01428.x
Schulz, A. (2018). *Efficient cognition: The evolution of representational decision making*. MIT Press.
Schulz, A. (2020). Enhancing thoughts: Culture, technology, and the evolution of human cognitive uniqueness. *Mind & Language*. https://doi.org/10.1111/mila.12320
Schulz, A. (2025). *It's only human: The evolution of distinctively human cognition*. Oxford University Press.
Schulz, A. (forthcoming). Core knowledge and its role in explaining uniquely human cognition: Some questions. *Behavioral and Brain* Sciences, *47*, e141. doi:10.1017/S0140525X23003199
Schulz, J. F., Bahrami-Rad, D., Beauchamp, J. P., & Henrich, J. (2019). The Church, intensive kinship, and global psychological variation. *Science*, *366*(6466), eaau5141. https://doi.org/doi:10.1126/science.aau5141
Shumaker, B. R. W., Walkup, K. R., & Beck, B. B. (2011). *Animal tool behavior: The use and manufacture of tools by animals* (rev. ed.). Johns Hopkins University Press.
Smith, K. M., & Apicella, C. L. (2020). Hadza hunter-gatherers disagree on perceptions of moral character. *Social Psychological and Personality Science*, *11*(5), 616–625. https://doi.org/10.1177/1948550619865051
Spelke, E. (2022). *What babies know*. Oxford University Press.
Stanford, P. K. (2018). The difference between ice cream and Nazis: Moral externalization and the evolution of human cooperation. *Behavioral and Brain Sciences*, *41*(e95).
Sterelny, K. (2003). *Thought in a hostile world: The evolution of human cognition*. Wiley-Blackwell.
Sterelny, K. (2012). *The evolved apprentice: How evolution made humans unique*. MIT Press.
Sterelny, K. (2017). Artifacts, symbols, thoughts. *Biological Theory*, *12*(4), 236–247.
Stich, S. (2018). Do we really externalize or objectivize moral demands? *Behavioral and Brain Sciences*, *41*, e113. https://doi.org/10.1017/S0140525X18000183
Suhler, C. L., & Churchland, P. (2011). Can innate, modular "Foundations" explain morality? Challenges for Haidt's moral foundations theory. *Journal of Cognitive Neuroscience*, *23*(9), 2103–2116. https://doi.org/10.1162/jocn.2011.21637
Tomasello, M. (1999). *The cultural origins of human cognition*. Harvard University Press.

Tooby, J., & Cosmides, L. (1992). The psychological foundations of culture. In J. Barkow, L. Cosmides, & J. Tooby (Eds.), *The adapted mind* (pp. 19–136). Oxford University Press.
Westra, E., & Andrews, K. (2022). A pluralistic framework for the psychology of norms. *Biology and Philosophy*, *37*(5), 1–30.
Westra, E., Fitzpatrick, S., Brosnan, S. F., Gruber, T., Hobaiter, C., Hopper, L. M., Kelly, D., Krupenye, C., Luncz, L. V., Theriault, J., & Andrews, K. (2024). In search of animal normativity: A framework for studying social norms in non-human animals. *Biological Reviews of the Cambridge Philosophical Society*, *99*(3), 1058–1074. https://doi.org/10.1111/brv.13056
Whiten, A., & Byrne, R. W. (Eds.). (1997). *Machiavellian intelligence II: Extensions and evaluations*. Cambridge University Press.
Wilson, E. O. (1971). *The insect societies*. Harvard University Press.
Witt, C. (2023). *Social goodness: The ontology of social norms*. Oxford University Press.
Zawidzki, T. W. (2018). Mindshaping. In S. Gallagher, A. Newen, & L. D. Bruin (Eds.), *The Oxford handbook of 4E cognition* (pp. 735–754). Oxford University Press.

2 The Cooperative Account of the Rise of Virtues

Brian Robinson

2.1 Introduction

Which traits count as virtues and why? Any account of the evolution of virtues must explain why some traits come to be regarded as virtues, why some are not, and why others cease to be so regarded. The importance of the question is for a simple reason. There is not (and never has been) a single agreed-upon list of virtues. Different virtue ethicists in different times and societies have produced different lists with at best limited overlap with the traits praised by other societies. To take one example, humility was not among Aristotle's list of virtues, though it has been much praised throughout the history of Christianity and is now receiving new attention and acclaim. To attribute these differences merely to cultural relativity is by itself not an explanation, for it fails to elucidate the function of virtues or to explain why a society prizes the traits it does.

This chapter offers an opening foray into this question. I will offer an account of the functions of virtues based on cooperation. This idea is not revolutionary. Others have tilled this ground before, and readers of this volume will likely encounter other chapters engaging with the link between cooperation and virtue. On this front, however, the specific contributions of this chapter are twofold. First, I advance the cooperation-virtues link by drawing on recent work in comparative anthropology that has produced the theory known as Morality as Cooperation (MAC) (Alfano et al., 2024; Curry, 2016; Curry, Mullins, et al., 2019; van Leeuwen et al., 2024). Second, I use humility as a test case, exploring why it has gone from ignored (or even disdained) to praised as a virtue.

2.2 Cooperation and Morality

Let's begin with cooperation, which requires a bit more explication than simply "working together." Cooperation is an end result, the product of

DOI: 10.4324/9781003632641-4

people acting cooperatively. So, instead of beginning with cooperation, it is better to begin with defining cooperative acts.[1]

> **Cooperative act:** One agent acting in a manner necessary for contributing to a common goal shared by another agent(s).

A few points to note about this definition. First, a cooperative act is performed by one person but involves at least one other. Two (or more) people must share the same goal and so shared intentions to contribute to that goal (Ritchie, 2020). Second, this definition does not require that the common goal be achieved. In a stag-hunt-esque scenario, you may act cooperatively by arriving on time to a meeting with me, while I forget and fail to show up at all. Cooperation was not achieved, but that fact does not entail that you did not act cooperatively. Third, it does not require that all agents benefit from the shared goal. I can cooperate with you toward the end of saving your life by sacrificing my own. That said, mutual benefit is very common.

With these preliminaries out of the way, we can turn to cooperation's connection with morality. Seeing some connection is nothing new. The relationship between morality and cooperation has been discussed by Aristotle, Augustine, Aquinas, Hume, Butler, Russell, Rawls, and Mackie, among others. (See Curry, 2016, p. 28 for details on this list.) More recently, Boehm (2012) develops an evolutionary account of the origin of morality, casting back to the prehistoric origins of virtue in early hunter-gatherer societies. Boehm argues for an evolutionary anchoring of morality in cooperation and punishment. Ridley (1996) spins a similar account, albeit with more attention to game theory. Tomasello and Vaish (2013) likewise argue for the origin of morality in cooperation, this time going back to our pre-human ancestors based on contemporary analyses of other non-human greater primates. Sterelny (2021) casts all the way back to the Pleistocene era to connect morality and cooperation.

We need not go back so far here.[2] I am not concerned at present with where the concept of virtues first emerged, but rather how ideas about which traits are virtues evolve over time. I take as a given that people across at least most cultures have some concept of morally good character traits and attribute those traits to people. But which traits they esteem are not static or uniform. The question is how we should account for these differences and changes.

The most recent foray into the examination of cooperation and morality is the theory MAC developed by Oliver Curry, which offers several compelling advancements. The first is that Curry (2016) presents a taxonomy of kinds of cooperation (or more accurately, cooperative acts). The different kinds are based on the different kinds of social interaction problems that societies face. The kinds of problems (which correspond to kinds of cooperation) are "(1) the allocation of resources to kin; (2) coordination to mutual advantage; (3) social exchange; and conflict resolution through contests featuring (4) hawkish displays of dominance and (5) dovish displays of deference; (6) division of disputed resources; and (7) recognition of prior possession" (Alfano et al., 2024, p. 2). Societies have faced these kinds of problems throughout history. One of the central claims of MAC is that societies will endorse cooperative responses to these problems. They do not have to be the same responses, however. Different societies will face different versions of these problems, have different ones more immediate than others, and have different cultural, technological, environmental, etc. resources available to them to respond to these problems. But whatever solutions a culture comes up with, MAC predicts that each culture will regard that solution as morally good. Consequently, MAC asserts that "each of these types of cooperation gives rise to a corresponding type of morality: (1) family values, (2) group loyalty, (3) reciprocity, (4) heroism, (5) deference, (6) fairness, and (7) property rights" (Alfano et al., 2024, p. 2).

The second advancement that MAC makes is in testability. MAC makes a clear prediction, namely, that societies will regard these seven kinds of cooperative acts as morally good. To test that prediction, Curry, Chesters, et al. (2019) examined an impressive 400 anthropological records from a random sample of 60 societies from across the globe spanning 300 years and diverse socioeconomic conditions. They found 962 observations of societies' moral evaluation of one or more of the seven kinds of cooperation. Of these, 961 (99.9%) of the cooperative acts were regarded as morally good. The only exception is where two kinds of cooperative acts were in conflict: the Chuuk people regarded hawkish acts of bravery in stealing to be good, despite violating property rights. Subsequent research extended these findings to all 256 societies in the Human Relations Area Files (eHRAF) using machine-reading analysis (Alfano et al., 2024).[3]

These results strongly indicate that people correlate morality with cooperation. Furthermore, they point to a fundamental universality to morality beneath surface-level cultural differences. Various societies have different moral standards because they found different cooperative solutions, because the problems of cooperation they faced were different (in kind or degree), because they had different resources available with which to solve them, or because they happened to have hit upon different viable solutions to the same problem.

While I appreciate the clarity on the nature of cooperation and find the empirical evidence impressive, it is important not to overstate the normative implications of this work. It would be incorrect to conclude that for any act, it is morally good if and only if it is cooperative. Morality and cooperation are not equivalent. MAC is not endorsing the claim that all cooperative acts are morally good. Cooperating with Nazis or oppressors is not; this claim is not what MAC's findings support. It would be a radical overreach to conclude either that all cooperation is moral or that all morality is cooperative.[4] Finally, we should not take these findings to support the maximally normative claim that we have a duty to do whatever is the most cooperative act we can. Absent a larger axiology that explains what goals are morally good that we should cooperate toward achieving (as well as explaining with whom to cooperate, how, and when), MAC does not provide a deontic formula to guide our everyday actions.

It is beyond the scope of this chapter to fully unpack all the mistakes of these claims. For now, it suffices to highlight the strong support MAC gives to a fundamental connection between morality and cooperation. This connection can be expressed in the first of the central claims of this chapter:

> **1 The Cooperation-Morality Connection**
>
> Generally speaking, cooperative acts are considered moral because they enable societies to get along (though they will not always redound to the agent's own benefit).

Admittedly, this is a ceteris paribus claim. There are exceptions. The idea, however, is that cooperative acts tend to help more often than not, which is why individuals and societies endorse them, even going so far as regarding them as moral (typically). That said, the connection between cooperation and morality is imprecise. Individuals or societies can forget this connection, retaining a moral endorsement of behaviors that long ago stopped being cooperatively useful. (This will be a point I return to in Section 2.4.)

2.3 Cooperation and Virtues

What MAC does allow for is an interesting and so far largely unexplored expansion in virtue ethics. Curry et al. (2022) offer a partial foray into virtue ethics, though with some limitations I will address below. Unlike more action-centered approaches to normative ethics, virtue ethics does not first and foremost attempt to explicate what we should do, focusing instead on what kind of person to be and what character traits we should cultivate. That said, just as it would be an oversimplification to simply equate morality with cooperation, so too would it be an exaggeration to claim simply that all virtues are cooperative.

Before proceeding, it is worthwhile to go over some basics about what virtues are. We need not settle all the details about the nature of virtues (regarding which there are many debates); a few relevant points will suffice. First, virtues are stable character traits. If someone is honest, generous, or humble, they are disposed to act honestly, generously, or humbly at least most of the time (Annas, 2011; Slote, 2001; Snow, 2011).[5]

Second, since traits are dispositions to act, to have a virtue does not mean that one acts a certain way all the time. To be generous does not mean one is always acting generously at all times, always trying to maximize their generosity. Rather, to have the virtue of generosity means that one will act generously at the right time, in the right way, to the right person(s), and for the right reason(s). For me to give my meager savings to a billionaire would not be a mark of my generosity. It would show either that I lack the understanding of who to be generous to or that my empathy is incorrectly focused. It would be evidence against me having the virtue. Traditionally, virtue ethicists have regarded it as impossible to explain precisely what constitutes the right time, in the right way, to the right person(s), and for the right reason(s). This is the non-codifiability thesis (Annas, 2011, p. 19).

Third, virtues are learned (or cultivated, if you prefer); they are not innate. What is most important here, however, is how they are learned. Traditionally, virtue ethics has held that one cannot be an autodidact of virtue. You cannot train yourself. The development of virtues requires other people. Accounts of how much one's virtue depends on others vary from merely requiring moral exemplars to emulate to a requirement of being brought up in a virtuous society populated (mostly) by virtuous people, or even that one's virtue is externalized to be partially constituted by one's company (Robinson & Alfano, 2016). This component of virtue theory is why in the next section I will focus on societal views about the moral status of traits.

Fourth, I define virtues as traits that serve a function.[6] They help people live well. For now, it is best to set aside the question of what constitutes living well, be it eudaimonia, a virtuous life, a life of pleasure, the realization of one's goals, etc. The point is that whatever living well amounts to, being virtuous is supposed to help. This is meant as a minimalist claim with respect to the connection between virtues and thriving.

The Stoics took an extreme view, maintaining that virtue is both necessary and sufficient for living well. Aristotle regarded virtue as merely necessary. My minimalist claim is committed to neither necessity nor sufficiency. It is possible to have virtue and not to live well (at least on some version of what that means). Likewise, this minimalist claim leaves open the possibility that one can live well without being virtuous.[7] That said, just because these are possibilities does not mean that they are likely or common. On this minimalist view, virtues are those traits that are a society's current best bets for how to live well. You are more likely than not to have life (or a portion of it) go well if you have a virtue. For example, with regard to when and how much of the truth to speak, being an honest person is more likely to work out in one's favor overall than any other alternative trait (or being inconsistent in one's behavior due to incontinence). This is not to say that being a liar never worked out; it has on occasion. There are no guarantees. Being virtuous does not ensure that life will go well, but it does load the dice in your favor. The minimalist claim is that virtuous people live well more often than not and that individual virtues are those specific traits that tend to help people live well. So, if one is deciding what sort of person to be, being virtuous is supposed to be the better bet.

Fifth, virtues are morally good traits. It is not just that virtues tend to help you do well in life; they are morally good.

The reason for emphasizing these points is that we are now in a position to connect MAC and virtues. Let's assume that MAC is generally correct: there is a fundamental connection between morality and cooperation. Since virtues are supposed to be morally good traits, then virtues too must be cooperative in some way. In a moment, I will propose an overview of a cooperative account of virtues to serve as this connection. But as a preliminary, it is worth discussing why focus on virtues, more specifically why it helps to convert MAC into a virtue-based format.

There are two motivations. First, attributing traits to others is very useful. We all know and interact with lots of people. So, we are constantly required to decide with whom to cooperate, when to cooperate, and how to cooperate. Trustworthiness is a major factor in making these decisions. Yet we cannot constantly and carefully assess each person's trustworthiness, not even in general to say nothing of their trustworthiness in a specific domain (e.g., telling the truth, empathy). So we attribute character traits (including virtues) as a way to handle a lot of this work (Snow, 2011). Attributing virtues to others is a useful heuristic to track trustworthiness.

The second motivation goes back to the non-codifiability thesis (Annas, 2011). Again, this is the idea that it is a fool's errand to attempt to specify all the conditions under which one should tell the truth or give one's time or money to others, for example. It is easier and more useful to people as they go about their daily lives to tell them to be honest and generous. A similar claim can be made about cooperation. Cooperation is good, in general. We should cooperate, but not all the time, with everyone, or toward any end. Linking cooperation and virtue will allow us to sidestep the problem of developing a cooperative algorithm for morality.

With these motivations clear, I can lay out what I take to be the connection between MAC and virtue ethics, which is the second main claim of this chapter:

> **2 Virtues as Cooperative Dispositions**
>
> Generally speaking, virtues are considered morally good because they are recognized as cooperative character traits.

(2) is a ceteris paribus claim. It is not a perfect, exceptionless rule. Sometimes, a trait is regarded as a virtue even though it is not cooperative. I will explain exceptions of this sort in the latter portion of the next section. (2) comes with two sub-claims, the first of which is

2a The Action-Disposition Link

> A virtue is a disposition to perform a particular type of cooperative action.

Honesty, for example, is a virtue. Someone with this trait is disposed to perform a particular type of cooperative action, namely, telling the truth. Truth telling is generally cooperative and the right thing to do. A key point regarding (2a) is that it provides a principled mechanism for two important distinctions. First, virtues are distinguished from one another based on what kind of cooperative act the virtue disposes a person to perform. Second, virtues are distinguished from other character traits (i.e., vices or amoral traits) by being a disposition to perform certain kinds of actions, i.e., cooperative actions, which are separately defined apart from the trait itself. This approach avoids the circular reasoning of defining a virtue in terms of an action type and that action type in terms of the virtue.

An advantage of connecting MAC and virtue theory is that doing so allows redress of a shortcoming of focusing on cooperative acts alone. Recall from (1) that cooperative acts often, but not always, redound to the agent's benefit. If the level of analysis is individual actions, this is the outcome we should expect. If, however, the level of analysis shifts to a more long-term one, we can consider the benefits of actions in aggregate. In other words, we can consider whether traits are in one's favor in the long run, smoothing over occasional disadvantageous outcomes. This leads to the second sub-claim:

2b The Long-Run Advantage Dispositions

> A virtue is a cooperative trait that is, in the long run, more to one's benefit than incontinency or other competing traits.

This claim needs unpacking. For any trait, consider what other traits one could have instead because they are mutually exclusive. For example, one cannot be both courageous and cowardly, both honest and dishonest, humble and arrogant. These are competing traits. Contrarily, one could be brave and dishonest. So, these are non-competing traits. For any set of competing traits, a person can have only one of them (or be incontinent, i.e., behave inconsistently).[8] The question becomes which of these competing traits (if any) will serve one best in life and help one live well. (2b) claims that whatever trait serves one the best is a virtue. As we will see, this determination does not have to remain fixed. It is not a necessary truth.

There is a final point of connection between MAC and virtue theory. Recall that Curry distinguishes (at least) seven different kinds of cooperative acts. Since virtues are defined in terms of cooperative behaviors, we should expect a similar seven-fold distinction for virtues, which is precisely what the third main claim of this chapter asserts:

3 The Seven Core Virtues

> The seven core types of cooperative behavior correspond to seven core virtues.

Identifying precisely which virtues are the core seven merits some attention. Curry et al. (2022) refer back to the seven types of morality (family values, group loyalty, reciprocity, heroism, deference, fairness, and property rights) and then identify two to four "virtues" corresponding to each type of morality (see their Table 1). For present purposes, there are, however, two problems with their list of virtues. First, not all of them are virtues. For example, they regard "duty of care" as one of the virtues corresponding to family values. As the name indicates, however, duty of care is a duty, not a virtue. The same can be said for their supposed virtues of unity, awe, and equality (among others). They are morally good, but they are not virtues. The second problem is that they put humility as a virtue corresponding to the moral category of deference. As I will discuss in the next section, the contemporary notion of humility that has received so much academic attention in the last 25 years is very different from this older conception of humility as a kind of deference or submitting to one's superiors. (All that said, the truly interesting claim of this chapter is that these seven types of morality are moral elements, which invokes a chemical metaphor meant to convey that they can be combined to create more complex moral compounds. Whether two virtues combine to create a more complex virtue is an intriguing notion meriting more consideration.)

For my part, I think that the list can be boiled down to seven virtues: filial piety, loyalty, trustworthiness, courage, deference, justice, and respect for property. Perhaps this is too simplistic and seven families of virtues will be better. Regardless, for present purposes, the key point is that we should expect near universal cultural endorsement of these seven virtues (albeit not always by these names or their cognates). The reason that these seven traits are regarded as virtues is that they are stable dispositions to behave in ways that consistently deploy cooperative responses to a society's problems of getting along.

2.4 How Virtues Change and Differ

By building on MAC (and in particular, by linking it with virtue theory), we are able to engage in "anthropological realism" (Flanagan, 2017), as opposed to armchair cultural relativism. In this section, I will sketch how that work can proceed, focusing on how cooperative virtues can be used to explain how views on which traits are virtues change over time and differ from one society to another.

The next major claim was introduced at the beginning of the chapter, namely, that there has never been a universal list of which traits are virtues.

> **4 The Variability Thesis**
>
> The moral status of traits varies from one society to another and can change over time within a society.

At first glance, (4) seems at odds with (3), which asserts cultural uniformity on seven virtues. In truth, they are not in tension, but to see why requires first explaining what I mean by 'moral status.' The moral status of a trait is based on three components: *valence*, *hierarchy*, and *expression*. The moral valence of a trait is whether it is a virtue, vice, or neither (i.e., amoral).

Hierarchy is about where a trait falls in the moral ordering of traits. Which traits are morally more important? For example, justice and modesty are both commonly regarded as virtues, but typically justice is much higher in the hierarchy than modesty. This point is not to suppose that an individual or a society has a definitive and precise ordering of how good every virtue is (and likewise how bad every vice is). But it does entail that not all virtues are equal. Some virtues clearly have a higher moral status than others within a society; i.e., they are regarded as more morally important to develop and exhibit.

Lastly, expression has to do with how a trait is expressed. Consider two generous people. One expresses their generosity primarily by giving their time, while the other gives money. Both are generous but express that trait differently, likely because they have different life circumstances affecting how they can be generous. At a social level, one society may focus on expressing a virtue one way, while another emphasizes expressing it quite differently. This difference may be a historical fluke or dependent on the differences in the cooperative problems the two societies faced or the resources available to each society. A difference in expression of a virtue does not entail that two societies regard the same trait with a different valence or have hierarchical differences.

With this picture of moral status in mind, we can turn to sub-claim (4a):

4a Core Virtue Consistency

Societal differences regarding the moral status of the seven core virtues will only relate to hierarchy and expression, not valence.

Some societies will rank humility above courage, for example, while other societies will have the reverse ranking. But they both regard humility and courage with the same valence, i.e., as virtues. We should expect all seven to be regarded as virtues. Likewise, (4a) tells us we should expect differences in how different societies express these same core virtues. What counts as an expression of filial piety in one society can be very different from another. The same virtue will be expressed with different behavior.

In a moment, we will come to consider why and how these differences evolve. For now, let's consider other, non-core virtues. It should be uncontroversial that there are more than seven virtues. Setting aside the moral molecules hypothesis, what should be said about these non-core virtues? Claim (2) still applies. Non-core virtues are cooperative. Just because one society employs one virtue as a cooperative response to its recurrent social interaction problems does not mean that another will do likewise. This is sub-claim (4b):

4b Non-Core Virtue Variability

The moral status of all other traits (besides the core seven virtues) will vary in terms of valence, hierarchy, and expression from one society to another and can change over time within a society.

This claim (4b) reiterates the observation at the start of the chapter. There is no universally agreed-upon list of virtues. Actually, it goes further, as the reader can see. It is not just that different societies have different lists of which traits are virtues; they rank and express traits differently as well. I do not take (4b) to be a controversial claim, but making it explicit is important for what comes next.

And what comes next is the central question, the one we have been building toward from the start of the chapter:

Q Why do different societies develop in such a way that the moral status of traits differs between societies and also within societies over time?

In other words, what accounts for the differences that emerge over time and between societies? A naïve cultural relativism simply answers that the differences are due to the fact that the societies are different. But that is putting the cart before the horse. A large part of what distinguishes societies from one another is that they have different moral codes and that they view the moral status of traits differently. This naïve relativism fails to actually answer the question.

Given what has been asserted so far—namely, (1)–(4)—I hope my answer to (Q) will be readily apparent by now.

5 The Problems-and-Resources Account

> The Variability Thesis (4) is true because different societies, as they develop, face different recurrent social interaction problems and have different resources available to them to develop cooperative responses to those problems.

This fifth, central claim of the chapter has two parts: *problems* and *resources*. Let's start with the problems. Recall that for MAC (along with many other thinkers such as Hobbes and Hume), morality is fundamentally about solving social problems. We regard as moral those responses we come up with over time to help us all get along (at least well enough). This chapter advances the claim that these responses are best understood as virtues, instead of distinct, individual behaviors. Someone with a virtue will behave that way consistently, while someone without that virtue will behave inconsistently, thereby lowering utility and diminishing others' trust that this person will be cooperative next time.

The catch is that the problems differ from one society to another (or over time for the same society). For example, a society with abundant water will face different problems of cooperation than a desert society with very limited access to water. To be clear, I do not mean that the problems are so different in that they do not conform to the seven, game-theoretical kinds of cooperative problems Curry identifies. Rather, the point is that different problems get emphasized more than others due to context or take different specific forms. One society may deal with scarcity of water while another deals with scarcity of labor. Furthermore, situations change. The situations and cooperative problems faced by the same society change over time. We should expect these differences to produce different moral codes, which leads us to the responses side of (5).

What kinds of cooperative responses a society can or does deploy to deal with a particular recurrent social interaction problem can vary. Just as a society's particular context shapes what recurrent social interaction problems they face, so too does their context shape what cooperative responses they can deploy. The factors affecting and shaping those possible responses are vast, too vast to survey here. In short, they are shaped by the material, social, technological, and intellectual resources available to a society. Wealthier societies can throw money at problems that other societies could not, for example. Such a response could permit or even endorse consumption and consumerism or alternatively generosity and philanthropy. The key point is that the resources available to a society will shape how they cooperatively respond to recurrent social interaction problems. These differences, in turn, will produce differing views on the moral status of different traits. And just as with problems, the difference of responses is not purely inter-societal. They can and do occur within the same society over time as a society's resources change. Lastly, it is interesting to note that changes in the size of a society's population can be classified as changes in both the kinds of recurrent social interaction problems and in the viable responses to them.

This brings us to the last major claim of this chapter. Change in a society's views about the moral status of traits does not occur overnight. There is often a significant lag (assuming views change at all).[9]

6 The Cultural Lag Principle

> Social views on the moral status of traits usually change slowly and are not always up to date.

There are really two claims here that are worth breaking apart.

6a The Delayed Adoption Effect

> There will typically be a significant delay between when changes either in terms of recurrent social interaction problems faced or in terms of what cooperative responses are viable on the one hand and when those solutions are adopted on the other hand (including when corresponding views on the moral status of relevant traits).

6b The Historical Residue Effect

> Just because a society views a trait as morally good does not mean that it is now a cooperative solution to a recurrent social interaction problem, but it does entail that it at least used to be.

Though separate, the reasons for (6a) and (6b) are intermingled. Though social contexts change—thereby altering what problems are faced, what solutions work, or what solutions are available—social views often do not track these changes in real time. Cooperative responses that used to work well might not any longer. Alternatively, newer, even better solutions may emerge.

All that said, however, there is often a significant lag time. It takes time for individuals to recognize that these changes have taken place. To see why, imagine an ideal scenario wherein everyone is (a) immediately aware of what sort of person their neighbors are, (b) immediately aware of how well they themselves are doing, (c) constantly comparing how well they are doing to their neighbors, and (d) willing to change very quickly to be like any neighbor that does better than them. Alexander and Skyrms (1999) model such a world and find that justice spreads rapidly through a society.

In the real world, however, none of (a)–(d) is true. People do make rapid assessments of other people's character, but there is good reason to be skeptical of their accuracy (Harman, 1999; Ross, 1977). We do not have ready access to how well people are doing, even financially, to say nothing of their psychological well-being or even eudaimonia. Though there is a degree of "keeping up with the Joneses," we are not comparing ourselves to our neighbors with anything like the constant vigilance in Alexander and Skyrms' idealized model. We all have a lot more going on in life. Lastly and most importantly, one's character cannot change as quickly as the model imagines. Character traits are deeply rooted and slow to change. An unjust person cannot inculcate the virtue of justice in a day, a week, or a month just because they recognize that their just neighbor benefits from being just.

None of these observations is meant to disparage the model or the conclusions of Alexander and Skyrms (and similar work in modeling norm adoption, such as Axelrod, 1984; Bicchieri, 2005), but rather to emphasize that the modeling presents changes in morality occurring far more rapidly than occurs in reality. The models correctly identify the reasons and mechanisms of changes in moral views; it is just a great deal more messy and slower in the real world. For a person to modify their views on the moral status of traits requires engaging in a costly re-examination of themselves and their deeply held beliefs. Because it is so costly in time and effort, people do not do this re-examination often (nor should they), even if they would benefit from doing so. Change in moral views at the social level emerges from lots of individuals engaging in such re-examination, so social change in views on the moral status of traits will be even slower.

In short, a simple way to view a system of morals—including views on the moral status of various traits—is that it is a long-running, multi-layered kludge (Stich, 2006). Societies keep adopting quick and dirty, inelegant but cooperative responses to recurrent social interaction problems. Those responses get baked into the structure of society over time. They are hard to change. When new problems emerge or an old kludge stops working, a society will more often than not slap a new kludge on their moral system (some new cooperative solution that works well enough for now given all that has come before). These new changes will be adopted slowly over time. And older, vestigial moral views often linger after they cease being useful.

2.5 Case Study: Humility

If this account of how virtues evolve is on the right track, it should be able to explain a test case. Humility serves as a great test case since views on its moral status have varied over time and across societies.

Let's start with a few basic facts. First, humility was not a virtue for the ancient Greeks. It was not among Aristotle's virtues in *Nicomachean Ethics*, wherein he instead praised the μεγαλόψυχος (*megalopsychos*), i.e., the magnanimous or proud person (Chappell, 2021; MacIntyre, 1981, pp. 182–184; Newman, 1982; Roberts & Cleveland, 2017).[10] Second, humility came to be seen as a virtue by many medieval Christian thinkers, such as Augustine and Aquinas (Nadelhoffer & Wright, 2017; Pinsent, 2021). Third, as Christianity became less common in Europe, humility ceased to be regarded as a virtue as evidenced by Hume, Nietzsche, and Sidgwick (Newman, 1982; Richards, 1988). Meanwhile, those who kept some sort of faith held onto their esteem of humility, with Kant and Kierkegaard serving as chief examples (Lippett, 2017; Newman, 1982). Fourth, in contemporary WEIRD societies (Western, educated, industrialized, rich, and democratic) (Henrich et al., 2010), humility is having something of a renaissance, with many secular and religious figures alike investigating and praising it (Alfano et al., 2021). Wright (2019) even goes so far as to regard humility as the first virtue, i.e., the virtue necessary for developing other virtues. This is a short history of the evolution of humility in Western societies, which Bellitto (2023) describes in far greater detail.

This history, however, encounters a problem that my account above skipped over. My cooperation-based account of the evolution of virtues took traits to be stable things. In other words, for any trait *T*, it is the same trait between societies and over time; it is merely views on the moral status of *T* that differ. This sort of analysis overlooks the issue of how views on what a trait is can shift over time. Humility certainly seems to exhibit such change, at least prima facie.

There is, however, another way of looking at this change in how people understand humility specifically (or any other virtue). Consider the case of the term 'Madagascar.' In the debate of the last century over competing theories of reference, Evans reminds us of the fact that the name 'Madagascar' once referred to an inland portion of Africa, but now (due to a confusion by Marco Polo) refers to an island off the eastern coast of Africa (Evans, 1973). This example produced a cottage industry of debating whether the name 'Madagascar' shifted its referent or whether a new name of identical spelling and pronunciation was created anew with a different referent. I will set that debate aside for simplicity's sake and adopt the reference switching view.

The lesson we can apply from Evans is that the term 'humility' has shifted over time, picking out different (albeit related) traits over time and in different cultures.[11] The contemporary notion of humility picks out a trait that is different from but related to the one praised by Aquinas, for example. The task of explaining the evolution of humility becomes the joint task of explaining why the trait each society praised worked and perhaps also why the modern, WEIRD notion of humility did not. Lamentably, such a task is too grand for this section to cover in detail (though I recommend Bellitto (2023), who provides plenty of material for this job), and so I offer a mere sketch here.

Recall the four basic facts about humility at the start of the section. Put together, they demonstrate that there has not been a consensus on the moral status of humility even in Western societies. This is precisely what we should expect based on the preceding cooperative account of the evolution of virtues, particularly given (3) and (4). Humility is not a core virtue. Therefore, we should not expect universal agreement between cultures that it is a virtue. Some should—and do—reject it.

By 'humility,' I mean roughly the trait of being inattentive to one's meritorious or praiseworthy qualities or accomplishments. It is not about being humbled (i.e., brought low) or putting oneself down or having a false sense of one's own worth. It is the trait of not being overly concerned with yourself (often because of focus on others). I take this trait to be the contemporary referent of 'humility' in WEIRD societies. It has received considerable attention and endorsement of late (Garcia, 2006; Nadelhoffer et al., 2017; Roberts & Wood, 2007; Robinson, 2021; Robinson & Alfano, 2016; Tangney, 2000). Of course, that is not what has always been meant by 'humility' (or *humilitas* in Latin; there was no Greek term).

As noted, the modern conception of humility I just mentioned was not a virtue for the ancient Greeks. Claim (5) provides a few plausible explanations. It might be that the modern notion of humility simply was not useful in that context. One possible reason that humility was not a useful concept could be that ancient Greek societies were honor-based. Aristotle called honor the greatest external good (*NE* 1123b18, 1124a17). In an honor-based society, being inattentive to one's worth is less useful than being keenly aware of it as the *megalopsychos* would be. The *megalopsychos* is better prepared to merit honor and get it. Alternatively, the modern, historically laden notion of humility may not have been possible in ancient Greek societies. Our modern notion of humility is built up from what came before. Aristotle and Plato (lacking the same history we have today) did not have the modern conception available to them.

What then of the medieval Christians? The cooperative account of the evolution of virtues would claim that if they have different views on humility (which they did), then it must be because they faced different problems or had different responses available. They definitely faced different problems than the ancient Greeks. Between them, views on the divine (and humanity's relation to the divine) shifted radically. A significant issue they confronted (especially the devout) was how humans, as wretched sinners, should relate to God. If humans were wretched and redeemable only by divine intervention, there is nothing to be proud of; there is no place for the *megalopsychos*. Thus, humility (or *humilitas*) becomes a virtue. But note that Aquinas' notion of virtue is not the same as the modern notion. For medieval Christians, humility was being humbled, being brought low before God Almighty (Nadelhoffer & Wright, 2017).

Centuries later, as significant portions of Europe lost their faith and shifted to post-Christian societies, views on the moral status of humility (or the medieval notion of it anyway) shifted again. As (6) would lead us to expect, the change in views was not immediate. The cooperative account of the evolution of virtues predicts the change to be slow and contentious. And that is precisely what we find. Atheists tended to find little value in humility. Hume contemptuously calls it a monkish virtue, and Nietzsche had not but disdain for it. Claim (5) explains that this disregard for humility was because of thinkers like Hume and Nietzsche; the medieval notion of humility was a response to a problem that no longer existed (at least in the minds of non-Christians) and therefore was unworthy of moral praise.

What then of modern WEIRD societies? Why would they now be reviving and revising humility? Analyzing one's own society, including understanding the problems faced and the responses available, is a tricky undertaking. Modern sociologists and anthropologists can do a far better job of this than I. Presumably, however, humility serves a useful purpose; it enables us to get along better either directly or indirectly. In fact, that is what many have recently asserted. Humility is the foundation that makes other virtues possible (Wright, 2019). An increasingly crowded world facing problems entirely new to humanity requires new responses. Humility (in the modern sense) appears to be one such cooperative response that we believe helps. This shift in views on the moral status of humility required something new, however. It required dusting off the medieval notion of humility and updating it to something related but new that will address the problems of today.

There are two last points to be mindful of in closing. First, our modern WEIRD notion of humility should not be regarded as the definitively right conception of humility for all time, but rather as the one that is well suited to our time and place. Future people in different contexts, facing different recurrent social interaction problems, may still use the term 'humility' (or a term etymologically derived from it) to pick out a slightly different trait, one that serves their needs. Second, this section is a gesture toward a fuller explanation of the evolution of humility. It is my hope that others—using the central claims highlighted in this chapter—can fill that story in more explicitly and more widely (especially in terms of non-Western societies) than I could do here.

2.6 Conclusion

This chapter has provided a sketch—hopefully, a plausible one—of the evolution of virtues. Virtues evolve as stable patterns of behavior that respond to persistent problems of cooperation faced by a society, using the material, technological, social, and intellectual resources that society has on hand. In short, virtues evolved because they work. Yet, since different societies face different problems and have different resources, different traits evolve to be virtues. What works in one society may not work (or work as well) elsewhere. That said, this is still merely a sketch. A great deal needs to be filled in and expanded upon. Nevertheless, it is my hope that this chapter serves as an aid for the chapters to come. Some chapters may fill in parts of this story. Others may point out its flaws and shortcomings. Either way, the reader of this volume will (I hope) be now more prepared to engage with the excellent and enthralling chapters that follow this one.

In conclusion, I would note that the old adage, "When in Rome, do as the Romans do," turns out to have some merit. When in Rome, one will face different problems than those faced at home. Different patterns of behavior will work (i.e., help one thrive) than back home. Some of the virtues of home can be different than the virtues in Rome. And not only is that good but it is also precisely what we should have expected all along.

Notes

1 Cooperation and cooperative acts have been conflated. For example, Bohn and Carter run them together, stating, "Cooperation is a behavior that is adaptive because it provides a benefit to another individual. Reciprocal altruism, or reciprocity, is cooperation that is conditional on receiving help from the recipient" (2021, p. 6504).

2 That said, I am drawing on these works (Boehm, 2012; Ridley, 1996; Sterelny, 2021), but will not generally make explicit connections with them for the sake of brevity.

3 MAC has also been compared to Haidt's Moral Foundations Theory (MFT). Curry, Chesters, et al. (2019) report four studies in which MAC's questionnaire was a better fit with participants' moral attitudes than MFT's questionnaire, thereby raising further problems for the ad hoc and poorly performing (Davis et al., 2016, 2017) Moral Foundations Theory.

4 In public presentations of MAC, Curry himself lamentably endorsed both that all morality is cooperative and that the right thing to do is whatever is the most cooperative (Curry, 2019).

5 I am only concerned with the concept of virtues, not whether people actually possess them. So, to say that virtues are stable traits is to say that when a virtue is attributed to a person, the claim is that this person is disposed to regularly and reliably behave in a certain way. Regardless of whether anyone actually possesses robust global character traits (Alfano, 2013; Doris, 2002; Harman, 1999a), people make these attributions and seem to believe that people have virtues. The focus of this chapter is on how the concept of virtue is anchored in cooperation and why people regard different traits as virtues in different contexts.

6 To be clear, this is not an appeal to Aristotle's function argument. That argument focuses on the function of an object so that when an object realizes its function, the object has reached a good state (or just is good). For humans, Aristotle claims, the function is rational activity, which he then uses to define well-being and virtue. My minimalist claim runs in the other direction. I am explicitly declining to comment on what constitutes the good for a human being or what its function is. I am defining virtues (not humans) in terms of a function, much like a functionalist in philosophy of mind defines a mental state in terms of its function.

7 It can and has been argued that neither of these is possible. I am not here arguing one way or the other. Rather, the point of the minimalist view is to minimize the theoretical commitments taken on board, at least for now. Ruling this out is not necessary to explain the evolution of virtues, the task before us in this chapter.

8 A reviewer asked this good question: "What if one is humble at home, but not at work?" The point is that it is unclear when someone *has* a trait. There are three options. First, one might follow the situationists (Doris, 2002) in rejecting global character traits. So, one could regard humble-at-home and arrogant-at-work as two separate, domain-specific traits. These are not competing traits. The second option is to claim that such a person, who is only humble at home, but not at work, lacks the more global trait of humility. Third (and perhaps less intuitive), one could claim that humility is a low-fidelity virtue so that being humble at home is often enough to qualify this person as humble. For the sake of this chapter, any of these options will work.

9 It is reasonable to expect that sometimes a society will retain elements of its moral code (including traits they regard as virtues), refusing to change even when the traits are no longer useful. The most common case is when a society defines itself in terms of its moral code. Having a particular trait becomes a signifier of social membership. Moral codes as sources of social identity will tend to be more intractable.

10 One might raise an objection that Aristotle did discuss humility in *NE* IV.3 and that Plato presents Socrates as humble. First, regarding Plato, I have no objection to taking contemporary notions of humility or intellectual humility and applying them back to Plato's presentation of Socrates. But that is a far cry from the anachronistic claim that Plato himself regarded Socrates as humble. There is no textual evidence for that claim. Second, regarding humility in *NE* IV.3, one must be careful of working in translation, as interpreting Aristotle here is a fraught enterprise. Some translations do render μεγαλόψυχος (*megalopsychos*) as 'pride' and μικρόψυχος (*micropsychos*) as 'undue humility'. Others hew

more closely to the literal translations of 'great-souled' and 'small-souled' or 'mean-spirited'. It is hard to argue from this portion of *Nicomachean Ethics* that Aristotle had anything like the modern concept of pride or humility. The Greeks did have the notion of hubris, but that was roughly what we could call excessive pride.

11 For discussion of how the modern concept of humility relates to similar concepts by different names, see Metz (2021) on *ubuntu* in African philosophy, McLeod (2021) on humility in early Confucianism, Vasalou (2021) on Islamic philosophy, and Bommarito (2021) on Buddhist humility. A consistent theme is that the modern concept does not precisely translate, but that highly related traits were valued as morally good.

Works Cited

Alexander, J., & Skyrms, B. (1999). Bargaining with neighbors: Is justice contagious? *The Journal of Philosophy*, *96*(11), 588–598.

Alfano, M., Cheong, M., & Curry, O. S. (2024). Moral universals: A machine-reading analysis of 256 societies. *Heliyon*, *10*(6), e25940. https://doi.org/10.1016/j.heliyon.2024.e25940

Alfano, M., Lynch, M. P., & Tanesini, A. (2021). Introduction. In M. Alfano, M. P. Lynch, & A. Tanesini (Eds.), *The Routledge handbook of philosophy of humility* (pp. 1–6). Routledge.

Annas, J. (2011). *Intelligent virtue*. Oxford University Press.

Axelrod, R. (1984). *The evolution of cooperation*. Basic Books.

Bellitto, C. M. (2023). *Humility: The secret history of a lost virtue*. Georgetown University Press.

Bicchieri, C. (2005). *The grammar of society: The nature and dynamics of social norms*. Cambridge University Press.

Boehm, C. (2012). *Moral origins: The evolution of virtue, altruism, and shame*. Basic Books.

Bohn, K., & Carter, G. (2021). Reciprocal altruism and cooperation for mutual benefit. In T. K. Shackelford & V. A. Weekes-Shackelford (Eds.), *Encyclopedia of evolutionary psychological science* (pp. 6504–6507). Springer International Publishing. https://doi.org/10.1007/978-3-319-19650-3_1207

Bommarito, N. (2021). Buddhist humility. In M. Alfano, M. P. Lynch, & A. Tanesini (Eds.), *The Routledge handbook of philosophy of humility* (pp. 236–244). Routledge.

Chappell, S. G. (2021). Humility among the ancient Greeks. In M. Alfano, M. P. Lynch, & A. Tanesini (Eds.), *The Routledge handbook of philosophy of humility* (pp. 187–201). Routledge.

Curry, O. S. (2016). *Morality as Cooperation: A Problem-Centred Approach* (pp. 27–51). https://doi.org/10.1007/978-3-319-19671-8_2

Curry, O. S. (2019, November). *How to Be Good* [Video]. TEDx Prague. https://www.youtube.com/watch?v=lO-VxCBOju8

Curry, O. S., Alfano, M., Brandt, M. J., & Pelican, C. (2022). Moral molecules: Morality as a combinatorial system. *Review of Philosophy and Psychology*, *13*(4), 1039–1058. https://doi.org/10.1007/s13164-021-00540-x

Curry, O. S., Chesters, M. J., & Lissa, C. J. V. (2019). Mapping morality with a compass: Testing the theory of 'morality-as-cooperation' with a new questionnaire. *Journal of Research in Personality*, *78*, 106–124. https://doi.org/10.1016/j.jrp.2018.10.008

Curry, O. S., Mullins, D. A., & Whitehouse, H. (2019). Is it good to cooperate? Testing the theory of morality-as-cooperation in 60 societies. *Current Anthropology*. https://doi.org/10.1086/701478

Davis, D. E., Dooley, M., Hook, J. N., Choe, E. J. Y., & Mcelroy, S. (2017). The purity/sanctity subscale of the Moral Foundations Questionnaire does not work similarly for religious versus non-religious individuals. *Psychology of Religion and Spirituality*, *9*, 124–130.

Davis, D. E., Rice, K. G., Van Tongeren, D. R., Hook, J. N., DeBlaere, C., Worthington, E. L., & Choe, E. J. Y. (2016). The moral foundations hypothesis does not replicate well in Black samples. *Journal of Personality and Social Psychology*, *110*(4), e23–e30. https://doi.org/10.1037/pspp0000056

Doris, J. (2002). *Lack of character: Personality and moral behavior*. Cambridge University Press.

Evans, G. (1973). The causal theory of names. *Aristotelian Society Supplementary Volume*, *47*(1), 187–208.

Flanagan, O. (2017). *The geography of morals: Varieties of moral possibility*. Oxford University Press.

Garcia, J. L. A. (2006). Being unimpressed with ourselves: Reconceiving humility. *Philosophia*, *34*(4).

Harman, G. (1999). Moral philosophy meets social psychology: Virtue ethics and the fundamental attribution error. *Proceedings of the Aristotelian Society*, *99*, 315–331.

Henrich, J., Heine, S. J., & Norenzayan, A. (2010). The weirdest people in the world? *Behavioral and Brain Sciences*, *33*(2–3), 61–135. https://doi.org/10.1017/S0140525X0999152X

Lippett, J. (2017). Kierkegaard's virtues? Humility and gratitude as the grounds of contentment, patience and hope in Kierkegaard's moral psychology. In S. Minister, J. A. Simmons, & M. Strawser (Eds.), *Kierkegaard's god and the good life* (pp. 95–113). Indiana University Press.

MacIntyre, A. (1981). *After virtue*. University of Notre Dame Press.

McLeod, A. (2021). Humility in early Confucianism. In M. Alfano, M. P. Lynch, & A. Tanesini (Eds.), *The Routledge handbook of philosophy of humility* (pp. 245–256). Routledge.

Metz, T. (2021). Humility and the African philosophy of ubuntu. In M. Alfano, M. P. Lynch, & A. Tanesini (Eds.), *The Routledge handbook of philosophy of humility* (pp. 257–267). Routledge.

Nadelhoffer, T., & Wright, J. C. (2017). The twin dimensions of the virtue of humility: Low self-focus and high other-focus. In W. Sinnott-Armstrong & C. B. Miller (Eds.), *Moral psychology, volume 5: Virtue and character* (pp. 309–342). The MIT Press.

Nadelhoffer, T., Wright, J. C., Echols, M., Perini, T., & Venezia, K. (2017). Some varieties of humility worth wanting. *Journal of Moral Philosophy*, *14*(2), 168–200. https://doi.org/10.1163/17455243-46810056

Newman, J. (1982). Humility and self-realization. *The Journal of Value Inquiry*, *16*, 275–285.

Pinsent, A. C. (2021). Aquinas on humility and relational greatness. In M. Alfano, M. P. Lynch, & A. Tanesini (Eds.), *The Routledge handbook of philosophy of humility* (pp. 202–211). Routledge.

Richards, N. (1988). Is humility a virtue? *American Philosophical Quarterly*, *25*(3), 253–259.

Ridley, M. (1996). *The origins of virtue: Human instincts and the evolution of cooperation*. Penguin Books.

Ritchie, K. (2020). Minimal cooperation and group roles. In A. Fiebich (Ed.), *Minimal cooperation and shared agency*. Springer.

Roberts, R. C., & Cleveland, W. S. (2017). Humility from a philosophical point of view. In E. L. Worthington Jr., D. E. Davis, & J. N. Hook (Eds.), *Handbook in humility: Theory, research, and applications* (pp. 33–46). Routledge.

Roberts, R. C., & Wood, W. J. (2007). *Intellectual virtues: An essay in regulative epistemology*. Oxford University Press UK.

Robinson, B. (2021). "I'm SO Humble!": On the paradox of humility. In *The Routledge handbook of philosophy of humility* (pp. 26–35). Routledge.

Robinson, B., & Alfano, M. (2016). I know you are, but what am I? Anti-individualism in the development of intellectual humility and Wu-wei. *Logos & Episteme*, *4*, 435–459.

Ross, L. (1977). The intuitive psychologist and his shortcomings: Distortions in the attribution process. *Advances in Experimental Social Psychology*, 10, 173–220. https://doi.org/10.1016/S0065-2601(08)60357-3

Slote, M. (2001). *Morals from motives*. Oxford University Press.

Snow, N. E. (2011). *Virtue as social intelligence: An empirically grounded theory*. Routledge.

Sterelny, K. (2021). *The Pleistocene social contract: Culture and cooperation in human evolution*. Oxford University Press.

Stich, S. (2006). Is morality an elegant machine or a kludge? *Journal of Cognition and Culture*, *6*(1–2), 181–189. https://doi.org/10.1163/156853706776931349

Tangney, J. P. (2000). Humility: Theoretical perspectives, empirical findings and directions for future research. *Journal of Social and Clinical Psychology*, *19*(1), 70–82. https://doi.org/10.1521/jscp.2000.19.1.70

Tomasello, M., & Vaish, A. (2013). Origins of human cooperation and morality. *Annual Review of Psychology*, *64*, 231–255. https://doi.org/10.1146/annurev-psych-113011-143812

van Leeuwen, F., van Lissa, C. J., Papakonstantinou, T., Petersen, M. B., & Curry, O. S. (2024). Morality as cooperation, politics as conflict. *Social Psychological Bulletin*, *19*. https://doi.org/10.32872/spb.10157

Vasalou, S. (2021). Humility in the Islamic tradition. In M. Alfano, M. P. Lynch, & A. Tanesini (Eds.), *The Routledge handbook of philosophy of humility* (pp. 225–235). Routledge.

Wright, J. (2019). *Humility as a Foundational Virtue* (pp. 146–174). https://doi.org/10.1093/oso/9780190864873.003.0007

3 The Evolution of Malice and Envy

Rory Smead and Patrick Forber

3.1 Introduction

Humans are impressive cooperators. Indeed, cooperation and altruism have received most of the attention from scholars working to understand the origins of our social behavior, and for good reason. Many think that cooperation creates the feedback loop that explains our peculiar mix of sociality and cognition. As we began to reap the benefits of cooperation, we developed more sophisticated cognition that enabled more complex and more rewarding cooperation (see, e.g., Sterelny, 2012). Some see morality as a specific adaptation for maintaining our cooperative alliances (Stanford, 2018; Tomasello, 2016). Yet we can be nasty as well as nice. Revenge and vendettas are as familiar to us as friendship and collaboration, and we routinely recognize behaviors as motivated by spite. Even if we have the character to avoid acting on our vicious impulses, it is difficult to deny that we occasionally experience the desire to spite others.

However, despite its colloquial familiarity, spite is considered an evolutionary rarity. Evolutionary investigations into social behavior define four broad categories based on the costs and benefits of a social interaction for the actor and recipient. Biological altruism is behavior that costs the actor but benefits the recipient, whereas biological spite is behavior that costs the actor and harms the recipient (Hamilton, 1964a). Behaviors that benefit the actor are mutually beneficial when they help the recipient and selfish when they harm the recipient. These costs and benefits are measured using evolutionary fitness. In standard selection dynamics, mutually beneficial and selfish behavior are easy to explain. Natural selection favors behaviors that benefit the actor, especially if they also cost the recipient. Altruism and spite, on the other hand, present more of an evolutionary puzzle because they cost the actor. Prima facie, natural selection should ruthlessly eliminate these behaviors from the population. Yet many consider altruism common and spite rare. Why?

The key to the evolution of altruism is correlated interaction (Hamilton, 1964a; Skyrms, 1996). If the benefits of cooperation are preferentially

DOI: 10.4324/9781003632641-5

directed toward fellow cooperators, these behaviors can evolve and be stable in a population by excluding the free riders from the rewards of altruistic behavior. There are a variety of mechanisms in the world that generate correlated interaction, including recognition, kinship, spatial structure, and partner choice. Given the proliferation of mechanisms and the theoretical vindication, biologists see cooperation across the tree of life. Spite, on the other hand, did not receive the same vindication. As Hamilton famously remarked, when discussing the impacts of evolved spite, "[s]uch trends of selection in small populations, if they occur at all, must act like a final infection that kills failing twigs of the evolutionary tree" (Hamilton, 1971, pp. 199–200). Even so, spite is theoretically possible, and there exist some apparent cases of spite in the natural world (West & Gardner, 2010). Perhaps some bacteria (Inglis et al., 2009) and birds (Robinson, 1988) engage in spite. If the behavior rarely evolves, what should we make of the human capacity to spite?

Vicious behaviors are sometimes characterized as cases where our prosocial tendencies were overwhelmed by selfish motives. For instance, Tomasello claims that "... people lying, cheating, and stealing are simply instances when, for whatever reason, the individual's self-interested motives have won out" (Tomasello, 2016, p. 161). This is representative of a general trend in the literature that sees human nature as part selfish and part prosocial. This dual-conception focuses on selfish exploitation as the central corrosive force undermining cooperation and neglects the possibility of spite. Partner choice, punishment, and moral signaling evolved to safeguard our hypercooperative behaviors from defection and riding free (Boyd et al., 2003; Stanford, 2018). Given the persistent threat selfish free riders pose for cooperative endeavors, and Hamilton's influential pronouncement that spite should be rare, there is a tendency to reinterpret spite as selfishness. We expect selfishness to occur all the time, so most cases that look like that rare sort of behavior called spite are probably selfish, or so the argument goes. While some cases of spite may indeed be misguided selfishness, given the empirical work on human social behaviors (Marcus et al., 2014; McAuliffe et al., 2014), it would be a mistake to write all acts of spite off as mere cases of selfishness. For instance, Hume believed our nasty behavior deserved as much attention as our nice behavior. The *Treatise of Human Nature* gives an account of envy and malice that follows from the general principles behind benevolence and love:

> ... envy is excited by some present enjoyment of another, which by comparison diminishes our idea of our own: Whereas malice is the unprovok'd desire of producing evil to another, in order to reap a pleasure from the comparison.
>
> (Hume, *Treatise*, 2.2.8.12)

We agree with Hume. Spite needs to be taken seriously as a phenomenon, and this presents a genuine puzzle for both human and animal behavior. First, as we will review, on strict evolutionary accounts, we should expect to see spite much more frequently than Hamilton supposed. But perhaps more important, in species capable of learning and cultural transmission—and humans are the cultural transmitters par excellence—spiteful behaviors can be reliably learned by a number of dynamical rules, and some of the learning rules that lead to spite can be evolutionarily successful despite producing stable and costly harm in populations.

3.2 The Evolution of Altruism and Spite

Existing evolutionary explanations for spiteful behavior involve cases in which spite, for whatever reason, generates a *relative* advantage for the spiteful strategy despite the fact that the actors pay an *absolute* cost. If spite causes other strategy types in the group to have fewer offspring (on average) than the spiteful actors, then it can generate a relative advantage for spite even if it decreases the total number of offspring the group produces. This can occur in a number of different ways including certain population structures (Hawlena et al., 2010), greenbeard effects (Gardner & West, 2010; Lehmann et al., 2009), signaling (Smead & Forber, 2016), or simply small populations (Hamilton, 1970; Smead & Forber, 2013). One of the central features of all of these explanations is *anti-correlation* of strategic behavior, or when spite is more likely to be directed at non-spiteful individuals than it would be if randomly employed (Hamilton, 1970; Skyrms, 2007). Perhaps the best way to understand the connection between anti-correlated interactions and spite is to start with the connection between correlated interactions and altruism.

The Prisoner's Dilemma (Table 3.1, left side) provides a natural way to represent altruism because players have an individual incentive to defect, yet the best joint outcome is achieved by mutual cooperation. Each player decides whether to pay a cost (c) to confer a benefit (b) on an interaction partner. Altruism is strictly dominated in the Dilemma; thus, if individuals interact randomly, altruism will quickly go extinct. However, if altruists

Table 3.1 The normal forms for the Prisoner's Dilemma and Delight games, where $c > 0$ represents the cost of altruism/spite, $b > 0$ represents the benefit in the Dilemma, and $h > 0$ represents the harm in the Delight. Payoffs are symmetric and are shown for the row player (assume $b > c$ and $h > c$).

Dilemma	*altruism*	*egoism*	*Delight*	*social*	*spiteful*
altruism	b-c	-c	social	0	-h
egoism	b	0	spiteful	-c	-h-c

can preferentially interact with other altruists—that is, if interactions are sufficiently correlated—altruism can evolve. Let r represent the degree of correlation in a population; individuals interact with a partner of the same type with probability r, otherwise they interact randomly. In large populations with a proportion p of altruists, the fitness of each type is:

$$F(altruism, p) = (bp - c)(1 - r) + (b - c)r, \tag{3.1}$$

$$F(egoism, p) = bp(1 - r). \tag{3.2}$$

Altruism will be favored when $F(altruism, p) > F(egoism, p)$ or:

$$r > c/b. \tag{3.3}$$

So, when the degree of correlated interaction is greater than the ratio of the cost (to the actor) to the benefit (for the recipient), altruism should evolve. This is a version of Hamilton's rule (Hamilton, 1964a,b).

To make the role of anti-correlation precise, consider the Prisoner's Delight (Table 3.1, right side). This game provides a simple way to represent spite. In this game, both players have an opportunity to pay a cost (c) in order to harm (h) the other player. The social strategy (not harming) is strictly dominant and the only rational solution to the game. Spite can evolve if it is disproportionally directed at non-spiteful types—if interactions are *anti-correlated.* Let r^* represent the degree of anti-correlation in a population. This can be understood as the probability of meeting an individual of a different type. In large populations with a proportion x of spiteful individuals, the fitness of each type is:

$$F(social, x) = -hx(1 - r^*) - hr^*, \tag{3.4}$$

$$F(spite, x) = -hx(1 - r^*) - c. \tag{3.5}$$

Spite will be favored whenever $F(spite, x) > F(social, x)$ or:

$$r^* > c/h. \tag{3.6}$$

There is a clear symmetry between the version of Hamilton's rule in equation (3.3) and the condition for the evolution of spite given in equation (3.6). Thus, just as correlation is the key to the evolution of altruism, anti-correlation is the key to the evolution of spite.

Although we have begun to develop an understanding of how spite can evolve, human spite arguably still presents a puzzle for several reasons. First, human behavior is not hard-wired as many models of biological evolution suppose. Second, humans are probably among the most spiteful of species and act spitefully even in cases where condition (3.6) is not obviously satisfied (see, e.g., McCarthy-Jones, 2021 for an entertaining review of the human propensity to spite). Third, humans are arguably far more rational than other animals, making spiteful behavior doubly puzzling as the social strategy is the uniquely rational solution to the Prisoner's Delight. While the mathematical models for altruism and spite originated in biological evolution, the dynamics can apply to cultural evolution as well (Skyrms, 1996), and Darwinian cultural evolution is driven by success-based learning dynamics (Godfrey-Smith, 2012). In what follows, we formulate an explanation for spiteful behavior in humans as learned behavior, and argue that the kinds of learning that generate spite are both common to humans and can be evolutionarily successful. There are three intertwined questions that must be answered. First, what kinds of learning might reliably produce spiteful behavior? Second, do humans learn in these ways? Third, if so, why do we learn in those ways when they lead to detrimental behaviors? We will consider these questions from the perspective of both social learning and individual learning. As we will show, a diverse and wide range of learning mechanisms can produce spiteful behavior and we have good reason to believe humans learn in some of these ways, some of the time.

3.3 Learning to Spite by Imitation

Given the nature of spiteful behavior, it is easy to see how no purely self-interested learner would reliably learn to do it. Doing so would involve taking on costs one could have avoided. Indeed, given that spite in the Prisoner's Delight is strictly dominated, any learning rule that produces behaviors that yield higher personal payoffs will learn not to spite others. Consequently, if individuals are going to learn spiteful behaviors, it *must* involve paying attention to the behavior or payoffs of others. Perhaps the most obvious form of this is social learning, which is exemplified by imitation.

Imitation learning comes in many forms, but the core idea is that imitative individuals learn from the behavior of others. Very simple varieties of imitation may disregard payoffs entirely and copy behaviors blindly, perhaps based on what is most common. Blind imitation of the popular strategies can lead to virtually any behavior (including spite) provided that there are enough other individuals already behaving in that way. Such blind imitation can be an effective way of cheaply exploiting information gained by others. However, if there are too many blind imitators, then it is less effective at identifying successful behaviors (Rogers, 1988).

More sophisticated forms of imitation, those commonly studied in evolutionary game theory, incorporate both the payoffs and behaviors of others, often in comparison with the learner's payoffs. While the exact details vary between imitation rules, all such rules work by comparing the learner's behavior and payoff with the behaviors and payoffs of others. This comparison is then used to inform whether to copy the observed strategy. There are close parallels between imitation rules and biologically derived evolutionary dynamics (Sandholm, 2010). To see this, consider the work of Traulsen et al. (2005) who studied the following imitation rule. When learning, an individual *b* picks a random member from the population *a* and observes that person's strategy choices and recent payoffs. Individual *b* copies the strategy of individual *a* with a probability proportional to the difference in payoffs:

$$p = \frac{1}{2} + \frac{1}{2}\frac{\pi_a - \pi_b}{\Delta\pi_{max}}, \tag{3.7}$$

where π_i is the payoff to player *i* and $\Delta\pi_{max}$ is the maximum possible payoff difference. Traulsen et al. (2005) go on to show that this imitation rule, if employed by members of an infinite randomly mixing population, is equivalent to the standard replicator dynamic derived from biological principles of population growth under competition (Hofbauer & Sigmund, 1998; Taylor & Jonker, 1978). Therefore, spite can be learned via this imitation rule under the same conditions that it would evolve via natural selection: when condition (3.6) is satisfied. The only difference is in the interpretation of the cost and harm, which should be understood as factors contributing to the propensity for a strategy to be imitated rather than as reproductive fitness.

Similar connections are also present in finite populations. Forber and Smead (2014) show that the frequency-dependent Moran process (Moran, 1962; Taylor et al., 2004), a finite population dynamic, can reliably lead to spite provided that the harm-to-cost ratio for spite is large relative to population size. The Moran process works as follows: in each "generation," an individual is selected to "reproduce" with a probability proportional to the ratio of their fitness and the average fitness of the population. The new individual replaces a randomly selected individual who "dies." Traulsen et al. (2005) also investigate the Moran process as an imitation rule, showing that, in the limit, this is equivalent to the adjusted replicator dynamic, which is based on biological principles (Maynard Smith, 1982). These processes can be interpreted as cultural rather than biological evolution: an individual is chosen to be imitated with probability proportional to their success, and they are imitated by a randomly selected individual in the population. Thus, even in finite populations, there are close parallels between evolutionary dynamics and imitation learning.

The close formal similarities between imitation and biological reproduction offer a key insight into how spite may be learned. As mentioned above, there are many studies that reveal the various mechanisms that can promote the biological evolution of spite. These same mechanisms, in the case of imitation, would also promote the cultural evolution of spite. In any case where condition (3.6) is satisfied and the payoffs can be plausibly re-interpreted in terms of cultural transmission, we should expect spite to spread via imitation. Likewise, any mechanisms that allow for anti-correlation, such as signaling (Smead & Forber, 2016), can produce the conditions necessary for the evolution and spread of spite via imitation.

3.3.1 Spite in Dynamic Networks

Many theorists have argued for the importance of partner choice for the evolution and elaboration of prosocial behavior, and this is thought to be especially important for thinking about the role of morality in supporting and stabilizing complex cooperation (Baumard et al., 2013; Stanford, 2018). If individuals can preferentially select their partners, then they can exclude corrosive free riders from benefiting from cooperation. Perhaps mechanisms of partner choice can provide a remedy to the way imitation dynamics can spread spiteful behavior. If spiteful individuals can be excluded from interactions with social individuals, the harmful behavior should go extinct. The ability to exclude individuals from social interaction should directly combat the anti-correlation necessary for the evolution of spite.

One way to model partner choice uses a dynamic interaction network where the interaction patterns between players can coevolve with player strategy choice. Fulker et al. (2021) investigate a dynamic network implementation of the Prisoner's Delight to examine how a partner choice mechanism affects the evolution of spite. Each individual in the population is connected to other individuals via a set of continuous network weights and starts with a randomly assigned strategy (*social* or *spiteful*). Both network ties and strategy choices update based on payoffs received from playing the Prisoner's Delight. Each round, each individual selects an interaction partner based on their interaction weights. Interaction weights are updated based on the payoffs of the game via a reinforcement learning dynamic. As one might expect, individuals quickly learn to avoid the spiteful types and interact preferentially with social types. Individuals also have a chance to update their strategy choice each round via imitation learning. If they decide to update, they select an individual probabilistically based on their current network weights; thus, they are more likely to compare their strategy to their preferred interaction partners. They compare their average per interaction payoff to the payoff of their selected partner.

If their partner has a higher average payoff over their interactions in the previous round, the individual will update their strategy to that of their selected partner. When network learning occurs much faster than imitation learning, individuals can settle on their preferred partners before deciding whether to update their strategy choice.

The findings of the study are dramatic: even when individuals can quickly update their network ties, and therefore choose their partners based on payoffs they receive in the game, spite reliably emerges and takes over the population in a broad range of circumstances. Spite is much more likely to evolve when the ratio of cost to harm is favorable, as one might expect from equation (3.6). The results are robust across variation in learning speeds. How does this happen? The population starts with a uniform interaction network and a random distribution of social and spiteful individuals. The network evolves such that social individuals are preferentially targeted by both strategy types. As this happens, a core-periphery structure emerges where social individuals form the core, receiving many more visits from interaction partners, and any remaining spiteful individuals are pushed to the periphery. The trouble with this structure is that the spiteful individuals tend to avoid each other and interact with social individuals. This means that while the spiteful individuals pay a cost to harm, they tend to receive the benefit from targeting social individuals. Social individuals, because they are targeted by the spiteful, accrue some amount of harm over their interactions. This can push their average per interaction payoff below that of their spiteful neighbors. The costly harm drives down everyone's payoffs, but it hits the social strategy harder. When social individuals imitate, they occasionally see spiteful individuals doing relatively better, and copy their strategy. Spite starts to spread, first among the social agents who happen to have more spiteful partners, then accelerates and eventually eliminates the social strategy by preferentially targeting those individuals. The remaining social individuals suffer more and more harm, driving down their average payoff and increasing the chance they switch to the spiteful strategy. The social strategy collapses and then the network starts to lose its structure and move toward more random interaction. If a new social individual enters the population, they will be quickly targeted, suffer serious harm, and then switch to spiteful.

3.3.2 *Spiteful Humans?*

Even in sophisticated models where partner choice is available, spite can still evolve when imitation learning rules are in play. Because imitation compares payoffs, the spiteful trade-off—lower absolute payoffs to achieve a relative advantage—leads imitation learners to adopt spiteful strategies.

Humans are probably the most prolific imitators and social learners in nature, and this is thought to play an important role in human evolution and cognition (Sterelny, 2012; Tomasello, 1999). There is some evidence that other animals imitate various behaviors and techniques (Heyes & Galef, 1996), but imitation learning in humans is ubiquitous and undeniable. There are some questions about when imitation develops in humans, what cognitive processes are required for imitation, and to what extent imitation is employed in humans. Humans clearly pay attention to what other people do and what the results of their actions are and, in many cases, attempt to copy them. It should be no wonder that humans learn to spite.

So, if imitation leads to detrimental behaviors, why imitate? The first answer is that, if there is sufficient anti-correlation as specified in condition (3.6), this behavior is relatively advantageous in the population even if it carries an absolute cost for actors. A second answer is that imitation learning carries clear and systematic benefits to organisms that have the cognitive capacity to learn in this way. It allows one to benefit from the learning of others in a fast and efficient way, providing a shortcut to acquire sophisticated techniques, and helps coordinate social behavior (Alós-Ferrer & Schlag, 2009). Additionally, imitation learning can lead to other beneficial social behaviors such as cooperation and fairness (Alexander, 2007). The benefits of imitation may easily outweigh any modestly detrimental side effects that come about from opening the door to acquiring spiteful behavior.

3.4 Spite via Individual Learning

Humans are not simply imitators; they also learn individually by trial-and-error or reinforcement learning. Can such individual learning also lead to spite? Trial-and-error learning is common throughout nature and humans are particularly adept at it. The most basic forms of trial-and-error learning that factor in only payoffs to the individual actor, such as reinforcement learning, will generally not learn to spite. For example, Roth-Erev reinforcement learning (Roth & Erev, 1995) will never converge on a dominated strategy (Beggs, 2005), meaning that the social behavior in the Prisoner's Delight is the only eventual outcome. For individual learning to reliably produce spiteful behavior, it needs to be other-regarding in some way. In particular, learners will need to be averse to falling below others in direct comparisons and so be willing to pay a cost to avoid being worse off in such rankings. In short, they need to be envious. As we will see, this can take a variety of forms.

Note that we are using "envy" in a minimal sense of agents having a negative or aversive reaction to the payoffs of others. This is an account that doesn't require a psychological interpretation beyond the effects of behavioral learning.

3.4.1 Other-regarding Utility Functions

Suppose that individuals do not only care about the payoff to themselves, but also care about the payoff to the other player. This could be a positive attitude where the other's material payoff is seen as a (partial) benefit to the actor, or a negative one where it is seen as a cost. Such an individual's utility function could be modeled as:

$$U_i(s_i, s_j) = \pi_i(s_i, s_j) + \rho\pi_j(s_j, s_i). \tag{3.8}$$

Here $\pi_i(s_i, s_j)$ is i's material payoff from the interaction with j when they play s_i and s_j, respectively. ρ is a factor representing an attitude toward others' payoffs, which could be positive, where others' payoffs are seen favorably, or negative, where others' payoffs are viewed with envy.

In this model, using the payoffs from the Prisoner's Delight, we can specify when the utility for spiteful behavior will outweigh that of social behavior. Let $U(s,x)$ represent the utility of playing a strategy s against an opponent who plays *social* with probability x and *spite* with probability $1-x$. Then $U(spite, x) > U(social, x)$ if and only if

$$x(-\rho h - c) + (1-x)(-\rho h - \rho c - h - c) > (1-x)(-\rho c - h). \tag{3.9}$$

which reduces to

$$-\frac{c}{h} > \rho. \tag{3.10}$$

The parallels between conditions (3.10) and (3.6) are apparent. If ρ is negative and c/h ratio is sufficiently small, then spiteful behavior will generate greater utility for the actor regardless of the other player's behavior. Hence, any learning rule that tracks utility in this way will tend to learn spite whenever there is sufficient negative regard for others' payoffs.

3.4.2 Inequity Aversion

Human utility functions may be other-regarding in more sophisticated ways than expressed by equation (3.8). One well-studied version of other-regarding utility functions comes from the investigations of inequity aversion (Fehr & Schmidt, 1999). Although not without dispute (Bergh, 2008; Binmore & Shaked, 2010), there is substantial empirical evidence suggesting humans are inequity averse and that humans are other-regarding when it comes to decision-making (Blake & McAuliffe, 2011; Fehr & Schmidt, 2006; Henrich et al., 2001). Adapting the inequity aversion utility function from Fehr and Schmidt (1999) to our two-player game yields:

$$U_i(\pi_i, \pi_j) = \pi_i - \alpha_i \, max(\pi_j - \pi_i, 0) - \beta_i \, max(\pi_i - \pi_j, 0). \tag{3.11}$$

Here π_i denotes i's material payoff from a given interaction, α_i denotes i's aversion to disadvantageous inequity, and β_i denotes i's aversion to advantageous inequity.

Using this framework, the utility of each strategy against an opponent that plays *social* with probability x and spite with probability $(1-x)$ is:

$$U_i(social, x) = (1-x)(-h - \alpha_i(h-c)), \tag{3.12}$$

$$U_i(spite, x) = x(-c - \beta_i(h-c)) + (1-x)(-h-c). \tag{3.13}$$

For $x = 1$, $U_i(spite,1) < U_i(social,1)$ regardless of inequity aversion. However, for $x < 1$ it is possible that spite will generate higher utility than the alternative provided x is sufficiently small and α is sufficiently large. $U_i(spite, x) > U_i(social, x)$ if and only if

$$(1-x)\alpha_i - x\beta_i > \frac{c}{h-c}. \tag{3.14}$$

With a large enough α_i value, there will be some values of x that mean spite will generate higher utility for i than the alternative strategy. However, the dependence on the x value indicates a different dynamic than simple negative regard for others' payoffs. While *spite* can beget *spite*, *social* will incentivize *social*. For individuals with this kind of utility function, they will learn to play *spite* if their opponents *spite* frequently enough, but they will learn to play *social* if exposed to relatively little *spite*. This suggests that if two individuals both have a large aversion to disadvantageous inequity, an accidental act of spite may cause a back-and-forth escalation of spiteful behavior. The core idea here will likely be familiar given the many depictions of vendettas and escalating cycles of revenge in history, literature, and cinema.

3.4.3 *Trial and Error Learning*

We can imagine even simpler variations of individual learning that do not involve such nuanced transformations of utility functions. Simple trial-and-error learning rules can provide an illustrative example. Consider one of the simplest learning rules studied: Win-Stay, Lose-Randomize (or WSLR).

For individuals using this learning rule, strategies are initially chosen at random. Then, if the result of the game-play is classified as a "win," behavior remains unchanged. If the result of game-play is classified as a "loss," a behavior is chosen at random for the next play of the game. This learning rule has been shown to be highly effective at reaching efficient solutions in common interest signaling games (Barrett & Zollman, 2009) and is closely related to Win-Stay, Lose-Shift due to Robbins (1985) which has been shown to be effective in games of cooperation (Nowak & Sigmund, 1993). The only difference is whether there is a chance to repeat the same play that resulted in a previous "loss."

We can consider different versions of WSLR by considering different ways to classify wins and losses. The number of possible WSLR rules is potentially infinite, but here we will consider a simple model that only focuses on a small subset. Let s_i represent the strategy of player i and let $\pi_i(s_i, s_j)$ represent the payoff for player i using strategy s_i against a partner playing s_j. In addition to a traditional best-response variety of WSLR, we will focus our attention on a learning rule investigated by Smead (2015) called "competitive-response," which is an instantiation of the minimal sense of envy, so let us call this rule "envious." Each type of learner can be specified in terms of how it classifies "wins" (assuming non-wins are losses):

- Best-response (BR): $\pi_i(s_i, s_j) \geq \pi_i(s, s_j)$ for all s,
  win: = individually optimal outcome given the opponent's strategy.
- Envious-response (ER): $\pi_i(s_i, s_j) \geq \pi_j(s_j, s_i)$,
  win: = did not receive lower payoffs than the opponent.

BR learners approximate a trial-and-error version of economically rational individuals. Two BR learners will eventually play a pure strategy Nash equilibrium (if one exists). In the Prisoner's Delight, it is straightforward to see that because spite is never a best response, a BR learner will never learn to spite.

ER learners, on the other hand, are very different. While ER is simpler than BR in that it does not need to consider counterfactuals, it is also more sophisticated by making a payoff comparison between players. These learners aim only at ensuring they are not outperformed by their opponent. As a consequence, they will learn economically irrational behavior in many contexts. They are willing to take on additional costs if it means they are not doing worse than their opponents. In the Prisoner's Delight, these learners will reliably acquire spiteful behavior with reasonable frequency and never "unlearn" such behavior. Although spite is never an optimal response, it guarantees that the player will never receive less than their opponent provided that $h > c$.

Are learning rules like ER, or those with utility functions which include negative regard for others and/or inequity aversion, plausible? This question can be addressed in two ways. First, we may wonder whether humans show such tendencies in their learning. Second, we may wonder whether non-traditional rules, such as ER, should be expected to evolve in the first place. We have a brief remark on the first point below before turning to the evolution of learning in the next section.

As mentioned in the section on inequity aversion, it is well known that humans are sensitive to the payoffs of others. We relate to the sentiment, "keeping up with the Joneses," because even if we try not to let comparisons with others drive our actions, we at least recognize that humans have this sort of bias and that it may drive behavior in some cases. In addition to the many empirical studies supporting inequity aversion, a study by McAuliffe et al. (2014) shows that inequity is rejected by human children when disadvantageous to the decision-maker but not when advantageous—behavior qualitatively consistent with ER. Indeed, the authors of that study describe this as rejecting inequity "out of spite." Human learning is undoubtedly far more complicated than any of the learning rules presented here. Nevertheless, if there are ER-like elements to our learning processes—as there seems to be—we should not expect completely rational behavior in all cases and perhaps should expect humans to occasionally learn spiteful tendencies.

3.5 The Evolution of Envy

Humans may act out of malice because we learn in ways that track relative payoffs. We are sensitive not just to how well we are doing, but also to how this compares to others. We are, at least sometimes, envious as well as malicious. But the observation that these tendencies can lead to costly, spiteful behavior raises a deeper question. If such envious learning rules lead to detrimental behavior, why would those rules evolve instead of others?

The evolution of learning itself can be studied in game-theoretic settings (see, e.g., Smead, 2012, 2015; Dridi & Lehmann, 2014, 2015), and some of these studies have produced results that help us understand the evolution of envy. In particular, Smead (2015) explores the game-theoretic conditions that will favor the evolution of learners over non-learners. Smead finds that learners most readily evolve in games that produce stable, diverse populations. For instance, selection dynamics in games of conflict (Forber & Smead, 2015), such as Chicken, Snowdrift, or Hawk-Dove, tend to produce stable polymorphic equilibria. However, many learning rules that are successful in these accounts do not behave in "traditional" ways: "...if learners are to dominate the population, it will most likely involve learning rules that do not lead to best responses..." (Smead, 2015, p. 173).

Smead then considers a model that includes both classic BR learners and our ER learners. He finds that there are interaction settings (e.g., games of conflict) in which ER learners evolve readily and are maintained by selection. In many cases, when the game allows, these learning rules do very well against traditional BR learners, learning to exploit them and never letting themselves be exploited. Yet they often do poorly against their own types, leading to polymorphisms of learning types. Smead goes on to identify other cases, such as the classic Prisoner's Dilemma, where ER-style rules perform very well and have significant evolutionary success compared to BR rules. These "non-traditional" learners, such as ER, may be expected to evolve and play a significant role in social interactions:

> ...there is reason to think that best-response-type learning rules are unlikely to fully evolve in the context of social interaction ... if learners are to dominate the population, some of them will probably learn in atypical ways.
>
> (Smead, 2015, p. 175)

The idea that polymorphisms of learning may be abundant in natural populations is consistent both with other theoretical models (Dridi & Lehmann, 2015; Rogers, 1988) and with empirical studies of both animals and humans. For example, van Bergen et al. (2004) claim that guppies show "genuine personality differences" with respect to how much individual exploration they do. Costantini (2005, p. 279) argues that such personality differences are common in animals:

> Individuals differ non-randomly in the way they deal with stressors and novelties and appear to vary along a behavioral continuum from shy to bold. These differences co-vary with other behavioral traits such as aggression, exploration, risk taking fearfulness and reactivity.

The research of Costa-Gomes et al. (2001, p. 1196) indirectly suggests the same may be true of humans: "Subjects' decisions [in games] are too heterogeneous to be adequately described by a single decision rule..." More generally, the results above are consistent with the intuition that different individuals often learn in different ways. Polymorphisms of learners may also have implications for common modeling methodology in the social or biological sciences. It is frequently assumed that populations are relatively homogeneous in terms of their adaptive strategies and goals in strategic interaction (e.g., that individuals will adjust their behavior in the direction of higher payoffs). However, as Smead (2015) shows, this may not be the case in strategic settings. If populations that are polymorphic with respect to learning rules generate very different behavior than monomorphic populations, we may have reason to question the soundness of traditional modeling assumptions that presume uniformity in learning. This will be especially important if the learning rules we should expect to evolve reliably generate non-traditional, apparently irrational behavior, such as spite.

The evolutionary success of rules like ER in some settings provides a how-possible explanation for the emergence of these despite their associated costs in games like the Prisoner's Delight. The real world is complicated and we do not always know what game we are playing. Learning rules that are successful in many areas may be selected even though they generate costs in some cases. Envy and malice might be evolutionarily valuable for avoiding exploitation, fitness leveling, and maybe even getting some extra cooperation. The spite that results from such learning rules could generate relative advantages directly (in the right circumstances), or even persist as a byproduct of the broader success these rules have in social interactions.

3.6 Conclusion

There are many ways that cooperation and altruism can evolve, including interaction between relatives, spatial structure, reciprocity, and partner choice. Cooperation can also be learned when similar factors are in play during social or individual learning. The literature on the evolution of cooperation contains many models and cases exploring the various ways in which prosocial traits can emerge.

What we have shown here is that many of the elements that allow for the evolution of altruism (assortment, partner choice, reciprocity, etc.) also allow for the emergence and stability of spite. Spite can also be learned. This can happen via imitation whenever there is sufficient anti-correlation to provide spite a relative advantage, a condition analogous to the condition for the (biological) evolution of spite. Spite can also be systematically learned when individuals are envious, which may be instantiated in a number of ways, such as inequity aversion or other-regarding preferences. Furthermore, such envious learners can outperform traditional BR learners in certain contexts.

These are situations where the learning dynamics lead to the evolution of spite. The spiteful strategy generates a relative advantage over other strategies despite imposing direct costs on the actors. If these learning rules reliably produce these harmful behaviors, should selection eliminate these rules? Perhaps. But these are effective rules in many circumstances, and so it is possible for evolution via natural selection to produce such learners even though they learn detrimental behaviors, such as spite, in some settings. This points toward an intriguing explanation for the existence of spiteful behavior in humans. Spite may not be directly selected for. Rather, the learning mechanisms are selected for because they produce adaptive behavior in many situations, yet these same patterns of learning have a propensity to learn spiteful behavior.

References

Alexander, J. M. (2007). *The structural evolution of morality*. Cambridge University Press.

Alós-Ferrer, C., & Schlag, K. H. (2009). Imitation and learning. In *The handbook of rational and social choice* (Chapter 11, pp. 271–298). Oxford Scholarship Online Monographs.

Barrett, J., & Zollman, K. J. (2009). The role of forgetting in the evolution and learning of language. *Journal of Experimental & Theoretical Artificial Intelligence*, *21*(4), 293–309.

Baumard, N., André, J. B., & Sperber, D. (2013). A mutualistic approach to morality: The evolution of fairness by partner choice. *Behavioral and Brain Sciences*, *36*, 59–122.

Beggs, A. W. (2005). On the convergence of reinforcement learning. *Journal of Economic Theory*, *122*, 1–36.

Bergh, A. (2008). A critical note on the theory of inequity aversion. *The Journal of Socio-Economics*, *37*(5), 1789–1796.

Binmore, K., & Shaked, A. (2010). Experimental economics: Where next? *Journal of Economic Behavior & Organization*, *73*(1), 87–100.

Blake, P. R., & McAuliffe, K. (2011). I had so much it didn't seem fair: Eight-year-olds reject two forms of inequity. *Cognition*, *120*(2), 215–224.

Boyd, R., Gintis, H., Bowles, S., & Richerson, P. J. (2003). The evolution of altruistic punishment. *Proceedings of the National Academy of Sciences of the United States of America*, *100*, 3531–3535.

Costa-Gomes, M., Crawford, V. P., & Broseta, B. (2001). Cognition and behavior in normal-form games: An experimental study. *Econometrica*, *69*(5), 1193–1235.

Costantini, D. (2005). Animal personalities in a competitive game theory context. *Ethology Ecology & Evolution*, *17*, 279–281.

Dridi, S., & Lehmann, L. (2014). On learning dynamics underlying the evolution of learning rules. *Theoretical Population Biology*, *91*, 20–36.

Dridi, S., & Lehmann, L. (2015). A model for the evolution of reinforcement learning in fluctuating games. *Animal Behaviour*, *104*, 87–114.

Fehr, E., & Schmidt, K. M. (1999). A theory of fairness, competition, and cooperation. *The Quarterly Journal of Economics*, *114*(3), 817–868.

Fehr, E., & Schmidt, K. M. (2006). The economics of fairness, reciprocity and altruism–experimental evidence and new theories. In S.-C. Kolm (Ed.), *Handbook of the Economics of Giving, Altruism and Reciprocity*, *1*, 615–691.

Forber, P., & Smead, R. (2014). An evolutionary paradox for prosocial behavior. *The Journal of Philosophy*, *111*(3), 151–166.

Forber, P., & Smead, R. (2015). Evolution and the classification of social behavior. *Biology & Philosophy*, *30*(3), 405–421.

Fulker, Z., Forber, P., Smead, R., & Riedl, C. (2021). Spite is contagious on dynamic networks. *Nature Communications*, *12*, 260.

Gardner, A., & West, S. A. (2010). Greenbeards. *Evolution*, *64*, 25–38.
Godfrey-Smith, P. (2012). Darwinism and cultural change. *Philosophical Transactions of the Royal Society B: Biological Sciences*, *367*, 2160–2170.
Hamilton, W. D. (1964a). The genetical evolution of social behaviour, I. *Journal of Theoretical Biology*, *7*, 1–16.
Hamilton, W. D. (1964b). The genetical evolution of social behaviour, II. *Journal of Theoretical Biology*, *7*, 17–52.
Hamilton, W. D. (1970). Selfish and spiteful behavior in an evolutionary model. *Nature*, *228*, 1218–1220.
Hamilton, W. D. (1971). Selection of selfish and altruistic behavior in some extreme models. In *Man and beast: Comparative social behavior* (pp. 57–91). Smithsonian Institution Press.
Hawlena, H., Bashley, F., & Lively, C. M. (2010). The evolution of spite: Population structure and bacteriocin-mediated antagonism in two natural populations of xenorhabdus bacteria. *Evolution*, *64*(11), 3198–3204.
Henrich, J., Boyd, R., Bowles, S., Camerer, C., Fehr, E., Gintis, H., & McElreath, R. (2001). In search of homo economicus: Behavioral experiments in 15 small-scale societies. *American Economic Review*, *91*(2), 73–78.
Heyes, C. M., & Galef, B. G. Jr. (Eds.) (1996). *Social learning in animals: The roots of culture*. Academic Press.
Hofbauer, J., & Sigmund, K. (1998). *Evolutionary games and population dynamics*. Cambridge University Press.
Inglis, R. F., Gardner, A., Cornelis, P., & Buckling, A. (2009). Spite and virulence in the bacterium Pseudomonas aeruginosa. *Proceedings of the National Academy of Sciences of the United States of America*, *106*(14), 5703–5707.
Lehmann, L., Feldman, M. W., & Rousset, F. (2009). On the evolution of harming and recognition in finite panmictic and infinite structured populations. *Evolution*, *63*, 2896–2913.
Marcus, D. K., Zeigler-Hill, V., Mercer, S. H., & Norris, A. L. (2014). The psychology of spite and the measurement of spitefulness. *Psychological Assessment*, *26*(2), 563–574.
Maynard Smith, J. (1982). *Evolution and the theory of games*. Cambridge University Press.
McAuliffe, K., Blake, P. R., & Warneken, F. (2014). Children reject inequity out of spite. *Biology Letters*, *10*(12), 20140743.
McCarthy-Jones, S. (2021). *Spite: The upside of your dark side*. Basic Books.
Moran, P. A. P. (1962). *The statistical processes of evolutionary theory*. Clarendon Press.
Nowak, M. A., & Sigmund, K. (1993). A strategy of win-stay, lose-shift that outperforms tit-for-tat in the prisoner's dilemma game. *Nature*, *364*, 56–58.
Robbins, H. (1985). Some aspects of the sequential design of experiments. In *Herbert Robbins selected papers* (pp. 169–177). Springer.
Robinson, S. K. (1988). Anti-social and social behaviour of adolescent yellow-rumped caciques (Icterinae: Cacicus cela). *Animal Behaviour*, *36*, 1482–1495.
Rogers, A. R. (1988). Does biology constrain culture? *American Anthropologist*, *90*, 819–831.
Roth, A., & Erev, I. (1995). Learning in extensive form games: Experimental data and simple dynamical models in the intermediate term. *Games and Economic Behavior*, *8*, 164–212.
Sandholm, W. H. (2010). *Population games and evolutionary dynamics*. MIT Press.

Skyrms, B. (1996). *Evolution of the social contract*. Cambridge University Press.
Skyrms, B. (2007). *Evolution and the social contract. The Tanner lecture on human values*. University of Michigan.
Smead, R. (2012). Game theoretic equilibria and the evolution of learning. *Journal of Experimental and Theoretical Artificial Intelligence*, *24*(3), 301–313.
Smead, R., & Forber, P. (2013). The evolutionary dynamics of spite in finite populations. *Evolution*, *67*, 698–707.
Smead, R. (2015). The role of social interaction in the evolution of learning. *British Journal for the Philosophy of Science*, *66*(1), 161–180.
Smead, R., & Forber, P. (2016). The coevolution of recognition and social behavior. *Scientific Reports*, *6*, 25813.
Stanford, P. K. (2018). The difference between ice cream and Nazis: Moral externalization and the evolution of human cooperation. *Behavioral and Brain Sciences*, *41*, e95.
Sterelny, K. (2012). *The evolved apprentice*. MIT Press.
Taylor, C., Fudenberg, D., Sasaki, A., & Nowak, M. A. (2004). Evolutionary game dynamics in finite populations. *Bulletin of Mathematical Biology*, *66*, 1621–1644.
Taylor, P., & Jonker, L. (1978). Evolutionary stable strategies and game dynamics. *Mathematical Biosciences*, *40*, 145–56.
Tomasello, M. (1999). *The cultural origins of human cognition*. Harvard University Press.
Tomasello, M. (2016). *A natural history of human morality*. Harvard University Press.
Traulsen, A., Claussen, J. C., & Hauert, C. (2005). Coevolutionary dynamics: From finite to infinite populations. *Physical Review Letters*, *95*, 238701.
van Bergen, Y., Laland, K. N., & Hoppitt, W. (2004). Social learning, innovation, and intelligence in fish. In L. J. Rogers and G. Kaplan (Eds.), *Comparative vertebrate cognition*. Kluwer Academic/Plenum Publishers.
West, S. A., & Gardner, A. (2010). Altruism, spite, and greenbeards. *Science*, *327*, 1341–1344.

4 Situationism and the Evolution of Spite

Isaac Wiegman and Michael T. Dale

The human species excels at cooperation. Even by comparison with humanity's closest living relatives, we spontaneously and proactively cooperate with others at a cost and in ways that have defied evolutionary explanation until the last few decades (Alexander, 2017; Boehm, 2012; Wrangham, 2019). Even commonplace occurrences like standing in line with strangers or sitting with them in close quarters on a bus or plane can be understood as forms of cooperation that chimpanzees are largely incapable of. Can and should this excellence be understood in terms of virtues—excellences of character—as philosophers understand them?

There are many forms this question could take. Besides virtues, what else could explain these dispositions to cooperate? In moral psychology specifically (as in psychology more generally), the two main explanations for individual behavioral differences are internal (e.g., personality and character) and external (e.g., situation). Virtues are thought to be internal (even though they may be externally scaffolded in important respects, e.g., Westra & Kelly, 2025). On the Aristotelian account, virtues are character traits that are developed through habituation (Aristotle, 2000). They are *consistent*, insofar as they reliably manifest across a wide array of trait-relevant situations, and they are *stable*, insofar as they reliably manifest over repeated trials in similar trait-relevant conditions (Doris, 2002).

So, when we pose the question of how to explain our species' distinctive excellence at cooperation, part of what we want to know is whether we cooperate because of the internal features of each member of our species (including the norms we internalize through learning and development) or because of external factors, such as situational factors. In evolutionary biology, there has been a strong tendency to attempt explanations of cooperation in terms of internal, genetic variations that were selected over millions of years. And if the chimpanzee (*Pan troglodytes*) is our contrast class, it does seem likely that there are some key internal differences (some of which involve our ability to internalize norms, etc.) between chimps and

DOI: 10.4324/9781003632641-6

humans that explain our abilities to tolerate strangers in close quarters, among other forms of cooperation. We cannot reliably achieve this level of cooperation in a chimpanzee, simply by changing its situation or social context.[1]

However, here as in any scientific enterprise, the question of causality is never a simple either/or. The question is not whether intrinsic or extrinsic variables explain variance (insofar as we can cleanly separate them), but rather how much variance each class of variables explains. And this is where contemporary, empirically informed debates about virtue ethics have settled. The situationist critique claims that situational variables explain so much variance in human behavior that there is little left to be explained by virtues (Doris, 2002, 2022).

Importantly, the variances in behavior that interest situationists concern variation among human beings. By contrast, Aristotle's main interest was to capture distinctively human functions. So, differences between apes and human beings would have been of primary interest to him. Moreover, one might be skeptical that the situationist critique has any degree of plausibility for explaining how our species differs from our closest living relatives in the animal kingdom.

Here, we take up the challenge of making the situationist critique plausible in an evolutionary context. Our central question is this: to what extent are our cooperative tendencies explained by robust, consistent, internal tendencies to cooperate, and to what extent are they explained by abilities to respond flexibly to changing situational factors? We can capture the tension between these competing explanations with a paradox:

1. Plausible evolutionary explanations of human cooperation predict the widespread existence of robust and consistent virtues.
2. Psychologists have observed situational variation in virtue-relevant behaviors.
3. Observed situational variation is inconsistent with the widespread existence of robust and consistent virtues.

Elsewhere, we have explored the possibility of rejecting the third claim (Dale & Wiegman, 2026), but another approach is simply to reject the first. In this chapter, we suggest that some evolutionary pressures on cooperation and other categories of social behavior—spite in particular—actually predict substantial situational variation in virtue-relevant behaviors.

4.1 Aristotle on Virtue

Before we argue for situational variation in spite, we need to paint a clear picture of the alternative explanation. To that end, we will briefly outline

Aristotle's theory of the virtues and then summarize how some neo-Aristotelians have sought to defend and update his theory in light of evolutionary biology.

To begin with, Aristotle is a perfectionist about human well-being. In the words of Lemos (2008), Aristotle "[believed] that every human being, indeed every living thing, strives toward its own perfection, flourishing, or *eudaimonia*" (p. 66). Moreover, Aristotle thought that humans have a telos (or purpose) just as a pair of scissors has a telos and that we flourish when we satisfy that telos. Knowing our telos comes about through understanding how humans are distinct from other organisms. So, what is good for human beings is to perfect their distinctively human functions. Virtues, then, are the traits that allow humans to perfect their capacities (or function well as humans, etc.).

Aristotle was deeply aware of humanity's distinctive sociality; people are political animals in ways that bees are not (*Politics*, book 1, chapter 2 in Ackrill, 1987, p. 509), and this is why many Aristotelian virtues are social in nature. For instance, friendliness concerns a person's agreeableness in social relationships, and an important part of being just is being fair in one's dealings with others. He also emphasized humanity's rationality relative to other animals. Thus, we live well by reasoning well, and "The person reasoning well will act so as to cultivate those states of being—the virtues—that enable him to function properly" (Casebeer, 2003, p. 43). Importantly, the virtues are not innate for Aristotle. Rather, they are acquired via experience, through reward, punishment, and perhaps most importantly, habituation.

As Mizzoni (2019) points out, one of the central challenges to Aristotelian virtue ethics is what Bernard Williams calls "the first and hardest lesson of Darwinism, that there is no such teleology at all, and that there is no orchestral score provided from anywhere according to which human beings have a special part to play" (Williams, 1983, p. 565). However, a few philosophers (Casebeer, 2003; Lemos, 2008)—we will call them the Darwinian neo-Aristotelians—have defended Aristotle's view, in large part by attempting to naturalize the Aristotelian notion of telos or function. For example, Casebeer (2003) adopts Godfrey-Smith's (1994) modern-history theory of functions according to which "functions are dispositions and powers which explain the recent maintenance of a trait in a selective context" (p. 52). In other words, the function of a trait is the purpose for which it was most recently selected. For example, the primary function of the opposable thumb in many primates is to grasp tree limbs for locomotion. In humans, the function is likely to be far more diffuse and general, involving the construction and use of tools and weapons rather than locomotion. Of course, in both cases, the distal function of the thumb is to promote survival and reproduction, but just as Godfrey-Smith (1994) argues that modern-history functions (rather than distal functions) are explanatorily relevant in modern biology, Casebeer argues that these are the relevant functions for defining human flourishing: "Morally speaking, it is not true that anything goes, but neither is it true that our *only* proper function is to breed like rabbits" (Casebeer, 2003, p. 53). While Casebeer does not attempt to list the functions that constitute a flourishing human life (aside from the distal functions of survival and reproduction), he does claim that "Aristotle's virtues are in fact virtuous ... they help us live a fully functional life" (p. 156).

At this point, it is worth pointing out that one could adopt a naturalized Aristotelian account of human flourishing while remaining skeptical that such flourishing is best achieved by cultivating virtues. One might think instead that human flourishing is actually achieved via a bricolage of capacities that allow us to flexibly adjust to situational changes and that our best efforts at cultivating human flourishing will be directed at shaping our environment so that we respond to it in functional ways. Just as our thumbs, fingers, and wrists need ergonomic handholds, we might think that our moral psychology requires carefully constructed situations for it to function best. The point is that you could follow the Darwinian neo-Aristotelians this far without committing yourself to the existence of virtues as Aristotle conceives of them. Naturalizing Aristotelian functions only gets you part of the way to the existence of virtues or their necessity for proper function or flourishing.

Some contemporary work in the social sciences may help fill in this lacuna. For example, Haidt and Joseph (2004, 2008) draw on moral foundation theory to suggest five different domains of virtue defined by adaptive problems that our species has faced: harm/care, fairness/reciprocity, in-group/loyalty, authority/respect, and purity/sanctity. They view virtues as skills, and they take them to be built on and refined by intuitions and emotional responses that were adapted for navigating these different domains. On this view, virtues exist as internally developed traits that help us flourish as human beings through practiced skills. In a similar vein, Curry et al. (2019) identify seven different domains of cooperation from literature on the evolution of cooperation, some of which are distinctively human. Moreover, they demonstrate that "these specific forms of cooperative behavior ... [are] considered morally good wherever they arise, in all cultures" (p. 48).[2] While they only refer to some traits within these domains as "virtues," it is easy to see how their theory of morality-as-cooperation could be extended to concepts of virtue in the same way that Haidt and Joseph attempt. This does not even scratch the surface of work by primatologists, anthropologists, and cultural psychologists whose work draws out other distinctive human functions (such as social learning and coordinated suppression of bullies and cheats), many of which further emphasize sociality (Boehm, 2012; Henrich, 2015; Tomasello, 2016; Wrangham, 2019). Nevertheless, it demonstrates how the existence of virtues might seem plausible in an evolutionary framework.

So far, we have outlined Aristotle's main theoretical commitments regarding the virtues and outlined some of the evolutionary considerations with which some neo-Aristotelians have updated the theory, but of course, this is a small fraction of what Aristotle and others have to say about virtues. In several places, Aristotle appears to commit himself to some fairly firm psychological claims about the virtues:

> According to Aristotle (1984: 1105a27–b1), genuinely virtuous action proceeds from "firm and unchangeable character" rather than from transient motives. The virtues are *hexeis* (1984: 1106a11–12), and a *hexis* is a state that is "permanent and hard to change" (1984: Categories, 8b25–9a9). Accordingly, while the good person may suffer misfortune that impairs his activities and diminishes happiness, he "will never [*oudepote*] do the acts that are hateful and mean" (1984: 1100b32–4; cf. 1128b29; cf. Cooper 1999: 299n14). The presence of virtue is supposed to provide assurance as to what will get done as well as what won't; for Aristotle (1984: 1101a1–8;1140a26–b30), the paradigmatically virtuous *phronimos*, or practically wise man, is characterized by his ability to choose the course of action appropriate to whatever circumstance he is in, whether it be easy or excruciating. *Arete* [roughly, "excellence" in Greek], Aristotle says, is "always concerned with what is harder" (1984: 1105a8–10); standing firm in the most terrifying crises, not just in any frightening situation, is diagnostic of the brave person (1984: 1115a24–6; 1117a15–22).
>
> (Doris, 2002, p. 17)

These claims enshrine a key commitment of most virtue ethicists in the Aristotelian tradition: virtues are robust and consistent (see, e.g., Doris, 2002, chapter 2, esp pp. 16–18 for an overview).[3]

4.2 Situational Variation and Effect Size

This is where the situationist critique enters into the picture: they assert that human behavior is not well predicted by robust and consistent character traits. It is far better predicted by situational influences. One of the foremost studies cited by situationists is Milgram's (1974) obedience study, which revealed that ordinary, and seemingly compassionate, people were willing to administer debilitating shocks to an unseen participant if they were told to do so by an experimenter. The list of experiments illustrating the powerful influence of situations is long and has been much discussed elsewhere (see Doris, 2002 for an overview), so we will only quickly mention two others. Darley and Batson (1973) carried out a study on Princeton Theological Seminary students and found that 90% of the students did not stop to help a man in obvious distress slumped in a doorway when they were running late. Finally, subjects in Isen and Levin (1972) who found a dime were significantly more likely to help a stranger pick up her dropped papers than those who did not find a dime.[4]

Situationists have used this evidence to argue that the kinds of character traits that Aristotelian virtue ethics calls for are extremely rare and difficult to cultivate (Alfano, 2013; Doris, 2002; Harman, 2009; Merritt, 2000; Merritt et al., 2010; Sarkissian, 2010; Vranas, 2005). John Doris (2002) admits that people exhibit local character traits (i.e., traits that reliably manifest in similar circumstances), but global character traits (i.e., traits that are stable and consistent) are rare to nonexistent.

In response, some have pointed out that these studies often fail to replicate (Alfano, 2018; Miller, 2003, 2017), while others have pointed out other deficiencies in the design and execution of the studies (Jayawickreme et al., 2014; Kamtekar, 2004; Kristjánsson, 2008; Sabini & Silver, 2005; Snow, 2010; Solomon, 2005; Sreenivasan, 2002). Situationists have followed up by acknowledging some of these deficiencies in the experimental literature but maintain that the empirical basis for situationism (which they now label *character skepticism*) remains "secure" (Doris, 2022, p. 235; Waggoner et al., 2022).

They claim this primarily due to the persistent effect size of 0.3 for psychological traits in the social and personality psychology literatures (Doris, 2022, pp. 215–235; Waggoner et al., 2022). As Doris explains, an effect of 0.3 indicates that character traits have an at best minimal effect on behavior. Specifically, "discovering someone is appropriately attributed a trait associated with an effect size of r = .3 yields an increase in predictive success of 15 percent over *knowing nothing about them at all*" (Doris, 2022, p. 222). As a result of this, "dispositional variables have weaker influence, and situational variables stronger influence, than one should expect if one understands dispositional differences in terms of robust traits issuing in cross-situationally consistent behavior" (Doris, 2022, p. 217).

4.3 Evolutionary Explanations of Social Behaviors

If we look at character traits from an evolutionary perspective, what do we find? Should we expect the evolution of robust traits or strong situational determinants of behavior? Game theory can give us some limited answers to these questions. From a simplified[5] game theoretic perspective, there are four categories of social behavior that evolutionary theory needs to explain, and all have to do with the payoffs of pairwise interactions. If we have an actor and a recipient, the actor's behavior is...

A ...cooperative when it benefits the actor and recipient.
B ...altruistic when it benefits the recipient at a cost to the actor.
C ...spiteful when it imposes a cost to the recipient at a cost to the actor.
D ...selfish when it imposes a cost on the recipient to the benefit of the actor.

Clearly, each of these behavior types is relevant to one or more categories of virtue. This is not to say that all cases of cooperation are virtuous or that all cases of spite are vicious. Spiteful behavior can be virtuous, and cooperative behavior can be vicious: Out of courage, you might impose a cost on a war criminal at a cost to yourself. Out of cowardice, you might cooperate with the operatives of an oppressive regime. Nevertheless, if one's tendency toward spite or cooperation in general were more strongly influenced by non-moral, situational variables rather than by one's character, then this would tell against virtue as the best explanation for these behaviors in many cases. If spiteful actions are more likely in conditions of scarcity (as we argue below), then perhaps it is the situation of scarcity, rather than your courage, that motivates your costly "punishment" of the war criminal.[6]

Our argument centers on Hamilton's Rule, a simple but powerful way to predict each of these categories of behavior across many species (see, e.g., Abbot et al., 2011): $RB > C$, where R is a measure of relatedness (which can be positive or negative), B is the benefit to the recipient (again, positive or negative), and C is the cost to the actor (also positive or negative). Put simply, selection will favor behaviors that cost the actor less than the benefits to the recipient, weighted by the relatedness of the recipient to the actor. Positive relatedness favors altruism; negative relatedness favors spite.

Hamilton's rule makes sense when fitness is understood inclusively:

> Inclusive fitness may be imagined as the personal fitness which an individual actually expresses in its production of adult offspring ... after it has been first stripped ... of all [social effects] ... then augmented by ... the quantities of harm and benefit which the individual himself causes to the fitnesses of his neighbours ... [weighted by] the coefficients of relationship appropriate to the neighbours whom he affects: unity for clonal individuals, one-half for sibs, one-quarter for half-sibs, one-eighth for cousins, ... and finally zero for all neighbours whose relationship can be considered negligibly small.
>
> (Hamilton, 1964, p. 8)

In other words, one's fitness is affected by relatedness.[7] Positive relatedness to another means that helping the other will promote the spread of your genes in a population. Negative relatedness means that helping will reduce the spread of your genes. If helping someone makes your genes less frequent in the population by comparison with competitors, then you are negatively related to them.

Grafen (1985) formalized these ideas in his geometrical theory of relatedness, showing that relatedness depends on gene frequency in a population, rather than shared ancestry. If your genes are rare, helping a relative boosts their spread (as in Figure 4.1a), and thus you are positively related in that circumstance. If your genes are common, helping that same relative may dilute your genetic advantage (as in Figure 4.1b), making you negatively related to the same individual only because the population differs. Figure 4.1 illustrates this shift: when your genes are widespread, aiding a recipient who shares only some of them can reduce your overall representation.

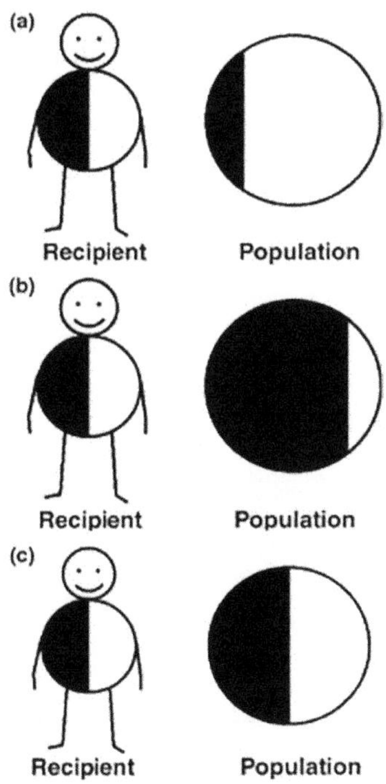

Figure 4.1 Illustration of Grafen's (1985) geometrical theory of relatedness, depicting high (a), low (b), and intermediate (c) degrees of relatedness to the same individual relative to different populations. From Gardner and West (2004) with permission.

Grafen's geometrical theory of relatedness (1985) is one of the simplest and best ways of capturing relatedness on a scale of 1 (identical twins) to –1 or even lower.[8] His formalization also explains why average relatedness in a population approaches zero as population size increases:

$$R = -1/(N - 1)$$

where R is average relatedness in a population and N is the population.

As you can see from Figure 4.2, as population size decreases, average relatedness becomes increasingly negative.

On this way of understanding fitness, altruism and spite can promote fitness, even if they lead to your death, at least insofar as they increase the proportion of your genes in the next generation. They can do this either directly, by contributing to your relatives' reproductive success, or indirectly, by decreasing the proportion of competing genes in the next generation. An instance of the latter would be if you were to sacrifice your life to also end that of someone who is negatively related to you, thereby reducing their reproductive success and so also the amount of competition for your kin's offspring (the number of which may be unaffected by your action).

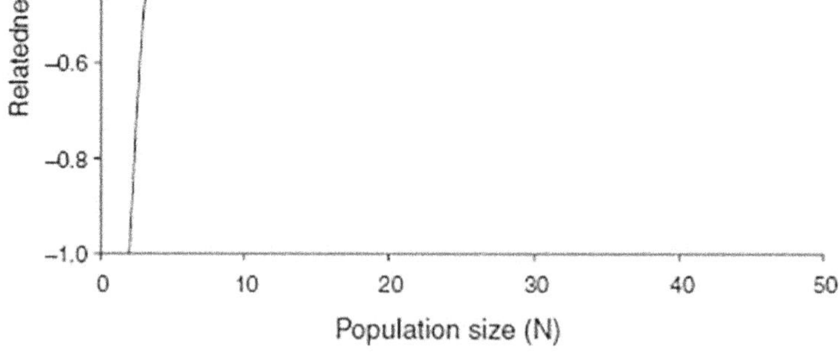

Figure 4.2 Relatedness as a function of population size according to Grafen's (1985) theory of relatedness. From Gardner and West (2004) with permission.

Of course, there are other processes aside from selection that affect evolutionary outcomes, like drift and gene flow. Nevertheless, selection is often a stronger influence than these other processes. So, many key evolutionary outcomes depend on replicator dynamics, and as we have just shown, Grafen's theory of relatedness shows that these dynamics depend on relatedness (similar alleles for genes that determine certain traits) relative to a population among which different alleles compete for copies in the next generation.

4.3.1 Applying Hamilton's Rule and Grafen's Theory of Relatedness

Evolutionary outcomes are not straightforwardly determined by the number of reproducing offspring (or copies of one's genes) in the next generation. Rather, they are more accurately determined by the *proportion* of offspring (or copies of genes) *relative to competing individuals* (or perhaps competing alleles) in the population. If you produce a large number of offspring and so do all your many competitors, the effects of selection may be negligible. However, if you produce a large number of offspring and your competitors do not, there is likely strong selection pressure favoring your phenotype.

It follows that in conditions of abundance, inclusive fitness reduces to personal fitness (fitness irrespective of social effects of behavior); the social environment does not exert any selection pressure. The net costs and benefits of the social environment get washed out by the abundance. In any but the smallest population, taking away resources from a competitor (or even killing them outright) makes no difference if the competitor can just go find more elsewhere or if there are plenty of other competitors in the population (which there are likely to be when abundance reigns). Abundance reduces competition for resources by definition.

Scarcity changes everything. Scarcity of resources limits the holding capacity of an environment, thus limiting the number of offspring a population can sustain to reproductive maturity. This is why taking away or destroying the resources of another makes more sense in conditions of scarcity. Your control over an absolute quantity of resources matters less than having a greater quantity than other individuals, because if you consistently have more than your competitors and this translates to relative fecundity, then your offspring will represent a higher proportion of the next generation. And so, you can increase your inclusive fitness, making things easier for your offspring (even if they are few) by hampering the reproductive potential of your competitors. If competitors have fewer offspring when the pie of resources is already small, then your offspring will not only be a larger proportion of the population but also will have fewer competitors to split the pie with in the next generation. Spite can clearly pay off in these conditions, insofar as one can harm a negatively related competitor at a smaller cost to oneself (modulated by relatedness, of course). Population size amplifies this effect. In small populations, negative relatedness is common (as demonstrated in Figure 4.2), and harming others can significantly boost your relative fitness.

So, both scarcity and population size change the payoffs for spite and selfishness. We can add that both formally and empirically, preferences for relative advantage are likely to drive spite and that these kinds of preferences tend to dominate in small populations. In psychology and behavioral economics, one of the best ways to measure spite, and to distinguish it from pure self-interest, is the Ultimatum Game. In this game, one player, the sender, receives a windfall sum of money, say $10, and then decides how to split it with another player, the receiver. The receiver can accept the offer or reject it. If the receiver rejects the offer, then neither party receives any of the money. Of course, the self-interested option for the receiver is to accept any nonzero amount of money, and that is also the Nash equilibrium for this game. However, most people will choose to reject low offers (of less than half), and they appear to do so because of an aversion to inequity (Camerer, 2003; Henrich, 2000; Henrich et al., 2004; Oosterbeek et al., 2004). In other words, most people prefer to receive a payoff of zero rather than for the sender to receive a greater share of the windfall. This is also spiteful behavior, because the receiver is paying a cost (losing the money offered) to impose a cost on the sender.

Besides scarcity and population size, there is yet a third situational variable that influences spiteful behavior: whether the scale of competition is global or local. That is, in many circumstances, it is not the entire population of a species that matters for the outcome of the gene replication game, but rather the portion of the population that defines the competitive arena. In their discussion of spite, Gardner and West (2004) put the point like this:

> Hamilton was correct when he stated that spite should be restricted to tiny populations; however, the 'population' of interest is that of the competitive arena … as competition becomes increasingly local, the reference population shrinks towards the size of the social arena, which may contain only a few individuals (small N) and/or a significant proportion of identifiable positively related kin, such that the negative relatedness towards the other potential recipients is nontrivial, enhancing the selective value of spite … Increasingly local competition means that fitness is increasingly dominated by success relative to the social group average, and so altruism is less favoured. Conversely, spiteful behaviour incurs a direct cost and reduces the success of social partners, so that more spiteful individuals can have higher success relative to the group average, but suffer a reduction in

> absolute success. When competition is global and fitness is proportional to absolute success, spite cannot be favoured, but as competition becomes increasingly local fitness is increasingly determined by success relative to social partners, so that spite can be a winning strategy.
>
> (pp. 1197–1198)

As Gardner and West hypothesize, humans do appear to be sensitive to the scale of competition. We are more likely to reject low offers in the ultimatum game if success is determined by results of interactions with other individuals, and less likely to reject low offers when success is determined by global rankings, where one is unlikely to actually interact with many of one's competitors (Barclay & Stoller, 2014). When one's success or failure depends on beating the people we interact with, it can often be advantageous to harm a competitor at a cost so that our local competitors don't beat us. In other words, when competition is local, we begin to care more about our payoff relative to our competitors than about the absolute value of our payoff. Earning a household income of 85 thousand dollars a year when the Joneses make 80 thousand actually feels better than making 90 thousand when the Joneses make 95 thousand dollars a year. Even though the scale of competition and population size are technically distinct variables, we will lump them together in what follows by assuming that the relevant population consists of one's competitors.

In an analytical model, Eaton and Eswaran (2003) showed that preferences for higher relative payoffs (rather than absolute payoffs) are evolutionarily stable in smaller populations on the scale of hunter-gatherer populations in the Pleistocene (where they seem to be referring to the population of one's competitors rather than the population of humans at the time). In other words, Eaton and Eswaran show that in circumstances where spiteful behavior is advantageous, so are the preferences that support this behavior.[9] We can conjecture that preferences for relative advantage (rather than absolute payoffs) also become more advantageous when resources are scarce, since this is another circumstance where spiteful behaviors are advantageous. Conversely, preferences for relative payoffs become less advantageous (relative to preferences for absolute payoffs) as population size increases and when resources are abundant. This is the beginning of a line of thought that supports situationism about virtue-relevant behaviors like spite and cooperation.

4.4 Evolution of Situational Cooperation, Selfishness, and Spite

The line of thought is this: first, as explained above, in smaller populations and also in conditions of scarcity, evolution tends to favor spiteful behaviors and preferences. That is, spite and selfishness, as well as the concerns for relative (rather than absolute) payoffs (e.g., Eaton and Eswaran, 2003), become more advantageous and cooperation becomes less so in conditions of scarcity, in smaller populations, and where the scale of competition is local (particularly when kin relations can be easily tracked). A corollary is that spiteful and selfish behaviors become less advantageous as population size increases and as resource scarcity decreases. Thus, we have already shown that the adaptiveness of selfishness and spite as opposed to cooperation varies according to two key situational variables: population size and resource scarcity. In this section, we want to show that there was strong variation in these situational variables over significant spans of human evolution, favoring individuals who could flexibly alternate between the pursuit of absolute and relative payoffs. Moreover, there is considerable evidence that humans do exhibit more selfish and spiteful behavior in extended conditions of scarcity. We will also show that there is a plausible neural mechanism for the modulation of spiteful behavior involving serotonin and dopamine, which appears to have some key genetic determinants. If humans evolved to be more spiteful in conditions of scarcity, genetically influenced mechanisms are exactly what we would expect to find. The best explanation for all of this is that humans evolved to have conditional preferences that strongly influence virtue-relevant behaviors depending on three key situational variables (i.e., scarcity, population size, and scale of competition).

4.4.1 *Evidence of Scarcity and Flexible Responses to Scarcity*

Having shown the advantages of spite in situations of scarcity (etc.), the next step is to notice that there has been significant variation in resource scarcity (and likely also population size) in human evolutionary history. The Pleistocene epoch—between 2.5 million years ago and 11,000 years ago—is when we evolved many of the traits that make us distinctive as a species, and in particular the later Pleistocene, beginning around 1 million years ago (Boehm, 2012). Importantly, this era of time was marked with extreme climate variability, especially in eastern Africa, the area of the world that our ancestors inhabited (Gould, 1982; Timmermann et al., 2022). This is when eastern Africa dried out and became mostly savannah and open grassland (Levin et al., 2004). Throughout this transition, there were rapidly changing weather patterns, almost certainly resulting in regular droughts and famines for our evolutionary ancestors.

As a result of this, there was pressure for our ancestors to evolve flexible responses to scarcity (which was frequent, but not pervasive) and, likely also, flexible responses to population size. In fact, decreasing the size of the population with which one competes for resources is a common response to famine:

> [T]o the extent that we have an ethnography of famine (see Boehm [2012], pp. 274–8), it suggests that foragers respond to famine with social atomization. Bands split into family units as the population spreads itself more thinly over landscapes in response to its lowered carrying capacity. Fission–fusion becomes fission–more–fusion. That response was probably particularly adaptive in the Pleistocene, with its lower population densities. But if the initial response to famine is to disaggregate, to fragment into even smaller daily units of association, the option of attempting to expropriate the neighbour's territory is taken off the table.
>
> (Sterelny, 2016, p. 41)

Sterelny is drawing on a well-established method: deriving an understanding of the behavior of our evolutionary ancestors from observations of extant hunter-gatherer groups. The comparison is, of course, far from perfect, but it is often the best we have to go on. To see how modern hunter-gatherers respond to scarcity, let's first consider how they behave during times of plenty. Balikci (1970) observed the Netsilik, an Inuit group in the far north of central Canada, as they hunt seals on the sea ice. After making a kill, a hunter divides the seal into 14 different parts. For seven of those parts, he has permanent "giving partners" to whom he will give the assigned body parts, no questions asked. For instance, he has a permanent "flipper partner," whom he will always give the flipper to. This ensures that everyone has at least some meat throughout the hunting months.

Thus, during times of plenty, there is usually some kind of system of sharing that requires hunters to give away parts of their kill to other clan members, and this ensures that they receive food in the future, when another hunter makes a kill. Indeed, among the Netsilik,

> [T]here is a general rule ... according to which all able bodied men should contribute to hunting, and the returns of the hunt should be shared according to established custom. Any activity in exception to this rule was bound to provoke criticism, various forms of conflict, and frequently social ostracism. Lazy hunters were barely tolerated by the community. They were the objects of backbiting and ostracism until the opportunity came for an open quarrel. Stingy men ... were treated similarly.
>
> (Balikci, 1970, pp. 176–177)

Similar sharing expectations are found among the Ache of Paraguay (Gurven et al., 2001) the Mbuti Pygmies of central Africa (Turnbull, 1961), and the !Kung of southern Africa (Lee, 1979).

Of course, there are sometimes the rare, extremely bountiful times, such as when an Inuit group coordinates an ambush on an entire herd of caribou at a narrow river crossing and slaughters scores of animals (Binford, 1978). During these events, there is no need for sharing norms, as there is so much meat going around and it often cannot be preserved for long. But for the most part, these events are rare, and sharing norms ensure that everyone at least gets some meat most of the time, regardless of their success in a particular hunt.

However, times of scarcity do occur, and they occur fairly often. Keeley (1988) surveyed 40 North American foraging bands about food scarcity. Of these 40, 13 regularly experienced famines, with members dying of starvation. Another 12 were subject to occasional famines.[10] Food shortages are also known to occur among the peoples in the interior of Australia, as well as the !Kung and !Ko bushmen of southern Africa (Gould, 1982).

When a famine approaches, there are usually two stages of response (Boehm, 2012). First, families continue to share small game and plant foods among themselves, but the consumption of large game diminishes, as hunters have less energy and large game becomes lean and scarce. The result is much less sharing among nonkin. The next stage is when sharing breaks down even among families, and egoistic tendencies overtake nepotistic ones. At this point, even cannibalism is not all that uncommon. During one famine, Netsilik parents resorted to eating their own children (Balikci, 1970).[11]

Finally, we can observe flexible responses to scarcity versus abundance in the outcomes of economic games like *Joy of Destruction*. In this game, participants decide whether to sacrifice income in order to lower another player's payoff. Importantly, there is no material gain for harming, no punishment for wrongdoing, and no reduction in inequality—the game is merely testing for a participant's willingness to harm others. Prediger et al. (2014) had modern-day pastoralists play Joy of Destruction, and the results of the game depended markedly on the quality of grazing pastures (which determines quality and quantity of resources produced): harmful behavior among villagers occurred twice as often in areas where resources were more scarce.

Of course, we need to keep in mind that these are reports of modern-day hunter-gatherer and pastoralist groups. If anything, the occurrence of droughts and famines was *even more common* for our evolutionary ancestors, due to the extremely variable climate patterns in eastern Africa during the late Pleistocene (Gould, 1982; Timmermann et al., 2022).

4.4.2 Neural Mechanisms That Modulate Spiteful Behavior

The final step in our line of thought concerns genetically influenced mechanisms for spiteful behavior. This step is important because the fact that a behavior can increase fitness is not actually sufficient for thinking that the behavior is an evolutionary product. Learned behaviors are a good example. If I happened to learn to extract nutrients from a poisonous plant by cooking it a certain way, this might increase my fitness in a particular environment. However, if this trait isn't reliably passed on to offspring such that it continues to exist because of this increase in fitness, then it's not a product of Darwinian evolution. By contrast, selfish and spiteful behaviors are influenced by specific genes, which clearly are heritable.[12] Now, we cannot prove beyond the shadow of a doubt that these genes were selected for these effects (perhaps these effects are byproducts or perhaps they are related to some more general-purpose trait). In fact, proof for or against this claim seems impossible to muster, even with all that is known about spiteful behaviors and their neural mechanisms. However, we can show that given what is known, it is quite possible that selfish and spiteful behaviors increase in conditions of scarcity because of heritable mechanisms.

There is a fairly wide range of evidence connecting dopamine signals to specifically social valuations or comparisons (Batten et al., 2024; Pedroni et al., 2014; Sáez et al., 2015). This goes beyond the well-known claim that dopamine signals track reward value, because it suggests further that these signals estimate reward in humans via social comparison. In other words, dopaminergic reward signals may very well be signals of relative payoffs, rather than absolute ones. Now, recall that Eaton and Eswaran (2003) showed that evolutionary pressures in small populations lead to the evolution of preferences for relative payoffs of exactly this kind. Moreover, the preference for relative payoffs is perhaps the primary motivation for spiteful behavior in humans, since spiteful action against a competitor is a way of reducing the difference between your rewards and theirs (McCarthy-Jones, 2021, ch. 3). So, dopaminergic reward signals may have been shaped to implement exactly these kinds of preferences.

Dopamine is also importantly connected to selfish, spiteful, and antisocial behavior, and a downregulation of human "hyperaltruism." That is, when experimental participants are given drugs that increase tonic dopamine concentrations (e.g., with drugs like tolcapone or L-dopa), they are more willing to inflict electric shocks on other people for money (Crockett et al., 2015) and also more willing to make unfair offers in the dictator game (Pedroni et al., 2014), which is similar to the ultimatum game without a punishment option for the receiver.

Finally, there are genetic modulators of dopamine tone. Consider, for example, the genes that code for the COMT enzyme. This enzyme breaks down dopamine in the intercellular spaces within the brain. When its action is more efficient, there are lower levels of dopamine in the intercellular spaces of the brain. Certain variants of the COMT enzyme appear to diminish the function of this enzyme in a way that likely increases tonic dopamine concentrations and also appears to increase rejection of low offers in the ultimatum game (Strobel et al., 2011).

Now, this is probably less than half of the story that we really need to prove that there are heritable mechanisms to increase spite in conditions of scarcity. All we have shown thus far is that certain genes can influence *some* spiteful behaviors, likely by influencing the reward signals of dopamine. We have not demonstrated the existence of any mechanism for increasing spiteful behaviors specifically in response to scarcity. And so far as we know, there is no solid evidence of this.

However, there are some hints and allegations. Serotonin appears to have an opposite influence on spiteful behavior compared to dopamine. For example, when Crockett and colleagues induced serotonin depletion in experimental participants, their rejections of unfair offers in the ultimatum game went up (Crockett et al., 2013). By contrast, enhancing serotonin with a reuptake inhibitor reduced unfair offers in the ultimatum game (Crockett et al., 2010). Now, serotonin is a notoriously complex neurotransmitter, with many different receptor types and possibly many different signaling roles in different parts of the brain. However, a recent hypothesis is that tonic levels of serotonin signal resource availability, with lower levels indicating scarcity (as hypothesized in Koolschijn et al., 2024). If serotonin levels do function to decrease in response to scarcity, then perhaps this would be a mechanism by which spiteful behaviors increase in those conditions.

Granted, there are many bones of likely contention in these lines of reasoning about the function of dopamine and serotonin, and there is room for debate about the likelihood that selection intervened on this system to modulate selfish and spiteful behavior in conditions of scarcity. Perhaps the observed effects of these neurotransmitters on spite and selfish behavior are a byproduct of some more general function for which they were selected. Nevertheless, given that proof (for or against) is far off, all we can do is demonstrate the possibility, given what we know. And we take ourselves to have shown that given the known effects of dopamine and serotonin on some spiteful behaviors, it is quite possible that these systems could modulate spiteful behavior in conditions of scarcity, as predicted by the evolutionary pressures we discussed above. The key idea is that if dopamine is a signal of relative reward value, then amplifying dopamine signals or reducing serotonin signals (which for the most part have opposing effects to dopamine) will make people care more about relative reward value and incentivize spiteful actions, and it is easy to see how nature could have accomplished this via genes that manipulate tonic concentrations of these neurotransmitters. There may also be heritable pathways influencing spite involving hormones like testosterone.[13]

To sum up the larger line of thought here, we can put the argument as an inference to the best explanation. Selfish and spiteful behavior (and more generally preferences for relative rather than absolute payoffs) may very well have heritable determinants (involving dopamine and serotonin function);[14] there is substantial evidence that humans are far more likely to engage in such behaviors in conditions of scarcity; and evolutionary considerations predict that flexibly switching toward these behaviors is advantageous when resources become scarce. So, the best explanation appears to be that humans have evolved to become more selfish and spiteful in conditions of scarcity.

Perhaps the predictions of this hypothesis seem mundane if we are focused only on selfish behaviors; of course, people will take care of themselves first when there is very little to go around. That is something you can easily predict on the hypothesis that humans act rationally by maximizing their utility. However, spiteful behaviors seem more surprising when resources are scarce; if there are few resources, why would you ever pay a cost to make someone worse off? It certainly does not maximize utility to do so.

Virtue theories are another class of hypotheses that likely make opposing predictions. As discussed above (and in great detail in Doris, 2002, ch. 2), many virtue theorists claim that, for example, generous people will have robust dispositions for generous action even in the most extreme situations. At the very least, we know of no theories of virtue that would predict that a more or less generous person might flip and become downright spiteful in conditions of scarcity.[15]

Evolutionary theories of cooperation might also be read as making opposing predictions (as we discussed above). In the expansive literature on the evolution of cooperation, one might easily conclude that cooperation is a fairly fixed behavioral strategy in human populations. Yet the foregoing arguments give the lie to this apparent conclusion. Evolved situational flexibility seems a far better explanation than robust and consistent virtues or cooperative tendencies.

Perhaps the models that explain cooperation are only aimed at explaining its prevalence in some limited range of conditions, which may exclude scarcity.[16] But if so, we are still in a position to reject the first premise of the paradox with which we began:

1 Plausible evolutionary explanations of human cooperation predict the widespread existence of robust and consistent virtues.
2 Psychologists have observed situational variation in virtue-relevant behaviors.
3 Observed situational variation is inconsistent with the widespread existence of robust and consistent virtues.

If models of cooperation only explain its prevalence in a certain range of conditions, then they do not really predict that cooperative virtues will be robust or consistent outside of that range. If explanations of cooperation are to be subsumed under a more general theory that takes into account situational variables like scarcity and population size, then such explanations do not actually predict robust and consistent virtues. By design, virtue is likely to be fragile in the face of scarcity.[17]

4.5 Concluding Thoughts

We have argued that the apparent fragility of virtue is a feature of human psychology shaped by evolutionary pressures. Conditions of scarcity, small population size, and local competition favored individuals capable of flexibly shifting between cooperative, selfish, and spiteful strategies rather than embodying fixed, cross-situational virtues. Neurobiological evidence—particularly involving dopaminergic and serotonergic modulation of social preferences—suggests that these adaptive shifts are at least partly heritable. This undermines the assumption, central to many virtue theories, that robust character traits underlie human cooperation. Instead, our species' remarkable social capacities appear to depend on situational sensitivity.

One might wish to avoid this conclusion precisely because it shows human goodness (understood in terms of individual virtue) to be fragile and situation-dependent by design. However, just as personal growth often requires sober acknowledgment of one's own current limitations, it seems we must acknowledge these facts about humanity to scaffold moral behavior and human flourishing. Strategies of generalized reciprocity in evolutionary game theory (see, e.g., Barta et al., 2011; Herne et al., 2013; Pfeiffer et al., 2005) provide another possible mechanism for the flexibility of spite and also a scaffold for generosity: when we receive generosity, we are more likely to pay it forward. This represents a flexible response to abundance that mirrors the putative flexibility of spite. Thus, if we see that gratitude and generosity promote human flourishing and that it is often the mere perception of abundance versus scarcity that influences them, then we have important reasons to promote perceptions of abundance over scarcity and to also promote conditions of recognizable abundance. Acknowledging the fragility of virtue in the face of specific challenges can give us the means to reinforce it.

Another reason for optimism about human goodness is that in some conditions, spite can actually stabilize outcomes that we deem to be good. As Forber and Smead (2014) have demonstrated, spite can stabilize fair or equal distributions of payoffs in social interactions like the ultimatum game, and it may have also played a role in the evolution of capacities to recognize worthy beneficiaries of cooperation, punishment, or avoidance (Forber & Smead, 2016). Perhaps even when our individual motivations are most vicious, they can lead us to moral achievements at higher levels of organization.[18]

Notes

1 Indeed, by comparison with chimpanzees, it appears humans have undergone hundreds of thousands of years of self-domestication, diminishing our reactiveness to each other and increasing our willingness to cooperate with one another (Wrangham, 2019).

2 See also Robinson, this volume.

3 To us, it is interesting that to some degree, these claims capture common sense, or folk psychology, about the virtues but are actually fairly divorced from the rest of Aristotle's theory. That is, nothing about Aristotle's broader theory of virtues, teleology, eudaimonism, or perfectionism about human well-being suggests to us that virtues must be robust and consistent. Certainly, Aristotle claims that being *consistently* virtuous is necessary for flourishing, but this seems an empirical claim on which he might very well be wrong, even if everything else he says is true. Perhaps this theoretical disconnect (and the corresponding difference in emphasis) explains why situationism has never been a concern for the Darwinian neo-Aristotelians and perhaps also why situationists have largely ignored evolutionary explanations of human cooperation: the Darwinian neo-Aristotelians are focused on the existence of virtues qua traits that enable human flourishing, whereas the situationists are focused on the nonexistence of virtues qua traits that are also robust and consistent (as the common sense view of virtues seems to paint them). As we point out below (in note 16), this may be an easy divide to reconcile: it is possible that virtues exist in the first sense but not the second: traits that help someone function well as a human being across a range of "ordinary" conditions might very well fail to be robust across every trait-relevant situation.

4 This is actually an instance of generalized reciprocity (see, e.g., Pfeiffer et al., 2005, Barta et al., 2011), where situational windfalls are ecological signals of resource abundance and thus prime the organism for prosocial behaviors. As we suggest in the conclusion, this kind of reactive generosity may be the mirror image of the situational flexibility of spite.

5 As Forber and Smead (2015) have argued, Hamilton's (1964) categorization assumes unilateral action, and so also elides interdependency of payoffs (as in the Stag Hunt), and does not track frequency-dependent effects, which are often significant for evolutionary outcomes. Forber and Smead's preferred categorization tracks interaction types (Coordination, Social Dilemma, Conflict, Social Delight), rather than action types. Our argument is more conceptual than dynamical, and as we are focused on virtue relevant actions, Hamilton's classification is a useful heuristic for our purposes.

6 Relatedly, Raihani and Bshary (2019) have argued that punishment often has a competitive function of imposing costs on competitors, rather than just a deterrent function. Similarly, Forber and Smead (2016) have suggested that punishment may have evolved as a co-opted form of spite.

7 There is an alternative to inclusive fitness, called neighbor-modulated fitness. When Hamilton's rule is interpreted in this way, it leads to many of the same

outcomes and predictions as inclusive fitness. However, as Okasha and Birch (2015, p. 26) suggest, it obscures causal relationships between the agent's actions and fitness outcomes and makes it difficult to say that an agent is acting to maximize its fitness, which is a main attraction of the concept of inclusive fitness.

8 See Madgwick (2020).

9 Relatedly, Hill and Buss (2006) draw out and test several predictions based on the importance of preferences for relative advantage in human evolution.

10 The famine free groups mostly lived in California and were sedentary, an indication that they were perhaps not as representative of our Pleistocene ancestors as the other groups were.

11 There is some evidence that in the very short term, low blood sugar and/or the hungry state it causes can actually make people more prosocial, in the sense that hungry people are more likely to favor welfare programs (Aarøe & Petersen, 2013). This may seem to contradict evidence concerning even the initial stages of response to famine. However, there are several key differences. One is that this research was conducted on WEIRD populations, so the findings may not generalize. Another is that low blood sugar indicates only short-term and non-urgent scarcity: you can live without the carbs that quickly raise blood sugar to more comfortable levels, but you cannot live without amino acids like tryptophan (which are building blocks for muscle and neurotransmitters like serotonin). Yet another difference is that support for welfare is generous only in a weak sense. In a state of low blood sugar, people who advocate government welfare programs might not feel as supportive of norms requiring them to share food with others. In other words, these experiments do not demonstrate costly forms of generosity or cooperation in the face of scarcity.

12 This is not to say that the behaviors themselves are genetically determined. The evidence we consider is more favorable toward the hypothesis that prepared learning mechanisms bias human minds toward selfish behaviors in certain conditions. And this also means that cultural evolution could play an important role in propagating the learning of such behaviors through a population.

13 Higher testosterone levels in men have been correlated with increased rejection of unfair offers in the ultimatum game (Burnham, 2007), and temporary increases in testosterone also increase men's focus on certain markers of status.

14 However, the question of whether we can identify genetic bases of behavior is somewhat fraught. For discussion of the conceptual and methodological challenges to the nature–nurture divide and genetic determinism, see Longino (2013), Downes (2016), and Kronfeldner (2018, ch. 4). Kronfeldner critiques genetic determinism and reductionism, emphasizing the developmental/interactionist consensus and the entanglement of biological, cultural, and environmental factors.

15 However, on a naturalized evolutionary account of human functioning like Casebeer's (2003, discussed above), it is difficult to see why spitefulness in conditions of scarcity would not be a part of what it means to be a well-functioning human. If so, this would seem to commit the Darwinian neo-Aristotelian to saying that one must act spitefully in conditions of scarcity to flourish as a human being, since our evolutionary history shaped us to so act in those conditions. In other words, if Aristotelian functions are naturalized, then perhaps the virtues will not actually conform to the folk psychological notions of virtues as robust and consistent traits. This may be an instance in which a naturalized account is running afoul of Hume's is-ought divide.

Just because human beings were designed to behave spitefully in conditions of scarcity doesn't mean that they should behave spitefully in conditions of scarcity or that it is virtuous to do so. We are unsure of how Aristotle or the Darwinian neo-Aristotelians should address this problem.

16 Indeed, this may be the seed of reconciliation between situationism and neo-Aristotelian virtue ethics. Surely there are some social functions, excellence at which is required most of the time and in most places for individual humans to succeed by surviving and reproducing. That these functions would have environmental requirements and "norms of reaction" to various environmental conditions does not detract from their existence as real functions in which humans can excel. For example, extreme generosity is characteristic of individuals with high prestige, and it conduces to the reproductive success of prestigious individuals (Henrich, 2015). Generosity is thus a plausible neo-Aristotelian virtue, and it remains so, even if the most prestigious of individuals would give in to stinginess in conditions of scarcity. In other words, the neo-Aristotelian picture may be more plausible if it relaxes the robustness requirement: even behaviors that fail to be robust to certain situational variations can help someone excel at the various functions relevant to survival and reproduction.

17 Importantly, vices may be far less situationally variable. Whereas most people are only likely to be spiteful or selfish in conditions of scarcity (etc.), about one third of people have fairly consistent tendencies toward spite in some experiments (e.g., Kimbrough and Reiss, 2012). People with dark triad personality traits (psychopathy, narcissism, and Machiavellianism) and personality disorders like sociopathy, psychopathy, or borderline personality disorder are also consistently spiteful (see, e.g., Moshagen et al., 2018). In other work, we have suggested that part of the function of virtue language is to track reputation so that cooperators can reliably avoid defectors and bullies. Such language would be useful even if virtues are not robust and consistent, so long as it helps us to avoid those who are robustly and consistently vicious.

18 We are especially grateful to John Doris whose ideas and encouragement provided the inspiration for this essay, and to Patrick Forber and Rory Smead for their invaluable feedback.

References

Aarøe, L., & Petersen, M. B. (2013). Hunger games: Fluctuations in blood glucose levels influence support for social welfare. *Psychological Science*, *24*(12), 2550–2556.

Abbot, P., Abe, J., Alcock, J., Alizon, S., Alpedrinha, J. A., Andersson, M., Andre, J.-B., van Baalen, M., Balloux, F., Balshine, S., Barton, N., Beukeboom, L. W., Biernaskie, J. M., Bilde, T., Borgia, G., Breed, M., Brown, S., Bshary, R., Buckling, A., … & Zink, A. (2011). Inclusive fitness theory and eusociality. *Nature*, *471*(7339), E1–E4.

Ackrill, J. L. (1987). *A new Aristotle reader*. Princeton University Press.

Alexander, R. (2017). *The biology of moral systems*. Routledge.

Alfano, M. (2013). *Character as moral fiction*. Cambridge University Press.

Alfano, M. (2018). A plague on both your houses: Virtue theory after situationism and repligate. *Teoria*, *38*(2), 115–122.

Aristotle (1984). *The complete works of Aristotle*. Princeton University Press.

Aristotle (2000). *Nicomachean ethics*. Cambridge University Press.

Balikci, A. (1970). *The Netsilik Eskimo*. Natural History Press.
Barclay, P., & Stoller, B. (2014). Local competition sparks concerns for fairness in the ultimatum game. *Biology Letters*, *10*(5), 20140213.
Barta, Z., McNamara, J. M., Huszar, D. B., & Taborsky, M. (2011). Cooperation among non-relatives evolves by state-dependent generalized reciprocity. *Proceedings of the Royal Society B: Biological Sciences*, *278*(1707), 843–848.
Batten, S. R., Bang, D., Kopell, B. H., Davis, A. N., Heflin, M., Fu, Q., Perl, O., Ziafat, K., Hashemi, A., Saez, I., Barbosa, L. S., Twomey, T., Lohrenz, T., White, J. P., Dayan, P., Charney, A. W., Figee, M., Mayberg, H. S., Kishida, K. T., Gu, X., & Montague, P. R. (2024). Dopamine and serotonin in human substantia nigra track social context and value signals during economic exchange. *Nature Human Behaviour*, *8*(4), 718–728.
Binford, L. R. (1978). *Nunamiut ethnoarchaeology*. Academic Press.
Boehm, C. (2012). *Moral origins: The evolution of virtue, altruism, and shame*. Basic Books.
Burnham, T. C. (2007). High-testosterone men reject low ultimatum game offers. *Proceedings of the Royal Society B: Biological Sciences*, *274*(1623), 2327–2330.
Camerer, C. F. (2003). *Behavioral game theory: Experiments in strategic interaction*. Princeton University Press.
Casebeer, W. D. (2003). *Natural ethical facts*. MIT Press.
Cooper, J. M. (1999). *Reason and emotion*. Princeton University Press.
Crockett, M. J., Apergis-Schoute, A., Herrmann, B., Lieberman, M. D., Müller, U., Robbins, T. W., & Clark, L. (2013). Serotonin modulates striatal responses to fairness and retaliation in humans. *Journal of Neuroscience*, *33*(8), 3505–3513.
Crockett, M. J., Clark, L., Hauser, M. D., & Robbins, T. W. (2010). Serotonin selectively influences moral judgment and behavior through effects on harm aversion. *Proceedings of the National Academy of Sciences*, *107*(40), 17433–17438.
Crockett, M. J., Siegel, J. Z., Kurth-Nelson, Z., Dayan, P., & Dolan, R. J. (2015). Dissociable effects of serotonin and dopamine on the valuation of harm in moral decision-making. *Current Biology*, *25*(8), 1055–1060.
Curry, O. S., Mullins, D. A., & Whitehouse, H. (2019). Is it good to cooperate? Testing the theory of morality-as-cooperation in 60 societies. *Current Anthropology*, *60*(1), 47–69.
Dale, M. T., & Wiegman, I. (2026). Evolution, situationism, and virtue ethics. *Philosophical Studies*. https://doi.org/10.1007/s11098-026-02527-9
Darley, J., & Batson, D. (1973). From Jerusalem to Jericho: A study of situational and dispositional variables in helping behavior. *Journal of Personality and Social Psychology*, *27*, 100–108.
Doris, J. (2002). *Lack of character*. Cambridge University Press.
Doris, J. (2022). *Character trouble*. Oxford University Press.
Downes, S. M. (2016). Confronting variation in the social and behavioral sciences. *Philosophy of Science*, *83*(5), 909–920.
Eaton, B. C., & Eswaran, M. (2003). The evolution of preferences and competition: A rationalization of Veblen's theory of invidious comparisons. *Canadian Journal of Economics*, *36*(4), 832–859.
Forber, P., & Smead, R. (2014). The evolution of fairness through spite. *Proceedings of the Royal Society B: Biological Sciences*, *281*(1780), 20132439.
Forber, P., & Smead, R. (2015). Evolution and the classification of social behavior. *Biology & Philosophy*, *30*(3), 405–421.
Forber, P., & Smead, R. (2016). The evolution of spite, recognition, and morality. *Philosophy of Science*, *83*(5), 884–896.

Gardner, A., & West, S. A. (2004). Spite and the scale of competition. *Journal of Evolutionary Biology*, *17*(6).
Godfrey-Smith, P. (1994). A modern history theory of functions. *Noûs*, *28*(3), 344–362.
Gould, R. A. (1982). *To have and have not: The ecology of sharing among hunter-gatherers*. In N. M. Williams & E. S. Hunn (Eds.), *Resource managers: North American and Australian hunter-gatherers* (pp. 69–91). AAAS Press.
Grafen, A. (1985). A geometric view of relatedness. In R. Dawkins & M. Ridley (Eds.), *Oxford surveys in evolutionary biology* (Vol. 2, pp. 28–89). Oxford University Press.
Gurven, M., Allen-Arave, W., Hill, K., & Hurtado, A. M. (2001). Reservation food sharing among the ache of Paraguay. *Human Nature*, *12*(4), 273–297.
Haidt, J., & Joseph, C. (2004). Intuitive ethics: How innately prepared intuitions generate culturally variable virtues. *Daedalus*, *133*(4), 55–66.
Haidt, J., & Joseph, C. (2008). The moral mind: How five sets of innate intuitions guide the development of many culture-specific virtues. In P. Carruthers, S. Laurence, & S. Stich (Eds.), *The innate mind: Structure and content* (Vol. 3, pp. 367–391). Oxford University Press.
Hamilton, W. D. (1964). The genetical evolution of social behaviour I & II. *Journal of Theoretical Biology*, *7*(1), 1–52.
Harman, G. (2009). Skepticism about character traits. *Journal of Ethics*, *13*(2/3), 235–242.
Henrich, J. (2000). Does culture matter in economic behavior? Ultimatum game bargaining among the Machiguenga of the Peruvian Amazon. *American Economic Review*, *90*(4), 973–979.
Henrich, J., et al. (2004). Overview and Synthesis. In Henrich, J., Boyd, R., Bowles, S., Camerer, C., Fehr, E., & Gintis, H. (Eds.), *Foundations of human sociality* (pp. 8–54). Oxford University Press.
Henrich, J. (2015). *The secret of our success*. Princeton University Press.
Herne, K., Lappalainen, O., & Kestilä-Kekkonen, E. (2013). Experimental comparison of direct, general, and indirect reciprocity. *The Journal of Socio-Economics*, *45*, 38–46.
Hill, S. E., & Buss, D. M. (2006). Envy and positional bias in the evolutionary psychology of management. *Managerial and Decision Economics*, *27*(2–3), 131–143.
Isen, A., & Levin, P. (1972). Effect of feeling good on helping: Cookies and kindness. *Journal of Personality and Social Psychology*, *21*, 384–388.
Jayawickreme, E., Meindl, P., Heizer, E., Furr, R., & Fleeson, W. (2014). Virtuous states and virtuous traits: How the empirical evidence regarding the existence of broad traits saves virtue ethics from the situationist critique. *Theory and Research in Education*, *12*(3), 282–308.
Kamtekar, R. (2004). Situationism and virtue ethics on the content of our character. *Ethics*, *114*(3), 458–491.
Keeley, L. H. (1988). Hunter-gatherer economic complexity and "population pressure". *Journal of Anthropological Archaeology*, *7*(4), 373–411.
Kimbrough, E. O., & Reiss, J. P. (2012). Measuring the distribution of spitefulness. *PLOS ONE*, *7*(8), e41812.
Koolschijn, R. S., Polner, B., Hoomans, J. M., Cools, R., Vassena, E., & den Ouden, H. E. M. (2024). Resources, costs and long-term value: An integrative perspective on serotonin and meta-decision making. *Current Opinion in Behavioral Sciences*, *60*, 101453.

Kristjánsson, K. (2008). An Aristotelian critique of situationism. *Philosophy*, *83*(1), 55–76.
Kronfeldner, M. (2018). *What's left of human nature? A post-essentialist, pluralist, and interactive account of a contested concept*. MIT press.
Lee, R. B. (1979). *The! Kung San: Men, women, and work in a foraging society*. Cambridge University Press.
Lemos, J. (2008). *Commonsense Darwinism: Evolution, morality, and the human condition*. Open Court.
Levin, N. E., Quade, J., Simpson, S. W., Semaw, S., & Rogers, M. (2004). Isotopic evidence for Plio-Pleistocene environmental change at Gona, Ethiopia. *Earth and Planetary Science Letters*, *219*(1–2), 93–110.
Longino, H. (2013). *Studying human behavior*. Brill.
Madgwick, P. G. (2020). Spite and the geometry of negative relatedness. *The American Naturalist*, *196*(5), E119–E126.
McCarthy-Jones, S. (2021). *Spite: The upside of your dark side*. Basic Books.
Merritt, M. (2000). Virtue ethics and situationist personality psychology. *Ethical Theory and Moral Practice*, *3*, 365–383.
Merritt, M., Doris, J. M., & Harman, G. (2010). Character. In J. M. Doris & The Moral Psychology Research Group (Eds.), *The moral psychology handbook* (pp. 354–400). Oxford University Press.
Milgram, S. (1974). *Obedience to authority*. Harper & Row.
Miller, C. (2003). Social psychology and virtue ethics. *The Journal of Ethics*, *7*(4), 365–392.
Miller, C. (2017). Character and situationism: New directions. *Ethical Theory and Moral Practice*, *20*(3), 459–471.
Mizzoni, J. (2019). Evolution and (Aristotelian) virtue ethics. *Human Affairs*, *29*(2), 199–206.
Moshagen, M., Hilbig, B. E., & Zettler, I. (2018). The dark core of personality. *Psychological Review*, *125*(5), 656–688.
Okasha, S., & Birch, J. (2015). Kin selection and its critics. *BioScience*, *65*(1), 22–32.
Oosterbeek, H., Sloof, R., & van de Kuilen, G. (2004). Cultural differences in ultimatum game experiments: Evidence from a meta-analysis. *Experimental Economics*, *7*(2), 171–188.
Pedroni, A., Eisenegger, C., Hartmann, M. N., Fischbacher, U., & Knoch, D. (2014). Dopaminergic stimulation increases selfish behavior in the absence of punishment threat. *Psychopharmacology*, *231*(1), 135–141.
Pfeiffer, T., Rutte, C., Killingback, T., Taborsky, M., & Bonhoeffer, S. (2005). Evolution of cooperation by generalized reciprocity. *Proceedings of the Royal Society B: Biological Sciences*, *272*(1568), 1115–1120.
Prediger, S., Vollan, B., & Herrmann, B. (2014). Resource scarcity and antisocial behavior. *Journal of Public Economics*, *119*, 1–9.
Raihani, N. J., & Bshary, R. (2019). Punishment: One tool, many uses. *Evolutionary Human Sciences*, *1*, e12, 1–26.
Sabini, J., & Silver, M. (2005). Lack of character? Situationism critiqued. *Ethics*, *115*(3), 535–562.
Sáez, I., Zhu, L., Set, E., Kayser, A., & Hsu, M. (2015). Dopamine modulates egalitarian behavior in humans. *Current Biology*, *25*(7), 912–919.
Sarkissian, H. (2010). Minor tweaks, major payoffs: The problems and promise of situationism in moral philosophy. *Philosophers' Imprint*, *10*(9), 1–15.
Snow, N. (2010). *Virtue as social intelligence*. Routledge.

Solomon, R. (2005). What's character got to do with it? *Philosophy and Phenomenological Research*, *71*(3), 648–655.
Sreenivasan, G. (2002). Errors about errors: Virtue theory and trait attribution. *Mind*, *111*(441), 47–68.
Sterelny, K. (2016). Cooperation, culture, and conflict. *British Journal for the Philosophy of Science*, *67*(1), 31–58.
Strobel, A., Zimmermann, J., Schmitz, A., Reuter, M., Lis, S., Windmann, S., & Kirsch, P. (2011). Beyond revenge: Neural and genetic bases of altruistic punishment. *Neuroimage*, *54*(1), 671–680.
Timmermann, A., Yun, K. S., Raia, P., Ruan, J., Mondanaro, A., Zeller, E., Zollikofer, C. P. E., Ponce de León, M. S., Lemmon, D., Willeit, M., & Ganopolski, A. (2022). Climate effects on archaic human habitats and species successions. *Nature*, *604*(7905), 495–501.
Tomasello, M. (2016). *A natural history of human morality*. Harvard University Press.
Turnbull, C. M. (1961). *The forest people*. Simon and Schuster.
Vranas, P. (2005). The indeterminacy paradox: Character evaluations and human psychology. *Nous*, *39*, 1–42.
Waggoner, M., Doris, J., & Vargas, M. (2022). Situationism, moral improvement, and moral responsibility. In M. Vargas & J. Doris (Eds.), *The Oxford handbook of moral psychology* (pp. 629–658). Oxford University Press.
Westra, E., & Kelly, D. (2025). Natural born jerks? Virtue signaling and the social scaffolding of human agency. In T. Zawidzki (Ed.), *Routledge handbook of mindshaping* (pp. 225–239). Routledge.
Williams, B. (1983). Evolution, ethics, and the representation problem. In *Evolution from molecules to men* (pp. 555–566). Cambridge University Press.
Wrangham, R. (2019). *The goodness paradox: The strange relationship between virtue and violence in human evolution*. Pantheon.

Part II

Sentimental Foundations

5 Evolution and Moral Sentimentalism

Michael Slote

Evolution is relevant to any consideration of the validity of moral sentimentalism in a number of ways, and so, of course, are all the arguments that ethical rationalists have (implicitly or explicitly) used to question the philosophical value of ethics and metaethics done in sentimentalist terms. In this chapter, I shall be dealing with both these topics.

1

Since Hume, moral sentimentalists have sought to ground normative morality in empathy-based motives/sentiments like benevolence and compassion. Hume was the first philosopher ever to discuss the relationship between empathy and ethics, but he brought in factors other than empathy in his overall explanations of moral phenomena. There are consequentialist and even Kant-anticipating deontological aspects to his normative thinking, but in my own work I have sought to show that Hume and other normative sentimentalists gave up too quickly on the idea that empathy is the most general underlying basis of morality. I have elsewhere argued that the ethical rationalist commitment to respect, autonomy, deontology, and justice can be honored in plausible, different terms within a virtue-ethical moral philosophy based on empathy, but there is no time or space for me to say much more about this here.[1] I refer the reader to my book discussion of this issue or these issues and only hope they will be willing to see my argument here in a conditionalizing way: *if* normative morality can basically be understood in empathic terms, then here is how evolution is relevant to our understanding of morality and here is why evolutionary debunking arguments that may work against various forms of ethical rationalism don't tend to undermine a sentimentalist morality based in empathy.

2

So first, so to speak, the positive contribution of evolution to normative morality. If morality works on an empathic basis, humans have to be able

DOI: 10.4324/9781003632641-8

to react empathically to situations that call out for moral attention, so it would be good to be able to say something, even if only sketchily, about the evolutionary development of empathy. I have elsewhere gone into this issue somewhat more fully than I propose to do now, but putting things in a summary way, one can say that empathy evolutionarily develops in (at least) three different stages.[2]

It is well known that empathy works best or more forcefully when the empathizer is near the person or animal they are empathizing with, and the instinct or desire for proximity goes very far back in the evolutionary history of this planet. Marine iguanas and crocodiles tend to sun themselves together in groups, and, of course, schools of fish show a similar kind of instinct/desire for proximity to conspecifics. It is not difficult to see why such proximity—such flocking, as we might call it—offers evolutionary advantages. Even where the animals who seek each other's proximity don't hunt or seek food together, their proximity to one another gives them some protection against outsiders.

But just as group proximity conveys evolutionary advantages, so too are there evolutionary advantages in intragroup imitation. It will be useful to a group if those who have acquired various skills or practical knowledge are imitated by those who so far lack those assets. But what if people with knowledge or skills were to imitate those who lack them? That certainly wouldn't be to the advantage of the group—far from it! But is such a thing possible in evolutionary terms? If the knowledgeable and skilled copy the tendencies of those who lack those qualities, the group as a whole will be at an evolutionary disadvantage, so evolutionary forces will tend to reinforce imitation by the less knowledgeable or skilled of those with greater knowledge or skill. We see a cute example of this with our own babies. They imitate us; we don't imitate them except for the occasional purpose of getting them to cutely imitate us. Imitation goes far back into evolutionary history: some toads learn to imitate what they see other toads have successfully done, and there are countless examples of imitative behavior to be found among the great apes.

Proximity is evolutionarily useful and is a necessary precondition for imitation and its species usefulness. But we can go further with this. Although we don't usually think of it in this way, empathy is a form of imitation carried out at a rather high level of mental ability. When I feel or take in the pain of someone who is suffering, my state in some sense is an imitation or copy of what the other person is going through. In typical or perhaps all cases, the imitation that occurs isn't something done deliberately. The pain of the other spreads, as Hume put it, by contagion into us. We don't seem to have any say in the matter.

However, in the *Treatise of Human Nature* (Book II, Part I, Section 11), Hume seems to allow that in at least some cases there is a volitional/

desiderative element behind our empathy with others. The instinct of imitation presupposes some sort of desire or instinct for being like other people (forget the animals now), and Hume says that when we find ourselves in attitudinal disagreement with someone we are spending time with, we tend to move in the direction of agreeing with them. We take in and assimilate their attitudes and thereby come to disagree less (or not at all) with them. This suggests or more than suggests that we humans don't like being different from others, and if empathy tends to diminish such difference, the imitation it involves has some sort of instinctual emotional basis.

Such empathy-based willing imitation clearly has evolutionary survival value. Our desire to be like others and thus to imitate others adds to, or grounds, group coherence, and a group that thinks and feels together will obviously have some sort of advantage over one where there is discord of opinion and attitude. For all these reasons, then, empathy takes imitation to a new level, a level beyond sheer imitation, and of course human beings are not the only creatures to which empathy can be attributed. Elephants, dolphins, dogs, and cats are said to be capable of empathy, and there may be other species where this capacity can be found. But I don't want to speak about the usefulness, evolutionarily, of empathy within non-human species. Let's focus on the human and see what benefits empathy confers on us. In particular, since empathy with the suffering of another can sometimes make us momentarily or for a longer period forsake our own self-interested purposes, we might suppose that empathy needn't represent an evolutionary advantage for the empathic individual considered as such.

But what somewhat compromises individual self-interest may serve the interests of the group, and a group of people lacking in empathy would be less likely to help one another than a group of empathic individuals. Interpersonal empathy (there is not thought to be such a thing as empathy for one's ongoing self) will help a group to flourish, and that obvious fact may help explain why empathy has developed among humans. Proximity, imitation, and empathy all offer evolutionary advantages, but so too, of course, does the development of cognitive capabilities. Our greater intellectual capacity as compared with other species (including those possessed of empathy) offers humans an evolutionary advantage that only the possibility of nuclear mass destruction can make us skeptical about.

So let me now raise a central issue for this chapter. Ethical rationalists think our capacity for morality is, to a large extent, independent of empathic factors. Reason or rationality, unaided by empathy, can tell us what is wrong and make us try to avoid doing what is wrong. Sentimentalists disagree, of course, and we cannot settle the issue between them here. I prefer, rather, to set the issue of sentimentalism vs. rationalism largely to one side and focus on what empathy can contribute to (human) morality.

3

A virtue-ethical sentimentalist like myself holds that valid moral distinctions are all anchored in empathy. There certainly is variety within our empathic reactions, but this sentimentalist wants to say that that variety can be intuitively ordered in a way that does justice to what most of us think or feel about the dictates of morality. However, the evolutionary sketch of developing empathy that has been offered here doesn't take us all the way to morality as we conceive it because it offers no explanation of why it should be to the advantage of one group to have empathic concern for other groups that it may compete with for territory or other resources. In fact, it is difficult to see why a combination of empathy for the in-group and aggressive non-empathic feelings toward those outside the group shouldn't be to the evolutionary advantage of any given group. Evolution doesn't by itself readily explain why we are empathically appalled by what some groups do to other groups.

So, for the moral sentimentalist, there is a kind of evolutionary mystery in the fact that we are and morally ought to be concerned with people outside our group (however broadly or narrowly we define it). But no matter! Our moral concepts may go beyond anything that can be explained in evolutionary terms, and if we pay attention to what our moral concepts actually are, we may be able to see how and why valid moral thinking doesn't have to depend on evolutionary considerations. That, at any rate, is what I hope to argue for in what follows.

The normative sentimentalist thinks that right and wrong can be explained in terms of the presence or absence of empathy in various intentional actions. But it is also possible to be a *metaethical* sentimentalist about moral terms or concepts, and Hume was (somewhat inconsistently) both of those things. Well, I think he was right to understand the meaning of moral concepts/vocabulary in basically sentimentalist terms and, in particular, in terms of empathy, but I shall try here to streamline and modernize his views in a way that allows us both to better understand what we mean when we speak of right and wrong and moral goodness and to see why evolutionary considerations pose no threat to sentimentalist ethics and metaethics.

4

Hume held that our understanding of moral concepts involves a kind of second-order application of empathy. The normatively virtuous person acts empathically or sympathetically toward others and does so in proportion as they are intimate with him or exemplify other empathy-related psychological mechanisms of association. I needn't go into that here, since

I am supposing that the reader has been willing to give me a conditional pass on the validity of normative sentimentalism along Hume's or more recent lines. The question now is with metaethics, and Hume's explanation of how we *understand* moral concepts and judgments makes use of empathy operating at a different level from what it occupies when someone *acts* virtuously. Hume offers one account, his main account, of what such understanding involves early on in the *Treatise*, but toward the end his discussion suggests another way of using second-order empathy to understand moral judgment, one I find more promising for present-day purposes. In Book III, Part III, Section III, he mentions the example of a friend whose warm-hearted actions on behalf of his friend warm the heart of the observer. We are warmed, Hume says, by the warmth the friend displays toward his friend, but the first-mentioned kind of warmth isn't agential but observational. It is second-order empathically felt warmth at the first-order warmth the acting friend displays toward his friend.

Now Hume ignores cases where we see someone do something harmful to another person, but he just as easily could have said that in such cases we are chilled by the cold-heartedness of the person who does the harming. Such chill would be a second-order empathic reaction to the lack of empathy shown by the person who (let us say gratuitously) harms another. Hume treats the second-order empathic reaction we feel toward the helpful friend as a kind of approval, and many philosophers during and after Hume's time have questioned whether this is the right word for such a sentimentally reactive state. But we needn't resolve this issue in order to use what Hume says about second-order empathy as a basis for understanding how we understand ordinary moral judgments.

Now, an ideal-observer theorist of the meaning of moral judgments might try to take advantage of what I, following Hume, have been saying about second-order empathy by claiming that we can define "x is morally right" as "x will tend to evoke a second-order reaction of warmth from any normal human observer acquainted with what the doing of x involves." Hume himself suggests such an ideal-observer account of moral concepts in his essay "Of the Standard of Taste." But this would make moral judgments relative to human nature in a way that ethical rationalists from Kant onward (and I myself) would find difficult to swallow. So, is there a better way for the metaethical sentimentalist to conceptualize our understanding of moral concepts/judgments and to do so in a way that evades criticism from an evolutionary standpoint?

I think there is, and to demonstrate that possibility, I would like to appeal to an analogy that has been harped upon for a long time by sentimentalists, the analogy between sensory qualities like color and value qualities like goodness. I think a proper account of how we understand color terms like "(objectively) red" can steer us toward an understanding of moral attributes and judgments that doesn't make them relative to human nature and allows them to have full objective validity. This is a tall order, but the first step of the way requires us to look at Saul Kripke's account of our understanding of "red" in *Naming and Necessity*.[3] Kripke points out that objective redness cannot be plausibly understood as a disposition to cause certain red experiences in us because things that are now red would or could continue to be red even if human beings became universally color-blind. So how can we say that those things could or would still count as red under such conditions? Kripke holds that we cannot analytically define "red" in dispositional or any other terms, but that the reference of "red" is fixed for us by the experiences we normally have of red objects (in good light, etc.). There is some reflectance property in red objects that causes our normal red experiences/perceptions, but ordinary folks don't know what that is, and science had to make a lot of progress before the nature of that property was properly understood.

But then, according to Kripke, the objects that color-blind humans cannot experience redly can still count, in such unfortunate circumstances, as red if they have, say, the reflectance property r that now causes red objects to be perceived by us as red. "The property in objects that causes normal red perceptions" fixes the reference of "red" for other possible worlds/situations without being analytically or conceptually equivalent to any objectual disposition to cause red experiences. So, for Kripke, we can understand judgments about redness via reference fixing and without being able to analytically define them. And it takes scientific experiment to determine that reflectance property r causes our ordinary red experiences, so the conclusion that objects with r in other possible scenarios of the world count as red (even if we cannot see them that way) is empirical, even if it holds for all worlds and is thus, according to Kripke, a metaphysically necessary truth. But how, you may well ask, can this help us to theoretically understand how we normally understand moral judgments?

The historically influential idea of analogy between sensory and value properties may make us try to see how something like Kripkean reference fixing might apply to moral concepts. And we can start with the idea, to some extent culled from Hume, that empathy of a second-order kind is essential to our understanding of moral judgments. I have suggested in my own previous work that such an approach might be made to work if we can fix the reference (say) of "morally good" via a reference to the kind of second-order empathy Hume mentions toward the end of the *Treatise*. Perhaps we can fix the reference of "moral goodness" via the description "the property of agential actions or motives that under veridical circumstances is empathically felt as warm by an/the observer." But think about it! The only property in an agent's actions/motives that can be *empathically* felt as warm is agential warm-heartedness. So, we can draw the conclusion that agential warm-heartedness (aka empathic caringness) is what moral goodness *is*. And we didn't have to rely on science, as in the case of redness, to draw this conclusion. The metaethical sentimentalist can say that it is a necessary a priori truth that moral goodness is nothing less or more than warm-hearted caringness, and this offers a sentimentalist metaethical reply to the many Kantians who have objected to the contingent human-relativity that is built into earlier forms of metaethical sentimentalism.

5

Let me now get back to issues concerning evolution, and I would like to speak first about the "debunking" evolutionary considerations that Sharon Street, in her influential article "A Darwinian Dilemma for Realist Theories of Value," directs at recent ethical-rationalist accounts of morality (especially T. M. Scanlon's).[4] She argues that certain facts about evolution give us strong reason to doubt whether such normative rationalism has any right to claim objectivity or truth for the ethical claims it makes. Many philosophers have sought to undermine her skeptical thesis, but I don't want to discuss their views or arguments here. Rather, I want to point to considerations that I think circumvent her arguments against ethical rationalism and allow us to maintain the objective validity of moral claims understood on a sentimentalist basis—a possibility Street never considers.

The ethical rationalists Street is thinking of follow Kant in refraining from making any attempt to analytically define moral terms, and, more generally, they don't offer us any basis (like reference fixing) for understanding how we ordinarily understand moral notions. I think this leaves them open to Street's type of objection in a way that a metaethical sentimentalism based on reference fixing is not. If our understanding of moral claims and concepts can be semantically grounded through reference fixing, then moral judgments made accordingly can be justified entirely in terms of the way we understand those judgments. And if the truth/objective validity of moral judgments is determined and assured by the way we understand such judgments, evolutionary facts cannot challenge or undermine them because those facts have nothing to do with the semantic considerations my defense of morality relies on. Because they offer no account of moral or meta-moral semantics, the ethical rationalists lack this kind of sure way out of the dilemma Street seeks to pose for them. Her arguments may or may not be successful, but it is at least clear that they have no force against the metaethical semantics of moral terms I (briefly) offered above or thus against the validity of moral claims made under the aegis of such metaethical semantics.

This still, however, leaves us with the problem I mentioned earlier about the way our understanding of morality could or could not have emerged via evolutionary processes. Our concern for those beyond our own group may be a spandrel of the concern for people in one's own group that gave our species evolutionary advantages. Who knows? But what is clear, what we have given ample reason to believe, is that our moral concepts take in a concern for people beyond our own group, however difficult it may be to account for that in evolutionary terms. And there are, in addition, particularities or peculiarities of our typical moral thinking and moral dispositions that may be difficult to explain in evolutionary terms. Why exactly, for example, are we moved more by clear and *present* danger than by equally great known danger in the future? It may be difficult or more than difficult at this point to know, but morality clearly judges it to be morally incumbent on us to help those presently in danger in preference to others we know will be in equal danger in the future. I say that that is no reason to abrogate the moral judgments we are inclined to make or the considerations of empathy that, in at least some measure, arguably underlie them. Our concepts of morality, our understanding of moral claims, insulate them from our inability to explain them in evolutionary terms. We cannot easily explain our capacity to understand infinity or real numbers in evolutionary terms, but that doesn't make us skeptical about those notions, and I think it can be seen, assuming only what I have been arguing for here, that evolution offers us no substantial reason to doubt what we or most of us think about right, wrong, and moral goodness.

6

But now and finally let us consider another objection to the kind of sentimentalist account of morality offered above, one once again coming from Sharon Street. In a less well-known but very interesting article called "Reply to Copp: Naturalism, Normativity, and the Varieties of Realism Worth Worrying About," Street argues that ethical naturalism cannot offer us a way out of her Darwinian dilemma because it cannot explain the normative/motivational force of moral judgments.[5] If we analyze "right" in naturalistic terms as "leading to the greatest happiness/pleasure of the greatest number," we will have no explanation of why the belief that something is wrong automatically motivates us (or gives us reason) to avoid doing that thing. It is not obvious that we have to *care* about what either does or does not advance the happiness of general humanity.

Now, many contemporary metaethicists reject the notion that moral claims sincerely advanced automatically have a certain amount of motivating power, but it seems to me that there may be a better way to respond to Street's criticisms of naturalism.[6] What if a sentimentalist view of morality can allow for a naturalistic understanding of moral claims that also attributes automatic motivating power to them and, more generally, to the moral considerations that are the focus of the sentimentalist account of normative morality? Empathy may be the key to making good on such theoretical aspirations.

Hume was the first person to discuss the various ways in which empathy can motivate virtuous actions, but partly, I think, because of his general commitment to subjective idealism, he seems never to have regarded empathy as a way of *knowing* the mental state of another person. When we see someone in distress and feel their distress as our own, we register their distress within ourselves in much the same way that we can register the existence of a chair when we have the perceptual experience of seeming to see a chair.

Empathy allows us to know the mind of another person, but of course the moral value of empathy consists in the presumed fact that empathy leads us to help them.

However, these different sides of empathy—the cognitive/epistemic and the motivational—are typically regarded as only contingently related, and if I can show you this is a mistake, we will be well on our way to showing that sentimentalist naturalism allows certain empirical considerations to be automatically motivating. Consider a father who is empathically infected by his daughter's enthusiasm for stamp collecting. This involves taking in more than some generalized state of enthusiasm; it means taking in her enthusiasm as directed toward the collecting of stamps. Genuine empathy involves taking in an emotion or feeling *along with its intentional object*, and if we apply this idea to empathy with the distress of some other person, we can see that empathy with that person's distress involves empathy with the distress they are feeling about, say, the shooting pain in their arm. In which case, the empathizer feels not just distress but distress about the given pain, a pain they are not feeling but that the other person *is* feeling. Distress with something automatically entails a desire to eliminate or lessen it, and this means that empathy with the distress or suffering of another automatically involves motivation to eliminate or lessen the pain *that the other person is distressed about.*

In that case, the knowledge of another person's (temporary) reality that is involved in feeling empathy for their distress automatically amounts to moral motivation to help them. Knowledge and motivation can be more necessarily intertwined than we have thought. Now, moral sentimentalism anchors its normative moral judgments in considerations about empathy: what shows empathy is morally desirable and what shows a lack of empathy is morally suspect or wrong. So, if our previous discussion shows that empathy has a necessary connection to helping motivation, the considerations of empathy that ground moral judgment and moral action have an intrinsic bearing on practical motivation. Morality has to do with empathy and empathy directly involves altruistic motivation, and the connection with naturalism is then visible through the fact that nothing in the above reference fixing account of concepts like moral goodness takes us outside the realm of natural fact(s). So, a metaethical naturalism based in empathy can also account for the motivating force of the considerations that favor the rightness or wrongness of an action.

That someone is distressed by the shooting pain in their arm motivates an empathic person to help them. Such empirical facts about another person make moral demands on us that automatically motivate someone who empathically knows them. The sentimentalist holds, in other words, that empathy-accessed considerations concerning what others are experiencing are both automatically motivating and play an essential role in our moral evaluations of what people do in the relevant circumstances. But what about moral judgments? The anti-naturalist ethical rationalist (like Scanlon) who believes in moral objectivity holds that a sincere belief that something is wrong automatically entails at least some motivation to avoid doing it. Can something similar be said about or for metaethical sentimentalist accounts of moral judgment?

Let me simplify matters a bit by noting how odd it would be to imagine someone capable of second-order empathy but incapable of the first-order kind. For this to occur, we have to imagine that someone, say, who is warmed by the warm-heartedness a friend shows to his friend, is himself incapable of any sort of agential warm-heartedness. And this doesn't correspond to anything we can readily imagine to be, in a given instance, the case. To be capable of second-order warmth at agential warm-heartedness, one has to at least know what agential warm-heartedness is, and is there any obvious path to such knowledge other than one's own (occasional) feelings of agential empathy for the plight of others? Someone who can make moral judgments has to have some sort of moral motivation as well. This means that a naturalistic metaethics along sentimentalist lines is capable of understanding and accounting for automatic moral motivation. Other forms of naturalism may not be able to accomplish this, but Street's idea that no form of ethical naturalism can allow for intrinsic moral motivation seems to be mistaken in the light of the arguments for naturalism and sentimentalism I have given you here.

However, when I earlier introduced the idea of inherent moral motivation, I also (somewhat in passing) raised the parallel issue of whether naturalistic sentimentalism can show that moral considerations and judgments are automatically *reason giving*. If it cannot, if it has nothing to say about morality-related practical reasons for action, then we may well wonder whether such sentimentalism offers a completely satisfying picture of the moral life. This might then be the sticking point that could undermine the kind of virtue-ethical sentimentalism I have been defending here.

But there is, in fact, a sentimentalist answer and solution to this issue that connects with what we have been saying about empathy. The sentimentalist holds that empathy allows for the self-interested motivation of someone in pain to transfer to the empathizer as an altruistic motive to help that someone. But if motives can be transferred by empathy, why can't this also happen with reasons? When I share the suffering person's desire for the shooting pain in their arm to end, may I not also share the reason they have to want that pain to end? Their reason is, of course, a self-interested one, but for the empathizer who shares it, it is as altruistic as the empathically derived motive to help them get rid of the pain. I have elsewhere discussed at some length how such a transfer of reasons can occur, but I believe that what I have just said makes it clear that moral sentimentalism can provide for automatic practical/altruistic/moral reasons in something like the way it provides for automatic practical/altruistic/moral motivation.[7] Naturalistic virtue-ethical sentimentalism seems to me, consequently, to be overall in pretty good shape.

Michael Slote, University of Miami

Notes

1 See my *Moral Sentimentalism,* Oxford University Press, 2010.
2 For a fuller account, see my "Empathy as an Instinct" in N. Roughley and T. Schramme, eds., *Forms of Fellow Feeling*, Cambridge University Press, 2017.
3 Oxford: Blackwell, 1980.
4 *Philosophical Review* 127, 2006, pp. 109–166.
5 *Philosophical Issues* 18, 2008, pp. 207–228.
6 See, for example, D. O. Brink, *Moral Realism and the Foundations of Ethics*, Cambridge University Press, 1989.
7 See my "Sentimentalist Practical Reason and Self-Sacrifice," *International Journal of Philosophical Studies* 26, 2019.

6 Emotional Plasticity and the Development of Virtue

Charlie Kurth

Early in the chapter of *On Virtue Ethics* devoted to discussing the place of emotion in virtue, Rosalind Hursthouse informs her audience that the details of an appropriate account of emotion "need not concern us" (2001, p. 111). Her thought seems to be that since emotions, as a class, have a uniformity—they are all belief-like evaluative appraisals—we don't need to get into the details about what they are or how they work. The real challenge, she maintains, lies in trying to understand how emotions can be educated so that they better contribute to virtuous thought and action. But Hursthouse's reasoning, while common within philosophy, is mistaken. Emotions do *not* have the uniformity that Hursthouse and many others presume. Moreover, appreciating that emotions differ—more specifically, that the cognitive architecture that undergirds emotions differs—has revisionary implications for our understanding of what virtue is and how we can become better people.

To develop my argument, I begin by noting an important ambiguity in talk of "emotion cultivation." While it is sometimes used to refer to our efforts to control or regulate what we feel, this is not what Hursthouse and the bulk of philosophers mean by the term. In talking about emotion cultivation, they're concerned with our ability to substantively *change* the emotion itself. But if change, not control, is what we want, then what does emotion science tell us about our ability to secure the needed change? To answer this, I turn to work in clinical psychology and psychiatry. Here we see that while it appears that some emotions can be substantively changed, others can only be controlled. This then raises a question, one that standard accounts of emotions' role in virtue are poorly positioned to answer: why do emotions differ in their susceptibility to being changed for the better? In particular, we'll see that the standard views that we get from Aristotelians and the Stoics struggle because they, like Hursthouse, see emotions as uniform in nature. To better understand both why emotions differ in their malleability and what this might entail for our understanding of virtue, I begin by turning

DOI: 10.4324/9781003632641-9

to work on the evolutionary origins of emotions. This research suggests that emotions differ in their susceptibility to being cultivated because their cognitive architecture differs. With this insight in hand, I then argue that if the cognitive architecture of emotions differs in the ways that evolutionary models suggest, this should lead us to revise our thinking about what virtue is and how moral education ought to proceed. In brief, the resulting account looks more like what Kant proposed and less like the accounts we get from Aristotle, the Stoics, or contemporary defenders of these views.

6.1 Emotion Cultivation: Substantive Change or Just Enhanced Control?

Before getting started, some background on "emotion cultivation" will be helpful. In what follows, I'll use this label to refer to our intentional efforts to bring lasting improvements to when and how we experience particular emotions. On this front, we find, for example, Aristotelian strategies of habituation, Stoic extirpation therapies, and Kantian techniques of moral education via catechistic instruction. But for what follows, we should also distinguish two distinct, but potentially overlapping, ways in which an emotion—anger, say—might be "cultivated." First, we can cultivate anger by changing the emotion itself. Alternatively, we can shape anger by changing the non-anger mechanisms that influence when and how one feels the emotion (e.g., general attentional or cognitive systems, display rules, motor control systems). Put slightly differently, if we assume emotions are dispositions, then the first model of cultivation is one that changes the dispositions themselves. It's akin to how the process of heat tempering changes the molecular structure of a glass vase and thereby the vase's fragility—its disposition to shatter if dropped. The second model, by contrast, doesn't change the dispositions, but rather shapes other cognitive systems that can mask, inhibit, or mimic the operation of the targeted disposition. Here, the analog is using bubble wrap to mask the vase's (unchanged) disposition to shatter.[1]

Moreover, and as we'll see in greater detail below, some cultivation strategies—like exposure and habituation—seem best understood as aiming for change. They often work from the bottom-up, making use of (e.g.) stimulus-based conditioning to change the emotional disposition. Others—like reappraisal and rule-based techniques—seem better understood as aiming to inhibit or mask the underlying disposition. They typically work from the top-down, drawing on high-cognitive mechanisms—e.g., metacognition, executive control—to impose constraints on how or when an emotion manifests itself. So, while there's a sense in which both ways of altering when and how one feels an emotion count as forms of "cultivation," it's also clear that only the former involves genuine change to the emotion itself.

With this background in hand, we can start to see why it matters in the context of emotions' place in moral development. Virtue, at least as it's generally understood within the Aristotelian and Stoic traditions, is about securing substantive emotional *change* and not merely enhancing our ability to *control* when and how we feel what we feel. Of course, what change amounts to differs radically between these two traditions—Aristotelians want to attune emotions, and Stoics want to eliminate them. But both are aligned in thinking that mere control is insufficient. While we will revisit the received view that change is always better than control when it comes to virtue, the distinction raises an important issue. If the dominant Aristotelian and Stoic accounts of virtue maintain that moral development demands substantive emotional change, then gauging our prospects for becoming more virtuous involves understanding the extent to which we can do more than just enhance our ability to control what we feel. So what can we do? As a start to answering this question, let's look at what therapists and clinicians working to help individuals with problematic feelings and emotional disorders have learned about change and control.

6.2 Insights from Clinical Psychology and Psychiatry

6.2.1 *Background*

The general presumption among philosophers is that emotions, as a class, exhibit a kind of uniformity with regard to our ability to shape them for the better. Of course, what this looks like changes depending on whose account we're looking at. For Aristotelians, the presumption is that emotions are uniform in their malleability: emotions are psychological states that we can substantively change for the better. So, for instance, Julia Annas, speaking in the context of how you and I should cultivate our emotions to promote virtue, notes that Aristotelians think "emotions can be so educated by our developing reason that they become wholly transformed" (2011, p. 68; also Sherman, 1989). Similarly, the Aristotelian Kristján Kristjánsson starts from the Aristotelian commitment to cognitivism about emotions, the idea that emotions are, or necessarily involve, evaluative beliefs (e.g., to feel anxious is to believe one's situation is worrisome). Given this, he argues that emotions as a class can be shaped for the better: "If emotions are cognitive, they are reason-responsive. If they are reason-responsive, they are educable" (2018, p. 167). While the Stoics' take on the role of emotion in virtue is very different, they too are committed to a uniformity claim by way of their cognitivism: emotions, as a class, are impervious to our efforts to shape them for the better because they systematically embed entrenched false evaluative beliefs. As such, all emotions must be extirpated if we're to become truly virtuous (e.g., Graver, 2008; Kamtekar, 2025).

But even a passing glance at the empirical record suggests the picture we get from Aristotelians and Stoics is unlikely to be true: some emotions are susceptible to change, some aren't. To highlight this, we'll look at three emotions that have received considerable theoretical and empirical attention: shame, anxiety, and disgust. For each emotion, I begin by drawing on work in philosophy and the affective sciences to characterize what I see as the best account of their functional roles. From here, I shift to discuss what work in clinical psychology and psychiatry tells us about our ability to shape these emotions for the better. Together, this groundwork will allow us to assess whether our efforts are likely to bring substantive change to these emotions or just enhance our ability to control them. A final point: in what follows, I will be following the philosophical mainstream in taking clinical findings to be applicable to our understanding of psychological functioning in general (e.g., Doris, 2015; Maibom, 2005; cf. Jackson, 2020); generally speaking, pathologies do not represent sui generis states but rather the tail ends of normally distributed psychological phenomena.

6.2.2 *Shame*

Shame is an emotion of self-assessment: one feels ashamed about one's infidelity or big ears because one feels one is not being seen (if only to oneself) as the person one takes oneself to be. Shame thus functions to both make individuals aware of (moral) reputation-damaging actions or traits and to prompt efforts to address the damage they have done. These efforts might take the form of making amends, self-improvement, or attempts to hide the shame-eliciting act or trait.[2]

Of course, this account of shame's function is not meant to suggest it's always a force for the good. As is all too familiar, shame can be a deeply problematic emotion. And while it's often experienced in response to moral failings, we can also be ashamed of morally benign things like our appearance. Alas, there are also shameless individuals, people who don't feel shame when they should. In light of shame misfires like these—false-negatives, false-positives, disproportionately experienced shame—what can we do to get more of the good, less of the bad?

Let's begin by considering strategies aimed at countering false-negatives, not feeling shame when we should. Here we see that efforts to encourage shame in the shameless have a spotted history. Much of this results from the morally and therapeutically dubious use of so-called "shame punishments"—sentencing a shoplifter, for instance, to stand on a street corner wearing a sandwich board with "I am a criminal" written on it. Some judges and legal scholars see these punishments as effective ways to harness shame to combat recidivism (Kahan, 1996). But this is, at best, a minority view. As critics like Martha Nussbaum (2004) argue, shame punishments are not only unlikely to actually work but also—worse—typically dehumanize their targets.

Setting shame punishments aside, focusing on how feelings of shame are elicited by challenges to *one's identity*—how one sees oneself—suggests there may be better ways to encourage shame in individuals who are insufficiently sensitive to it. Rather than directly shaming individuals by calling them out for their bad behavior, this more sophisticated work takes an indirect route: it aims to leverage shame by changing the substance of the underlying identities. So, for instance, we see that anti-violence against women campaigns are having a positive impact in their efforts to change what it means to be a man: hitting or verbally abusing one's girlfriend is shameful, not something to be proud of (Raymond et al., 2014).

Moving to interventions aimed at addressing false-positives and disproportionate shame, we find that both Acceptance and Commitment Theory (ACT) based interventions and Internal Family Systems (IFS) therapy show promise. ACT has many similarities with more traditional treatment protocols like Cognitive Behavioral Therapy (CBT). As the name suggests, ACT interventions aim to help individuals both *accept* that (strong) negative feelings are an inevitable part of life and *commit* to developing ways to work through (rather than fight) these feelings so that they can act in ways that better accord with their values.

So, for instance, consider a woman whose demanding job as a lawyer leaves her feeling that she's a failure—and so ashamed—as a mother. In a case like this, ACT interventions aim to, first, help the woman both understand the drivers of her self-criticism and appreciate that these shame-inducing critiques are part of a natural self-protective mechanism. So, for instance, the mother might come to see that her tendency to be self-critical is grounded in an underlying perfectionism that, while intended to help her be a better mom, more often just leaves her feeling ashamed of her "failings." The aim of this initial work is to help her gain non-judgmental distance from her (perceived) flaws. From here, the ACT strategy makes use of exercises that can help her both take a self-compassionate perspective on her (perceived) deficiencies and commit to acting in accordance with what she sees as most important.

It's important to understand how these two skills work together: the capacity for self-compassion is crucial in allowing the woman to detach herself from her identity as a lawyer so that she can better focus on being the kind of mother she wants to be. While this ability to detach is unlikely to eliminate the shame she feels, it does allow her to better align her identities with what she cares most about. Moreover, the emphasis on treating problematic shame by helping individuals better reconcile competing identities is even more prominent in IFS therapy, which explicitly theorizes shame as the upshot of clashes in the roles we take ourselves to occupy.

While the effectiveness of ACT and IFS interventions in moderating misplaced and excessive shame has only recently started to be assessed, the studies we have suggest that both can be effective (Hodgdon et al., 2022; Lear & Luoma, 2025) and that they can even outperform standard treatments in randomized controlled trials (Luoma et al., 2023).

6.2.3 *Anxiety*

Anxiety, like fear, is a threat-oriented emotion. But while fear sensitizes us to imminent and significant dangers, anxiety attunes us to situations where the nature and probability of the threat are unclear. Put differently, anxiety functions as an alarm that tracks problematic uncertainty, be it the dangers that might lurk in a dark alley, the possibility of being criticized about what one wears, or the complications that might arise from a poor medical decision.[3] In situations like these, anxiety works to both make individuals aware of the potential threat or challenge they face and prompt a range of risk minimization and risk assessment behaviors (e.g., information gathering, reflection, deliberation) aimed at helping the individual better address the uncertainty at hand. While an appreciation of shame's moral significance goes back to ancient times (Mun, 2024), awareness of anxiety's moral relevance is more recent. Kant, for instance, speaks of anxiety as an instrumentally valuable emotion—an alarm of sorts—in the context of moral judgment (1797/1996, pp. 559–562; Kurth, 2024a). And there's now a growing recognition of anxiety as an emotion that signals danger in our moral decision-making and that prompts us to reflect on—even change our mind about—what we ought to do (Kurth, 2015, 2018; Lacewing, 2005; Maibom, 2022).

This is not to deny that anxiety can go badly awry—we know all too well that it can. So, like shame, anxiety's standing as a morally important emotion improves to the extent that we can cultivate it. To assess this, let's begin by looking at interventions aimed at addressing false-negatives. Here, research on the effectiveness of "fear appeals"—a type of public safety advertisement (PSAs)—shows that, if properly designed, individuals can come to be anxious about things that should, but don't currently, worry them. These ads use graphic images, audio, and text to elicit anxiety in an effort to curb harmful behaviors like drunk driving. Not only are these PSAs effective in combating the risky behaviors they target, but the research suggests they're effective, in part, because the feelings of anxiety and unease that they provoke become associated with the recognition that the viewer is vulnerable to being harmed by the risky behavior depicted in the advertisement (e.g., Lewis et al., 2007).

Switching to work aimed at false-positives and disproportionate anxiety, we see that a range of treatments have proven effective. This includes well-established methods like CBT and more recent strategies like ACT. Taking these in turn, CBT works by, first, on the cognitive side, helping individuals recognize and understand the problematic thought patterns—e.g., catastrophizing, overgeneralizing—that tend to bring intense bouts of anxiety. This then allows individuals to better assess the evidence that affirms and challenges their worries. Second, and more on the behavioral side, CBT makes use of guided exposure to anxiety-provoking situations in order to help individuals better calibrate their responses to threats. A range of meta-analyses have found strong support for the effectiveness of CBT as a treatment for problematic anxiety in comparison to both wait-list and placebo controls (Carpenter et al., 2018; Olatunji et al., 2010). While the evidence speaking to ACT's effectiveness in treating problematic anxiety is not as robust as what we see for CBT, meta-studies and systematic reviews speak to its promise (e.g., Ferreria et al., 2022; Swain et al., 2013).

6.2.4 Disgust

Disgust functions as a guard against contamination, broadly construed. Unlike mere taste aversion—a reaction to things we deem unpalatable—disgust is a response to things we see as "contaminated." While the list of elicitors of disgust is strikingly diverse, what is often called "core disgust" is a sensitivity to things like disease, decay, feces, insects, bodily fluids, and other things that can make us sick. Developmentally, a sensitivity to core disgust elicitors tends to appear around 3 years of age, broadening to include sensitivities to social/moral "contaminants" (e.g., food taboos, cheating) around age 7. And regardless of what we're disgusted by, we tend to respond ballistically—quickly and automatically rejecting, purging, or otherwise seeking to avoid whatever it is that disgusts us.[4]

While there is some controversy about whether there is a distinctly moral form of disgust (Bollard, 2022), the emotion is uniformly seen as morally relevant—though controversy reigns about whether it's a force for good or ill. On the plus side, many take disgust to be a uniquely well-equipped response to certain wrongs. As William Miller explains, "there are those vices and offenses for which notions of ugliness, smelliness, sliminess readily apply and those for which they do not. Hypocrisy, betrayal, cruelty put us in the swamp of the disgusting, and no other moral sentiment seems as well qualified to express our disapprobation" (1997, p. 205; also Plakias, 2013). Others disagree. The breadth of things that we can become disgusted by, and the force of the emotion's reject/purge response, make all too likely that we will end up demonizing morally benign persons and behaviors—Jews, women, interracial marriage—that we've come to see as disgusting (Kelly & Morar, 2014; Nussbaum, 2004). But regardless of which side one is on, all agree that becoming a better person demands getting a handle on one's disgust.

Turning to what we might do to shape disgust for the better, we've seen that disgust sensitivities are very easy to acquire. In fact, it sometimes takes just one unfortunate experience with, say, tuna salad gone bad to become disgusted by anything fishy. But this flexibility is both an asset and a liability when it comes to addressing false-negatives. More specifically, given both the flexibility of our disgust acquisition mechanisms and their susceptibility to "one-shot" learning, it will often be very easy to develop disgust sensitivities to things we're not disgusted by but should be (factory farming, perhaps). But those same features mean it will be just as easy to acquire disgust sensitivities to things we're not disgusted by and should *not* be (e.g., inter-racial marriage).[5] So, putting this together, we get a recipe for skepticism: what we become disgusted by seems more likely to be the result of chance and fortune than our explicit efforts to educate the emotion (Kurth, 2021).

In sharp contrast to what we see in shame and anxiety interventions, disgust is largely *unresponsive* to the dominant strategies that one might deploy to change what one is disgusted by or alter how one responds when disgusted. As we've seen, exposure and habituation-based interventions are central tools underlying CBT, ACT, and IFS protocols that have proven effective in addressing shame and anxiety. But empirical work studying the effectiveness of these methods for disgust disappoints: disgust sensitivities are "little affected" by efforts to lessen or eliminate them through habituation and exposure (Meunier & Tolin, 2009; see also Engelhard et al., 2014, Study 1). Moreover, and in addition to these clinical findings, the work of atrocity studies scholars indicates that even those with very strong incentives to lose their disgust sensitivities—the dutiful Nazi's deep desire to not be repulsed by killing Jewish women and children—find that more killing doesn't make their disgust go away (see Munch-Jurisic, 2023 for review).

But while exposure and habituation interventions show little promise, efforts that make use of implementation intentions—a strategy that's proven effective in combating implicit bias—do better when deployed against problematic disgust. Implementation intention-based interventions aim to change the "if, then" rules that guide many of our actions. In the context of implicit-bias mitigation, they might, for instance, get an individual to engage a bias-mitigating "if Black, then approach" rule rather than a bias-oriented "if Black, then avoid" rule. For disgust, implementation intentions have been used to help nurses with problematic disgust sensitivities toward blood control both how they appraise these disgust-eliciting stimuli and how they subsequently respond. More specifically, we find that implementation intentions targeting appraisal (e.g., if I see blood, I'll take the perspective of a physician) are more effective in reducing how unpleasant one takes a stimulus to be compared to both controls and goal-oriented strategies where one just endorses the aim, for example, of not getting disgusted. Additionally, implementation intentions targeting response tendencies (e.g., if I see blood, I'll stay calm and relaxed) are also more effective in reducing how repulsed one feels (Schweiger Gallo et al., 2012; Webb et al., 2012).

6.2.5 Assessing What We've Learned

What can we say about how these emotion cultivation strategies work? Do they bring substantive change or merely enhance our control over what we feel? In some cases, the answer is straightforward. For instance, as we've seen, interventions for anxiety that make use of fear appeals aim to condition individuals to feel anxiety toward things they're not currently disposed to worry about. So, when successful, it seems we've brought substantive change to the emotion itself. In the other direction, as generally understood in the literature, implementation intentions don't aim to eliminate problematic disgust, but rather enhance control. As we've seen, they deploy a top-down approach, aiming to bolster an individual's higher-order capacity to both recognize situations likely to bring problematic disgust and engage rules that will help them better respond (Brownstein, 2015).

Other interventions, however, are more difficult to cleanly categorize. This is in part because treatment protocols like CBT, ACT, and IFS bring a range of distinct interventions together under a unifying theoretical framework. So, for instance, some parts of CBT, like its use of guided exposure to address false-positives and excessively strong responses, appear to secure genuine change in the emotion: this type of intervention, as we've seen, works to desensitize individuals to dubious elicitors and encourage dispositions to respond that are more proportionate to the threat or challenge at hand. But other aspects of CBT appear more focused on enhancing control. For example, the cognitive reappraisal work that aims to help individuals see that there's little reason to think that what they're anxious about really is worrisome. This suggests that the intervention works not by targeting anxiety itself, but rather by helping individuals bolster top-down rational control over how they appraise their situation.

All told, then, our look at clinical work reveals that there are significant differences in our ability to shape emotions for the better. For shame and anxiety, it appears that fear appeals, CBT, and ACT can substantively change these emotions so that we're better able to feel them at the right time and in the right way. But disgust is different. While we may have some ability to shape which disgust sensitivities we come to have, the tremendous ease with which we acquire these sensitivities suggests that what we end up being disgusted by is more the result of chance and fortune than any disgust-educating blueprint we might try to deploy. What's more, once we have developed a disgust sensitivity, there appears to be little we can do to change it. Rather, our best hope seems to be to enhance our ability to control when we feel disgust and how we subsequently respond.[6] Here, implementation intentions look promising.

But these insights leave a big question unanswered: Why do emotions differ in their susceptibility to cultivation? If we go in for a standard Aristotelian or Stoic account of virtue, this question is surprisingly difficult to answer. As we've seen, these views are committed to seeing emotions as a unified class insofar as both views build from a cognitive theory of emotion. And it's this commitment that leads defenders of the view to think that emotions, as a class, can be shaped for the better (Aristotelians) or not (Stoics). And that's the rub. After all, the data we reviewed above suggests that while some emotions can be educated, others cannot—and differences like these are *not* what these accounts, and their cognitivist commitments, would lead us to expect.[7]

This might seem too quick. After all, both Aristotelians and Stoics can emphasize that even if all emotions are (or involve) beliefs, this doesn't mean that all emotions will be alike in the degree to which they are (not) educatable. Situational factors—specifically, one's experiences and educational environment—matter. And on this front, some of us will be luckier than others. Appreciating this explains why some of us will be able to effectively shape (eliminate) our emotions, but others won't be. Or so the Aristotelians and Stoics might argue. Now this response is good as far as it goes. The trouble is that it just doesn't go very far. For the crucial questions—e.g., why is disgust resistant to change, but shame not—do *not* get answered. Rather, they're buried in appeals to luck and generic claims about variations in the educational environment.

Importantly, this isn't just an issue for views that endorse cognitivism. Psychological constructivism, while not wedded to understanding emotions in terms of evaluative beliefs, does take emotions to be the upshot of our social practices (e.g., Barrett, 2017; LeDoux, 2015). For you to experience anxiety or disgust just is for you to project your socially created ANXIETY or DISGUST concept on to the physiological sensations you're having. The key thing in this for our purposes is that constructivists take emotion concepts to be amalgamations of things like our cultural beliefs about an emotion (e.g., what elicits it, how it feels), our memories of experiences of it, and the emotion-specific behavioral scripts we've acquired via socialization. But again, we should ask why are some emotions more resistant to change than others? What is it about, for instance, the socially constructed nature of DISGUST (or our projection of it) that explains why it resists change in a way that ANXIETY does not?[8]

By contrast, a biological theory of emotion where (some) emotions are understood as evolutionary adaptations does better. These proposals don't presume that there's a single adaptive account that explains the emergence of all (or many) emotions. Rather, they aim to give evolutionary accounts of particular emotions, and don't presume that these various proposals will be the same. In fact, they aren't. For our purposes, what matters is that the best biological accounts of shame, anxiety, and disgust not only maintain that the adaptive scenarios from which these emotions emerged are different but also that these differences suggest that the cognitive architecture underlying these emotions differs as well. So, looking here might provide an answer to our question: perhaps the ways that evolution has shaped our various emotions help explain why some are more resistant to change than others.

6.3 Evolution's Impact on the Cognitive Architecture of Emotion

Suppose we go in, at least provisionally, for a biological theory of emotion on which at least some emotions—anxiety, shame, and disgust in particular—are understood as evolutionary adaptations. Might understanding the evolutionary origins of these emotions shed light on our question: why does an emotion like disgust resist being changed in ways that anxiety and shame don't? I think it does. But in making my case, I will not be able to defend the claim that disgust, shame, and anxiety are adaptations. The evidence and arguments for that have been made elsewhere.[9] Rather, I'll try to show how mining insights from this work can help explain both the differences that we see in these emotions' susceptibility to cultivation and why the interventions that work are able to secure their success.

We can start with disgust, where the evolutionary model is likely to be most familiar. Here, we want to understand what an account of disgust's evolutionary origins might say about why the emotion's learning mechanisms exhibit so much flexibility that we have little chance to outdo luck's influence on what we become disgusted by. We also want to understand why the disgust sensitivities we develop ossify in ways that leave efforts to enhance our control over them as the most promising strategy for addressing problems.

On this front, work examining disgust's evolutionary origins offers insight. As standardly understood, disgust is a mechanism that was originally selected for in order to help protect individuals against biological contaminants, but that was subsequently expanded in scope to also guard against social and moral "contaminants" (e.g., Kelly, 2011; Kurth, 2021; Rozin et al., 2008; Tybur et al., 2013). Moreover, this evolutionary account is generally taken to explain the distinctive features of disgust that are of interest to us. For instance, if disgust began as a pathogen defense mechanism, then that helps explain the flexibility of its learning mechanisms—such flexibility would allow an individual to guard against the distinctive pathogens of its environment. Similarly, a ballistic and rigid reject/purge response would allow for immediate protection against perceived contaminants. And the fact that disgust sensitivities ossify—resisting efforts to change them—also makes sense. After all, this would help prevent individuals from again eating something that had made them sick.

Shifting to anxiety, here too we want to understand why its learning mechanisms allow for so much flexibility. But we also want to know why, unlike disgust, the emotion promotes more flexible, teachable epistemic behaviors. Again, evolutionary models help. As suggested above, anxiety and fear are generally viewed as evolutionarily honed sensitivities to threats. Fear attunes individuals to *imminent* dangers and tends to bring a stereotyped pattern of defensive behavior—the proverbial fight/flight/freeze response. Anxiety, by contrast, cues one to threats and challenges whose imminence and significance are *uncertain*. And given this uncertainty, anxiety's behavioral routine tends to be more tempered and flexible: efforts to both learn more about the risk faced and identify ways to minimize exposure to it.

Moreover, and like disgust, given the diversity of potential threats that might lurk in an individual's environment, there would be adaptive advantages both in anxiety coming with a flexible learning mechanism (alas, there are few limits on what we can become anxious about) and in anxiety's learning mechanisms nonetheless being weighted toward things like snakes and spiders that are dangerous from an evolutionary perspective. Unsurprisingly, this is what we see (Biondi et al., 2025; Öhman, 2008). That said, while ingesting poisons can quickly bring death, what anxiety guards against—namely, *potential* threats—don't tend to have such rapid consequences. Appreciating this sheds light on why, in comparison to disgust, anxiety is more susceptible to our efforts to change it for the better. In contrast to the enduring protection that we get when disgust sensitivities ossify, anxiety's adaptive advantage is more epistemic in nature. When we're anxious, we start trying to make sense of, and strategize about, the challenge or uncertainty that we face. So here, a flexible response is advantageous: it helps us sort the things we should really be worried about from the ones we can safely ignore.

Finally, let's consider the distinctive features of shame that are of interest to us, namely, its susceptibility to being changed for the better and our observation that some of the most effective ways of doing this focus on changing how we see ourselves. On the dominant account of shame's evolutionary origins, the emotion began as a way of forestalling physical aggression in dominance hierarchies, but was subsequently co-opted to serve the more complex function of protecting one's status as a cooperative partner in the wake of serious norm violations (Maibom, 2010; Kurth, 2024b, 2025b; Fessler, 2004). This evolutionary model begins with the observation that submission responses tend to forestall aggression by dominants both because they blunt the aggressive motivations of the subordinate animal and because they signal that the subordinate will refrain from whatever behavior prompted the dominant's ire. From here, the account adds that, as our ancestors transitioned from dominance-based living to norm-governed social life, a new cooperative challenge emerged: how to sustain cooperation in the wake of serious norm violations. Shame—understood as a co-opted and augmented form of the earlier submission response—is then pointed to as a tool that seems particularly well equipped to address this challenge. As Heidi Maibom explains,

> the person who is ashamed shows to others—through the shame display—not just a recognition that they have failed to live up to public expectations, but also that they have an adverse emotional reaction to it ... Her shame indicates she can be counted on to live a life with others within the constraints set by the community.
> (2010, pp. 587–588)

Moreover, on this account, a key aspect of shame's transition from a basic appeasement response lies in the incorporation of a distinctive form of self-assessment into the mechanism that became shame (Kurth, 2025a; Maibom, 2010; Tracy & Robins, 2004). As we noted above, we feel ashamed when we're not seen as the person we take ourselves to be. And for this to happen, we need both a distinctive form of self-awareness—a sense of the identities and roles we occupy—and a distinctive perspective-taking capacity—a sensitivity to how we are seen by others (or ourselves at a distance). Support for this comes from various sources. For instance, animals that have self-oriented capacities like these (humans, chimps, orangutans) appear capable of self-conscious emotions like pride and shame, while those that lack them cannot (Tracy & Robins, 2004). Further evidence of shame's dependence on self-oriented capacities comes from work indicating that shame and other self-conscious emotions appear later in development than do emotions that do not require these capacities (Kurth, 2024b; Tracy & Robins, 2004).

In short, shame's evolutionary origins lie in its role as a tool that helps us protect our standing by helping us to both recognize occasions where we're not seen as the person we take (want) ourselves to be and engage in efforts to repair (or hide) the features of our selves that are the source of our shame. Given this, it's not surprising that shame is susceptible to being substantively changed, nor is it surprising that some of the most promising ways of shaping shame for the better work by way of our identities.

6.4 Implications for Our Understanding of Virtue

We've already looked at the standard Aristotelian and Stoic accounts of emotions' role in virtue and its development. But a bit more detail will be helpful. As we've seen, Aristotelians take emotions, as a class, to be malleable. Virtue, then, is a matter of bringing one's feelings into harmony with one's beliefs and motivations. Thus, one manifests the virtue of benevolence when one acts to help a person in need from a belief that they are in trouble and feelings of sympathy for their plight. By contrast, if one helps, but feels resentment for having to do it, one is merely continent—though one has done the right thing, one's emotions are out of tune with the situation. Given all this, Aristotelians see moral development as, among other things, a process where one changes one's emotions so that they're felt at the right time and in the right way.

The Stoics offer a very different picture. For them, emotions are inherently pernicious. For one, they systematically misevaluate the moral significance of one's situations. So, for instance, anger says the value of my reputation has been besmirched. But this is false since reputations are not things that have value, at least according to the Stoic axiology. But emotions are also highly prone to excess—hence, Seneca's claim, for example, that anger is an emotion "so hungry for vengeance that it brings down the avenger with it" (*De Ira*, I.1). What's more, Stoics take these flaws to be *inherent* features of emotions, not things that can be corrected (be it via emotional change or control). Thus, a central aspect of moral development is learning to extirpate one's emotions so that one's knowledge of and motivation toward right action is untainted (Graver, 2008; Kamtekar, 2025).

But in both cases, the view of emotions—as inherently malleable or completely impervious to correction—fits poorly with what emotion science tells us. Moreover, given the central place that these claims about emotions have in the Aristotelian and Stoic views, the trouble spreads (Kurth, 2022, Chapter 8). Consider: if emotions aren't malleable in the ways that Aristotelians maintain, then why think moral development should aim at changing emotions (as a class) or that virtue should be understood as the attunement of thoughts, motivations, and feelings? Similarly, if contrary to the Stoic line, at least some emotions can be cultivated, then why should moral development aim to root *all* of them out?

Were the Aristotelian and Stoic proposals the only options, it seems we'd be up a creek—pushed toward skepticism about whether there's space for virtue in an (empirically realistic) understanding of the human mind. This is where a less well appreciated way of thinking about emotion, virtue, and moral development becomes interesting. Call it the Kantian Account, given its roots in Kant and Kantians' thinking about emotion and virtue.

We can start with the Kantian Account's understanding of emotion, a view of the malleability of emotion that occupies the middle ground between the extremes staked out by the Aristotelians and Stoics. Specifically, the Kantian Account maintains that while some emotions can be changed for the better, others can only be controlled. Developing this, we can turn to Kant himself. He does not believe that the set of affective states that we call "emotions" forms a common category, as we see in the Aristotelian and Stoic view of emotions as all being evaluative beliefs. Rather, for him, what we call "emotions" fall into at least three different types of mental states—what he labels affects, passions, and moral feelings. The details of his view aren't important for us.[10] Rather, what matters is that Kant believes we need these different affective categories both because the "emotions" in these sets have different underlying cognitive architectures (though, of course, he doesn't put it that way) and because appreciating these differences in how emotion types work matters for how we should approach moral development.

To appreciate the Kantian Account's understanding of virtue and moral development, we can begin by noting that while Kant famously maintains that the only thing that has moral worth is action from respect for the moral law, he's also aware that, given the frailties of human psychology, our emotions and inclinations will often undermine our ability to make decisions, and so act, in ways that reflect our moral obligations. Given this, Kant takes virtue to be a unique form of reasoned *self-control*: it's the strength to overcome the emotions and inclinations that otherwise would lead us to act contrary to the moral law. Moral development, then, is a process where one learns how to "comba[t] natural impulses sufficiently to be able to master them when a situation comes up in which they threaten morality" (Kant 6, pp. 485, 598). What's more, Kant's account of moral development is informed by his associated account of emotion.

In particular, the Kantian Account proposes that the emotion cultivation techniques we adopt for moral education should reflect the kind of affective state we're looking to shape: only certain tools are appropriate for a given "emotion" type. So, for instance, Kant takes an affect like anger to be an intense and sudden feeling that generates beliefs and actions that bypass reason. Because these beliefs and actions have not been reflected on, they display a lack of virtue: they are more like animalistic responses than products of our higher, rational faculties. Thus, we see Kant recommending that individuals in the grip of anger should be distracted; they will then relax and calm down, giving reason the opportunity to reassert itself. Generalizing this, we might say that when it comes to affects, becoming more virtuous involves developing certain top-down, metacognitive capacities, including the ability to recognize situations where one's anger (say) will be piqued so that one can take the steps needed to neutralize its effects on one's subsequent thoughts and actions.

In contrast to the control-focused approach that Kant advocates for taming affects, we see him calling for a more change-focused approach for moral feelings. So, for instance, Kant holds that moral feelings like sympathy are to be actively cultivated (Baxley, 2010; Eran, 2020; cf. Thomason, 2017). More specifically, for Kant, sympathy is a feeling that we experience in response to the suffering of others—an emotion that draws our attention away from our own needs and toward the needs of others. Although sympathy itself does not have moral worth on Kant's account, it does contribute to one's ability to recognize what is morally relevant (i.e., the suffering of others) and does so in a manner that shapes our deliberations (e.g., shifting our focus from self to other). Given that sympathy is instrumentally valuable in these ways, Kant maintains that we have a duty to put ourselves in situations where our sympathy will be aroused and so changed for the better. In particular, in the *Doctrine of Virtue*, he advocates for visiting hospitals and soup kitchens in order to foster associations that enhance our sensitivity to human suffering. What we have, then, is a bottom-up exposure-based strategy.

Now we don't need to accept all of Kant's picture to see why it's relevant for our purposes. That is, even if we set aside Kant's austere account of moral worth or disagree with his typology of affective states, the rest provides us with a foundation for an account of emotion and virtue that's more empirically plausible than what we see in the Aristotelians and Stoic proposals. So here's my Kantian Account. The first four claims articulate the core of the Kantian Account's empirically informed understanding of emotion and human moral psychology, doing so in a way that answers the question that bedevils traditional Aristotelian and Stoic accounts—namely, why do emotions differ in their malleability?

1 "Emotion" does not pick out a natural kind, though some of the things we call emotions are natural kinds.
2 The emotions that are natural kinds are best understood on a biological model: as traits that have been shaped in significant ways by evolutionary forces.
3 Given evolution's shaping influence, we should expect that there will be (significant) differences in the cognitive architecture underlying natural kind emotions.

4 The differences in the cognitive architecture of natural kind emotions affect what we can do to shape them for the better: some we will be able to substantively change, others we may only hope to control.

The final pair of claims articulate the Kantian Account's resulting understanding of virtue and its development.

5 Virtue is a distinctive combination of self-knowledge and self-control: being virtuous is a matter of having the ability to recognize and overcome contrary emotions and inclinations in order to do the right thing. Importantly, this overcoming can manifest in various ways—in some cases, it will amount to substantively changing problematic emotions and inclinations, but in others it will involve enhancing one's ability to control what one feels and wants.
6 At its core, moral education aims to help individuals and communities develop the moral understanding and discipline needed to manifest the reasoned self-control characteristic of virtue.

6.5 Conclusion

Returning to where we began, we can better locate the error in Hursthouse's claim that an understanding of the nature of emotion is of little relevance to our appreciation of the place of emotion in virtue. As we've seen, the presumption that underlies this sentiment—that there's a significant degree of uniformity to the cognitive architecture underlying emotions—fits poorly with what we learn from emotion science. Not only does the cognitive architecture of emotions like shame, anxiety, and disgust differ but also these differences have significant consequences for our ability to shape them for the better. There's some irony here. After all, Kant's moral theory is often derided for being too theoretical—too out of touch with the reality of human moral psychology. And, on this point, Aristotelian and Stoic proposals are thought to do better. But from what we've seen here, the opposite is true. Not only does Kant's moral psychology fit better with what we learn from emotion science, but we also get an equally empirically realistic understanding of virtue and moral education.[11]

Notes

1 Exactly how the distinction between these two types of "cultivation" is drawn will turn on one's theory of emotion—specifically, how one distinguishes between emotions and the non-emotion mechanisms that shape them. What's important for our purposes is that the standard emotion theories all accept some form of this distinction.
2 I take this sketch to be common ground among shame theorists (e.g., in philosophy, see Deonna et al., 2011; Thomason, 2018; Williams, 1993; in psychology, see Tracy & Robins, 2004; de Hooge, 2013). That said, shame

is sometimes viewed as an emotion that predominantly brings hiding and withdrawal (e.g., Tangney & Dearing, 2002). However, meta-studies suggest otherwise: typically, shame brings repair and self-improvement efforts when the ashamed individual thinks she can make things better; it's only when she believes that she can't that we see hiding and the like (Leach & Cidam, 2015). For more, see de Hooge (2013), Maibom (2019), and Kurth (2025a).

3 For defenders of this view in psychology, see Barlow (2001) and Öhman (2008) and, in philosophy, Kurth (2018) and Vazard (2024).

4 Support for this picture is widespread in philosophy and psychology, e.g., Kelly (2011), Nussbaum (2004), and Tybur et al. (2013).

5 Understanding why exposure will lead to the quick acquisition of a disgust sensitivity is a complicated matter, one that turns on things like one's antecedent beliefs and the details of the exposure (May, 2018).

6 Disgust does not seem alone in its resistance to our efforts to shape it for the better. Hate, for instance, appears to have a similar profile. See Brudholm (2010) for discussion.

7 Of course, Aristotelians and Stoics could abandon their commitment to cognitivism. But it's unclear how helpful this would be. For one, excising the cognitivist commitment would require severe changes to the overall moral psychological picture that motivates these views—an issue that is especially acute for Stoics since so much of their view is grounded in a cognitivist theory of emotions (e.g., Kamtekar, 2025). Second, there's the task of providing an alternative understanding of emotions that not only avoids cognitivism's difficulties but also allows the Aristotelians and Stoics to hold on to their larger account of virtue and its development. As we'll see, it's not clear how this can be done.

8 In talking to prominent psychological constructivists about this, I received two answers. The first is a kind of skepticism about whether there really are differences in emotions' susceptibility to change. But I have a hard time squaring this suggestion with what I see in the empirical record. The second reply makes an analogy with three-dimensional state-space maps: an emotion like disgust resists change because it's "stuck in a trough." However, while there's certainly something right about this analogy, here too, the hard work—explaining *why* disgust is "stuck"—remains to be done.

9 For discussion of shame, see Tracy and Robins (2004), Maibom (2010), and Kurth (2025a). On anxiety, consult Öhman (2008) and Kurth (2016, 2018). And for disgust, the places to look are Kelly (2011), Tybur et al. (2013), and Rozin et al. (2008).

10 For more, see Baxley (2010), Eran (2020), and Kurth (2022: Chapter 8).

11 In writing this chapter, I've benefited from input from Isaac Wiegman, Ellen Fridland, and two anonymous referees.

References

Annas, J. (2011). *Intelligent virtue*. Oxford University Press.

Barlow, D. (2001). *Anxiety and its disorders*. Guilford Press.

Barrett, L. F. (2017). *How emotions are made*. Houghton Mifflin Harcourt.

Baxley, A. M. (2010). *Kant's theory of virtue*. Cambridge University Press.

Biondi, L., Gomes, N., Maior, R. S., & Soares, S. C. (2025). Revisiting "The malicious serpent": Phylogenetically threatening stimulus marked in the human brain. *Emotion Review*, *17*(2), 81–91.

Bollard, M. (2022). Is there such a thing as genuinely moral disgust? *Review of Philosophy and Psychology*, *13*(2), 501–522.
Brownstein, M. (2015). Implicit bias. In E. Zalta (Ed.), *The Stanford encyclopedia of philosophy* (Spring 2017 ed.). Stanford University. https://plato.stanford.edu/archives/spr2017/entries/implicit-bias/
Brudholm, T. (2010). Hatred as an attitude. *Philosophical Papers*, *39*(3), 289–313.
Carpenter, J. K., Andrews, L. A., Witcraft, S. M., Powers, M. B., Smits, J. A., & Hofmann, S. G. (2018). Cognitive behavioral therapy for anxiety and related disorders. *Depression and Anxiety*, *35*(6), 502–514.
de Hooge, I. (2013). Moral emotions and prosocial behaviour: It may be time to change our view of shame and guilt. In C. Mohiyeddini, M. Eysenck, & S. Bauer (Eds.), *Handbook of psychology of emotions* (Vol. 2): Recent theoretical perspectives and novel empirical findings (pp. 255–276). Nova.
Deonna, J., Rodogno, R., & Teroni, F. (2011). *In defense of shame*. Oxford University Press.
Doris, J. (2015). *Talking to ourselves*. Oxford University Press.
Engelhard, I., Leer, A., Lange, E., & Olatunji, B. (2014). Shaking that icky feeling. *Behavior Therapy*, *45*(5), 708–719.
Eran, U. (2020). Which emotions should Kantians cultivate (and which ones should they discipline)? *Kantian Review*, *25*, 53–76.
Ferreira, M. G., Mariano, L. I., de Rezende, J. V., Caramelli, P., & Kishita, N. (2022). Effects of group Acceptance and Commitment Therapy (ACT) on anxiety and depressive symptoms in adults: A meta-analysis. *Journal of Affective Disorders*, *309*, 297–308.
Fessler, D. (2004). Shame in two cultures. *Journal of Cognition and Culture*, *4*, 207–262.
Graver, M. (2008). *Stoicism and emotion*. University of Chicago Press.
Hodgdon, H. B., Anderson, F. G., Southwell, E., Hrubec, W., & Schwartz, R. (2022). Internal family systems (IFS) therapy for posttraumatic stress disorder (PTSD) among survivors of multiple childhood trauma. *Journal of Aggression, Maltreatment & Trauma*, *31*(1), 22–43.
Hursthouse, R. (2001). *On virtue ethics*. Oxford University Press.
Jackson, L. (2020). *Beyond virtue*. Cambridge University Press.
Kahan, D. M. (1996). What do alternative sanctions mean? *The University of Chicago Law Review*, *63*(2), 591–653.
Kamtekar, R. (2025). Stoic Emotion. In J. Klein & N. Powers (Eds.), *The Oxford handbook of Hellenistic philosophy* (pp. 426–446). Oxford University Press.
Kant, I. (1797/1996). Metaphysics of morals. In M. Gregor & A. Wood (Eds.), *Practical philosophy* (pp. 353–604). Cambridge University Press.
Kelly, D. (2011). *Yuck!* MIT Press.
Kelly, D., & Morar, N. (2014). Against the yuck factor. *Utilitas*, *26*, 153–177.
Kristjánsson, K. (2018). *Virtuous emotions*. Oxford University Press.
Kurth, C. (2015). Moral anxiety and moral agency. In M. Timmons (Ed.). Oxford studies in Normative Ethics (Vol. 5, 171–195). Oxford University Press.
Kurth, C. (2016). Anxiety, normative uncertainty, and social regulation. *Biology & Philosophy*, *31*(1), 1–21.
Kurth, C. (2018). *The anxious mind*. MIT Press.
Kurth, C. (2021). Cultivating disgust. *Emotion Review*, *13*(2), 101–112.
Kurth, C. (2022). *Emotion*. Routledge.
Kurth, C. (2024a). Moral anxiety: A Kantian perspective. In D. Rondel & S. Chopra (Eds.), *The moral psychology of anxiety*. Rowman & Littlefield.

Kurth, C. (2024b). An evolutionary account of guilt? *Philosophy of Science*, *91*, 510–518.
Kurth, C. (2025a). Shames and selves. *British Journal for the Philosophy of Science*. https://doi.org/10.1086/735280.
Kurth, C. (2025b). Emotion, adaptation, and natural kinds. *Passion*, *3*, 1–22.
Lacewing, M. (2005). Emotional self-awareness and ethical deliberation. *Ratio*, *18*(1), 65–81.
Leach, C. W., & Cidam, A. (2015). When is shame linked to constructive approach orientation? *Journal of Personality and Social Psychology*, *109*, 983.
Lear, M. K., & Luoma, J. B. (2025). A case study on transforming shame. *Journal of Clinical Psychology*, *81*, 878–887.
LeDoux, J. E. (2015). *Anxious*. Penguin.
Lewis, I., Watson, B., Tay, R., & White, K. (2007). The role of fear appeals in driver safety. *International Journal of Behavioral and Consultation Therapy*, *3*, 203–222.
Luoma, J. B., Rossi, S. L., Sereda, Y., Pavlov, N., Toussova, O., Vetrova, M., Bendiks, S., Kiriazova, T., Krupitsky, E., Lioznov, D., Blokhina, E., Lodi, S., & Lunze, K. (2023). An acceptance-based, intersectional stigma coping intervention for people with HIV who inject drugs—A randomized clinical trial. *The Lancet Regional Health–Europe*, *28*, 100611. https://doi.org/10.1016/j.lanepe.2023.100611.
Maibom, H. (2005). Moral unreason: The case of psychopathy. *Mind & Language*, *20*, 237–257.
Maibom, H. (2019). On the distinction between shame and guilt. In B. Cokelet and C. Maley (Eds.), *The moral psychology of guilt* (pp. 37–51). Rowman and Littlefield.
Maibom, H. L. (2010). The descent of shame. *Philosophy and Phenomenological Research*, *80*(3), 566–594.
Maibom, M. (2022). Don't worry, be happy? *Synthese*, *200*, 67.
May, J. (2018). "The limits of appealing to disgust." In N. Strohminger and V. Kumar (Eds.), *The moral psychology of disgust* (pp. 151–169). Rowman and Littlefield.
Meunier, S., & Tolin, D. (2009). The treatment of disgust. In B. Olatunji & D. McKay (Eds.), *Disgust and its disorders* (pp. 271–283). American Psychological Association.
Miller, W. I. (1997). *The anatomy of disgust*. Harvard University Press.
Mun, C. (2024). *Cultural perspectives on shame*. Routledge.
Munch-Jurisic, D. (2023). *Perpetrator disgust*. Oxford University Press.
Nussbaum, M. (2004). *Hiding from humanity*. Princeton University Press.
Öhman, A. (2008). Fear and anxiety. In M. Lewis, J. M. Haviland-Jones, & L. F. Barrett (Eds.), *Handbook of emotions* (pp. 127–156). Guilford Press.
Olatunji, B. O., Cisler, J. M., & Deacon, B. J. (2010). Efficacy of cognitive behavioral therapy for anxiety disorders: A review of meta-analytic findings. *Psychiatric Clinics*, *33*(3), 557–577.
Plakias, A. (2013). The good and the gross. *Ethical Theory and Moral Practice*, *16*, 261–278.
Raymond, L., Weldon, S., Kelly, D., Arriaga, X., & Clark, A. (2014). Making change. *Political Research Quarterly*, *67*(1), 197–211.
Rozin, P., Haidt, J., & McCauley, C. (2008). Disgust. In M. Lewis, J. Haviland-Jones, & L. Barrett (Eds.), *Handbook of emotions* (pp. 757–776). Guilford Press.

Schweiger Gallo, I., McCulloch, K. C., & Gollwitzer, P. M. (2012). Differential effects of various types of implementation intentions on the regulation of disgust. *Social Cognition*, *30*(1), 1–17.
Sherman, N. (1989). *The fabric of character*. Oxford University Press.
Swain, J., Hancock, K., Hainsworth, C., & Bowman, J. (2013). Acceptance and commitment therapy in the treatment of anxiety: A systematic review. *Clinical Psychology Review*, *33*(8), 965–978.
Tangney, J., & Dearing, R. (2002). *Shame and guilt*. The Guilford Press.
Thomason, K. (2017). A good enough heart. *Kantian Review*, *22*, 441–62.
Thomason, K. (2018). *Naked*. Oxford University Press.
Tracy, J., & Robins, R. (2004). Putting the self into self-conscious emotions. *Psychological Inquiry*, *15*, 103–125.
Tybur, J. M., Lieberman, D., Kurzban, R., & DeScioli, P. (2013). Disgust: Evolved function and structure. *Psychological Review*, *120*(1), 65–84.
Vazard, J. (2024). Feeling the unknown. *Erkenntnis*, *89*, 1275–1294.
Webb, T., Miles, E., & Sheeran, P. (2012). Dealing with feeling. *Psychological Bulletin*, *138*, 775–808.
Williams, B. (1993). *Shame and necessity*. University of California Press.

7 From Appeasement to Virtue

A Confucian Response to the Evolutionary Origins of Shame

Qiannan Li

7.1 Introduction

As Heidi Maibom notes, the dominant account of shame in the Anglo-American analytic tradition characterizes it as a painful emotion concerned with the failure to live up to certain standards, norms, and ideals (Maibom, 2010, p. 566). Because having a standard to live up to is central to this view, I refer to it as the standard account of shame. Within this framework, shame is typically understood either as an agent-centered response to one's own endorsed ideals or as a group-centered response to socially constructed norms (Thomason, 2018, p. 19). Despite their differences, both approaches treat shame primarily as a reaction to evaluative failure, whether internal or external.[1]

Recently, Maibom has offered a powerful challenge to this picture by turning to the evolutionary origins of shame. Drawing on work in evolutionary psychology, she argues that shame evolved from appeasement and submission behaviors observed in non-human animals, functioning to regulate social hierarchies and reduce the risk of aggression in group living (Maibom, 2010). On this appeasement interpretation, shame is a deeply heteronomous emotion: it is uniquely sensitive to the opinions of others, responsive to status differentials, and capable of arising even when individuals reject the standards by which they are judged. This evolutionary perspective has significant explanatory advantages, particularly in accounting for phenomena such as recalcitrant shame and persecution shame, which strain agent-centered accounts.

At the same time, appeasement-based interpretations face persistent difficulties. They struggle to explain why shame so often gives rise to anger or aggression rather than submission, why individuals experience second-hand shame on behalf of others, and why reflective, morally evaluative forms of shame—those tied to one's own ideals—appear normatively central rather than merely derivative. These challenges suggest that while

DOI: 10.4324/9781003632641-10

evolutionary explanations illuminate important aspects of shame, they do not by themselves capture the full moral and psychological significance of the emotion in human life.

This chapter proposes a Confucian-inspired alternative that both builds upon and revises the evolutionary story. Rather than rejecting evolutionary explanations outright, I argue that Confucian ethics can be understood as a culturally evolved framework for managing, shaping, and normatively evaluating shame. On this view, shame is not merely an inherited biological mechanism for appeasement, nor simply a reaction to violated standards. Instead, it functions as an affective alarm signaling threats to one's standing as a relational self—a self whose identity and worth are constituted through social roles, relationships, and practices of moral cultivation.

By interpreting shame through the lens of the relational self, the Confucian tradition offers a distinctive account of how biologically evolved emotional sensitivities can be redirected through cultural practices such as ritual propriety (*li* 禮), moral reflection, and character cultivation. In this sense, Confucianism exemplifies a form of cultural evolution: it transforms shame from a primarily defensive response to external threat into a moral resource that can support virtue development, relational harmony, and ethical self-improvement.

Framing shame in this way allows us to move beyond the dichotomy between autonomous and heteronomous accounts. Both internal moral reflection and sensitivity to others' judgments can be understood as responses to the same underlying concern: the preservation of one's relational integrity within a shared moral world. Shame's moral significance, on this view, lies not primarily in the standards it enforces, but in the care individuals take toward their own relational worth and toward the relationships that sustain a flourishing human life.

The argument proceeds as follows. Section 7.2 reconstructs Maibom's appeasement-based interpretation of the evolutionary origins of shame and highlights its theoretical strengths and limitations. Section 7.3 develops a Confucian account of shame as an affective response to threats to the relational self, situating this view within a broader framework of cultural evolution and moral cultivation. Section 7.4 shows how this Confucian perspective better accommodates aggression, second-hand shame, and the moral role of reflective shame, while also acknowledging shame's potential for vice. I conclude by suggesting that Confucian ethics offers a model for understanding how shame, as an evolved emotion, can be ethically transformed rather than merely endured or suppressed.

7.2 Heidi Maibom's Appeasement Interpretation of the Evolutionary Origin of Shame

7.2.1 The Core Argument and Theoretical Merits of the Appeasement Interpretation

Traditional theories of shame typically hold that it arises from an individual's failure to live up to their own internalized standards. Against this view, Maibom offers a powerful critique by drawing attention to phenomena such as *recalcitrant shame*—cases in which individuals feel shame despite rejecting the standards by which they are judged—and *persecution shame*, reported by survivors of genocide or systemic oppression (Maibom, 2010, pp. 571–572). These cases are difficult to explain if shame is primarily the result of failing to meet self-endorsed ideals.

Instead, Maibom turns to the evolutionary origins of shame. Many evolutionary psychologists argue that shame evolved from an innate disposition serving functions such as threat regulation and rank recognition—both crucial for survival in social species. On this view, shame operates as an appeasement strategy, reducing the risk of violence or retaliation following encounters with dominant individuals.[2] By signaling submission and a willingness to accept a lower status, shame helps maintain social stability.

Across species such as primates, wolves, and birds, subordinates display appeasement behaviors—lowering the head, avoiding eye contact, or retreating—to defuse tension and preserve group cohesion. Maibom argues that these behaviors exhibit evolutionary continuity with human shame, both in their physical expression and social function (Maibom, 2010, pp. 579–581). On this basis, she contends that shame is a heteronomous and deeply social emotion, shaped less by autonomous moral reasoning than by sensitivity to others' judgments. As she puts it, shame is "a profoundly social emotion uniquely sensitive to the opinion of social others" (Maibom, 2010, p. 567).

Maibom further argues that this account offers significant explanatory advantages over agent-centered theories, particularly in addressing three features of shame that such theories struggle to explain: its sensitivity to social status, its dependence on an audience, and its ability to arise without endorsement of external norms.

First, shame is closely tied to social hierarchy. Individuals in lower-status positions are typically more vulnerable to shame, while those with higher status are often more insulated from it. This pattern poses a challenge for agent-centered views, which explain shame in terms of failing one's own moral ideals (Maibom, 2010, p. 583). If shame were fully autonomous, it is unclear why social rank should so strongly influence one's susceptibility to it.

Maibom explains this pattern by appealing to the evolutionary role of submission in non-human animals. In many species, subordinates instinctively display appeasement behaviors—such as gaze aversion or physical withdrawal—to signal deference and reduce the risk of aggression. These responses are directed toward dominant individuals and help stabilize social hierarchies. On this account, brute dominance alone can be sufficient to elicit shame-like responses (Maibom, 2010).

This evolutionary background helps explain why status continues to shape the experience of shame in humans. Although human social systems are far more complex, the underlying emotional mechanisms persist. Even today, proximity to high-status individuals can elicit feelings of inferiority or shame, independently of any moral failing, and dominance plausibly plays a central role in persecution shame as well (Maibom, 2010). By conceiving shame as an evolutionarily conserved response to hierarchy, Maibom's theory illuminates why marginalized or subordinate individuals are disproportionately affected by shame in unequal societies—something agent-centered accounts struggle to explain.

Second, shame is almost always linked to a real or imagined audience. People typically feel shame when they are being observed, or when they imagine others judging them. This distinguishes shame from guilt, which can arise privately and without reference to others' opinions. Agent-centered theories have difficulty accounting for this audience-dependent feature.

Maibom again appeals to shame's evolutionary roots in submission behavior (Maibom, 2010, pp. 583–584). In non-human animals, appeasement displays are effective only in the presence of a social other—especially one capable of enforcing norms or posing a threat. Over time, this sensitivity to being seen became internalized in the human experience of shame.

In modern humans, the audience is often imagined rather than literal. Shame can arise in response to anticipated judgment from parents, peers, or abstract cultural ideals. The lingering influence of shame's original social function explains why imagined disapproval continues to exert such powerful emotional effects. This gives Maibom's account a distinctive advantage: it explains why social visibility remains central to shame even in private contexts.

Third, agent-centered accounts struggle to explain why shame frequently arises in cases where individuals explicitly reject the standards by which they are judged. Survivors of rape, genocide, or systemic abuse often report intense shame despite recognizing their lack of fault and rejecting the legitimacy of the imposed values. Similarly, individuals raised in repressive cultures may feel ashamed for violating taboos they no longer endorse.

Maibom argues that these cases reveal the fundamentally heteronomous nature of shame (Maibom, 2010, p. 583). Shame does not require endorsement of the relevant standards; it arises from being perceived—or perceiving oneself—as falling short of communal expectations. Because shame evolved as a mechanism of appeasement and reintegration, it remains highly responsive to external judgment, even when those judgments are rejected. This explains both recalcitrant shame, in which emotional responses conflict with personal beliefs, and persecution shame, in which shame results from being degraded or devalued by others.

Here, Maibom's theory shows its greatest strength. It accommodates the painful reality of shame without presupposing moral failure and avoids the strained strategies often employed by agent-centered views, such as positing unconscious norm endorsement or "deep down" acceptance of values. Instead, Maibom presents shame as a product of vulnerability and exposure to others' evaluations, whether fair or unjust.

By grounding shame in its evolutionary origins as a mechanism of social submission and appeasement, Maibom offers a compelling alternative to agent-centered models. Her account not only explains central features of shame—its status sensitivity, audience dependence, and capacity to arise without norm endorsement—but also deepens our understanding of its moral ambiguity. Shame, on this view, is not merely an internal reckoning with one's ideals, but a deeply heteronomous social emotion shaped by one's position in the social world and one's susceptibility to the gaze of others.

7.2.2 Challenges Facing the Appeasement Interpretation

While Maibom's appeasement-based account of shame offers valuable insight into the emotion's evolutionary roots and social dynamics, it also faces significant limitations. Several phenomena commonly associated with shame resist explanation within this framework, casting doubt on its overall comprehensiveness. Three challenges are particularly salient: the tendency for shame to elicit aggressive responses, which conflicts with the action orientation implied by appeasement; the phenomenon of second-hand shame, which calls into question the assumption that shame is always self-directed; and the conceptual gap between autonomous and heteronomous shame. Although Maibom acknowledges the existence of autonomous shame, she maintains that such forms should be regarded as secondary without offering a fully satisfactory explanation (Maibom, 2010, p. 588).

Aggressive responses to shame present a pressing challenge. On Maibom's account, shame evolved as a submissive response—a form of emotional retreat designed to de-escalate social tension and signal deference. Yet both psychological research and everyday experience suggest that shame often produces the opposite reaction: anger, defensiveness, and even aggression toward others (Tangney et al., 1992; Tangney et al., 2014; Thomason, 2018).[3] Individuals who feel shamed may lash out at those who caused or witnessed their shame, displace their feelings onto others, or engage in confrontational behavior.

These reactions sit uneasily with the appeasement model, which presupposes that shame primarily motivates withdrawal and self-effacement. If shame reliably functioned as a submissive response, retaliatory or aggressive behavior would be exceptional rather than common. The prevalence of such responses—especially in cases of humiliation—suggests that shame may serve multiple motivational roles, some of which conflict with appeasement. As a result, the evolutionary lineage Maibom emphasizes appears to capture only part of the emotional and behavioral complexity of shame in human life.

In response to this objection, Maibom and other proponents of group-centered accounts might appeal to Gabriele Taylor's (1985) suggestion that shame-induced violence is a product of honor cultures rather than a direct expression of shame itself. This reply, however, is inadequate for two reasons. First, it fails to explain shame-related violence outside the context of honor cultures (Kurth & Nelson, 2022). Second, it cannot account for cases in which shame-driven aggression is directed at strangers or innocent bystanders rather than at the shamer. While honor cultures often valorize revenge, they typically emphasize direct and public confrontation with the offender, not indiscriminate retaliation. Appealing to honor culture, therefore, does not sufficiently resolve the problem for group-centered theories of shame.

Second-hand shame raises a further difficulty. Individuals frequently report feeling ashamed on behalf of others—for example, when a family member, close friend, or public representative behaves in a norm-violating way without displaying any shame themselves. As Gail Weiss observes, "This 'secondhand' experience of shame on the part of another when there is no 'firsthand' experience of shame by the original agent has its own unique ethical dynamic that deserves further exploration" (Weiss, 2018, p. 538). In such cases, the person experiencing shame has not committed the relevant act but nonetheless bears the emotional burden in response to another's apparent shamelessness. This phenomenon challenges the idea that shame is strictly a self-directed response to one's own failure to appease others.

Within Maibom's framework, shame functions as a self-protective mechanism aimed at maintaining or repairing one's own status within a social hierarchy. Second-hand shame does not easily fit this model. It suggests that shame cannot be reduced to individual strategies of hierarchical survival but must also be understood as embedded in affiliative and communal relationships. Although Maibom's account is social in orientation, its focus on individual dynamics of submission limits its ability to capture this broader, relational dimension of shame.[4]

A further challenge concerns the distinction between autonomous and heteronomous shame. Maibom acknowledges that there are reflective, morally evaluative forms of shame arising from failures to meet one's own ideals. For example, an elite athlete may feel ashamed of falling short of a standard he alone has set for himself (Maibom, 2010, p. 588). Such experiences are central to moral self-assessment and cannot be reduced to internalized public disapproval; they involve a personal reckoning with one's ethical commitments. Nevertheless, Maibom asserts—without further argument—that these autonomous forms of shame are secondary to the socially embedded forms she prioritizes (Maibom, 2010).

This claim raises several unresolved questions. In what sense is autonomous shame "secondary"? Does this mean it is less frequent, less authentic, or normatively less significant? The lack of clarification becomes especially problematic when considering the moral force of accusations of shamelessness. In many Western philosophical traditions, to call someone shameless is to attribute a moral defect—to fault them for failing to recognize when they ought to feel shame. Shamelessness is often understood as a failure of moral self-evaluation (Mason, 2009) or as an inability to acknowledge the limits of one's self-conception and the authority of external perspectives (Thomason, 2018, p. 173). If shame is fundamentally a heteronomous response rooted in submission or appeasement, it becomes unclear why the absence of such a response should count as a moral failing. The normative weight carried by judgments of shamelessness, therefore, suggests that shame, at least in some of its forms, plays an essential role in moral judgment and self-evaluation that cannot be reduced to social compliance.

Taken together, these considerations indicate that while Maibom's appeasement-based interpretation captures important aspects of shame's social and evolutionary background, it leaves several central phenomena insufficiently explained. Aggressive responses to shame, second-hand shame, and the moral authority of reflective shame resist reduction to a model centered on submission and status de-escalation. What is missing, then, is not an evolutionary account as such, but a richer understanding of what is at stake when shame is experienced in human life. The following section develops a Confucian-inspired framework to address this gap by reconceiving shame as an affective response to threats to the relational self—an approach that preserves shame's social character while expanding its moral and psychological explanatory power.

7.3 A Confucian-Inspired Interpretation of the Evolutionary Origin of Shame

The Confucian account of shame offers an alternative perspective on its evolutionary origin, diverging from standard Western approaches in two fundamental respects. First, shame is not best understood as a signal of failure—whether with respect to personal ideals or social norms—but as an affective alarm that signals a threat to one's self-worth. Second, this self-worth is not the property of an independent, autonomous individual, but of a *relational self*—one whose identity and value are constituted through social relationships.

7.3.1 Shame as an Affective Alarm Signaling Threats to the Relational Self

In this section, I develop a Confucian-inspired account of shame that reconceptualizes the emotion not as a response to violated standards, but as an affective alarm signaling threats to one's standing as a relational self. Drawing on the *Analects* and the *Mencius*, this interpretation extends early Confucian insights to contemporary debates concerning the evolutionary origins and functions of moral emotions.

A central departure from many Western accounts lies in how shame is functionally characterized. Rather than treating shame primarily as a verdict of failure—whether failure to meet personal ideals or social norms—the Confucian tradition understands shame as an alerting mechanism. Shame arises when one perceives a disturbance in the conditions that sustain one's relational integrity: one's standing as a child, parent, friend, citizen, or moral agent embedded within a network of social roles.

This functional shift has important implications for the evolutionary significance of shame. If shame is understood as an affective alarm rather than a definitive judgment, its role cannot be reduced to submission or appeasement alone. Instead, shame signals that one's relational standing is endangered and demands a response—though what counts as an appropriate response varies with context, available resources, and the agent's level of moral cultivation.

Early Confucian thinkers emphasize that the disposition to feel shame (*xiu* 羞) is innate yet educable. Mencius famously identifies the "sense of shame" as one of the four moral sprouts: an incipient capacity that must be cultivated in order to develop into virtue. What matters most on this view is not the episodic experience of shame itself, but the underlying sensitivity to what is *unbearable*—to forms of treatment or conduct that threaten one's self-worth as a relational being. As Mencius observes, people naturally recoil from being treated with contempt, even when survival is at stake (*Mencius* 6A10).[5] This recoil reveals an evaluative sensitivity that precedes explicit moral reasoning.

Mencius illustrates this point with a well-known example:

> A basket of food and a bowl of soup—if one gets them, then one will live; if one does not get them, then one will die. But if they are given with contempt, then even a homeless person will not accept them. If they are trampled upon, then even a beggar will not take them.
> (*Mencius* 6A10; Van Norden, 2008, 154)

Here, shame is not a self-critical response to failed standards but an affective sensitivity to being demeaned. The sense of shame motivates resistance to, or withdrawal from, situations that undermine one's self-worth. On this reading, Confucians understand shame as a self-conscious disposition grounded in concern for relational value—a sensitivity to whether one is being treated appropriately within a given context and relationship.[6]

On the Confucian account, shame combines two elements: a valuing attitude toward one's own self-worth and a perception-like recognition of threats to that worth. A person is disposed to feel shame when two conditions are met. First, they possess a caring attitude toward their self-worth, which may be minimal or implicit. Even a child's intuitive sense of being loved by their parents suffices to ground this attitude. Second, they perceive a threat to that worth, either in their own conduct or in the behavior of others who contribute to their relational self. On this view, a person is shameless when they either lack a valuing attitude toward their own worth or fail to recognize threats to it. Shamelessness thus reflects a deficiency in one's sense of appropriateness. I return to this point in the next section, where I argue that this account better captures the moral criticism implicit in judgments of shamelessness than Maibom's appeasement model.

A second major divergence between Confucian and many Western accounts of shame concerns their underlying conception of the self. For Confucian thinkers, the self is intrinsically relational—not a bounded, autonomous individual, but a being "shaped, characterized, and identified in and through their relations" (Thompson, 2017, p. 889). As David Wong emphasizes, recognizing our relational nature allows us to see that "we are at least much less the sole authors of our actions" (Wong, 2004, p. 423). Virtue, for such beings, requires continual harmonization between one's own interests and those of others to whom one is related (Wong, 2004, p. 427).

This relational conception of the self contrasts with identity-based accounts of shame, which ground the emotion in conflicts among different facets of one's identity. For example, Thomason characterizes shame as "a tension between our identities and our self-conceptions," arising when some aspect of one's identity overshadows one's self-understanding (Thomason, 2018, p. 149). Similarly, Charlie Kurth defines shame as "the feeling that your ability to project a public self that corresponds to your self-conception has been substantively frustrated" (Kurth, 2025).

While the Confucian account shares with identity-based views the idea that shame involves a perceived threat to the self, it diverges in two key respects. First, the relational self does not presuppose sophisticated self-reflective capacities or a developed conception of identity. Even young children can experience threats to their relational standing without possessing such reflective resources. Second, Confucian shame is not defined in terms of coherence among identities. Instead, it understands the self as always embedded in a dynamic web of moral relationships, within which different aspects of the relational self become salient in different contexts. This fluidity reflects the situational and relational nature of Confucian selfhood.

Understanding shame as an affective alarm for the relational self also clarifies its place within cultural evolution.[7] While evolutionary theories plausibly explain why humans are disposed to experience shame in response to social threat or loss of standing, they do not determine how shame is interpreted, regulated, or normatively evaluated. Confucian practices—such as ritual propriety (*li* 禮), moral exemplars, and reflective self-cultivation—function as cultural mechanisms that reshape the triggers, meanings, and appropriate responses associated with shame.

From this perspective, Confucianism does not deny the biological roots of shame. Rather, it represents a stage of cultural evolution in which shame is redirected away from mere status preservation toward the cultivation of virtue and relational harmony. Cultural norms determine which threats to self-worth are legitimate, which audiences matter, and which responses—withdrawal, confrontation, apology, or reform—are appropriate. In this way, Confucian ethics illustrates how moral emotions are not only products of evolution but also tools refined by culture to support cooperative social life.

The culturally specific articulation of shame in Chinese further illustrates this point. Chinese distinguishes multiple forms and degrees of shame that are often collapsed in other languages. Terms such as *haixiu* (害羞, shyness) and *nankan* (难堪, embarrassment) describe relatively mild and easily reparable threats to relational standing. *xiukui* (羞愧, shameful guilt) captures cases in which shame is accompanied by self-endorsed moral failure, while *xiuchi* (羞耻, disgrace or dishonor) denotes severe and potentially irreparable damage to one's relational identity.

These linguistic distinctions underscore the role of cultural refinement in shaping emotional life. Differences in how cultures name, evaluate, and manage shame suggest that the emotion is shaped not only by biological evolution but also by cultural evolution, through which emotional responses are calibrated to sustain increasingly complex forms of social cooperation. Shame thus functions as a socially refined tool that helps individuals navigate relational expectations, repair social bonds, and regulate participation in shared practices. This cultural variability challenges accounts that treat shame as a fixed, universally biological response, while fitting naturally with the Confucian understanding of shame as grounded in the relational self.

Within this framework, shame arises not from the failure to meet self-endorsed or social standards, but from a perceived threat to one's standing as a relational being. This raises a further question: what does it mean to have proper standing as a relational self? Answering this requires situating shame within the broader Confucian ethical project, which centers on cultivating virtues such as *li* (ritual propriety), *ren* (humaneness), *zhi* (wisdom), and ultimately *yi* (righteousness), all directed toward harmonizing the self with others.

First, a relational self must exhibit sensitivity to the needs, reactions, and expectations of others. As Confucius remarks, "I will not be afflicted at men's not knowing me; I will be afflicted that I do not know men" (*Analects* 1.16).[8] To be humane (*ren*), as Thompson explains, is to have overcome immature self-centeredness and to have developed the discernment needed to navigate complex relational situations (Thompson, 2017, p. 891).

Second, such a person is able to take moral responsibility not only for their actions but also for how those actions are perceived and affect close others. According to Confucius, "He who aims to be a man of complete virtue in his food does not seek to gratify his appetite, nor in his dwelling place does he seek the appliances of ease; he is earnest in what he is doing, and careful in his speech; he frequents the company of men of principle that he may be rectified – such a person may be said indeed to love to learn" (*Analects* 1.14).

Third, they also demonstrate integrity across relationships by maintaining consistency and sincerity in moral behavior, regardless of social context. Confucius says, "The man of perfect virtue is cautious and slow in his speech" (*Analects* 12.3), and that one should feel shame if their words exceed their deeds (*Analects* 4.22). Thus, to properly function as a relational being, a person should embody sincerity and consistency in both values and actions.

Finally, they must have an appropriate understanding of the legitimate sources of self-worth to differentiate between proper and improper threats to their self-worth. Confucius provides multiple examples to illustrate the proper and improper factors for evaluating a person's self-worth. According to Confucius, people should not feel shame about their attire or meager sustenance (*Analects*, 4.9), nor should they be ashamed of inquiring from social inferiors (*Analects*, 5.15). Instead, shame should arise from not being true to one's word (*Analects*, 4.22) or from possessing affluence and esteem while residing in an ill-ordered state (*Analects*, 8.13).

Together, these capacities—sensitivity to others, moral responsibility, integrity, and proper self-understanding—constitute one's relational standing. When this standing is threatened, shame arises as an affective signal of disruption in the moral fabric of one's roles as a parent, child, friend, or community member.[9]

This Confucian account also differs from evolutionary theories that link shame to social attractiveness and prestige. According to the Social Attention Holding Potential (SAHP) model, human status depends on one's ability to attract positive social attention, and shame functions to protect a favorable self-image in order to secure social approval (Gilbert, 2001; Gilbert & McGuire, 1998). On this view, shame motivates the maintenance of external appearance rather than the management of threat.

The key difference lies in what shame is understood to protect. While the SAHP account emphasizes image and social recognition, the Confucian view insists that moral standing cannot be reduced to appearance alone. Shame is tied to the relational self, whose value depends not only on how one is seen but also on self-understanding and the fulfillment of ethical obligations within one's relationships.

Section 7.3.1 has articulated a Confucian account of shame as an affective alarm signaling threats to one's standing as a relational being, situating this account within a broader framework of cultural evolution and moral cultivation. The next section turns to the explanatory payoff of this reconceptualization, showing how the Confucian framework successfully addresses the very phenomena—aggressive responses to shame, secondhand shame, and the continuity between autonomous and heteronomous shame—that challenge appeasement-based interpretations.

7.3.2 *Responding to Challenges*

In this section, I will demonstrate that the Confucian account of shame not only addresses the three puzzles discussed in Section 7.2 that confront agent-centered views of shame but also avoids the challenges facing the appeasement interpretation of the evolutionary origins of shame.

7.3.2.1 *Confucian Responses to Three Key Challenges in Agent-Centered Accounts of Shame*

Confucianism conceives of the self not as an isolated, autonomous individual but as a relational being whose identity and value are constituted through social roles and interpersonal relationships. Because selfhood is inseparable from relational standing, threats to that standing naturally give rise to shame. On this view, shame is inherently social: it arises in response to disruptions within the network of relationships that define who one is.

Because the relational self is shaped by power dynamics within relationships, shame is necessarily sensitive to status. Importantly, the Confucian understanding of status extends beyond hierarchical or economic position to include one's moral standing within families, communities, and society at large. Shame thus tracks not only disparities in power but also failures or threats concerning how one is positioned within shared moral practices.

Maibom may likewise endorse a broadened notion of status. The key difference, however, lies in the Confucian emphasis on relational selves, which more clearly captures the shift from an individualistic perspective—one shared with non-human animals—to a collective perspective that is distinctive of human social life. For Confucians, status and power look different when individuals are considered in isolation than when they are understood as members of a collective moral community.

This difference is especially evident in cases of systemic injustice such as genocide, child abuse, or spousal abuse. When victims confront perpetrators as isolated individuals, asymmetries of physical, economic, or institutional power often prevent meaningful moral reckoning. By contrast, when victims organize and speak collectively—as in movements such as #MeToo—they can assert moral authority, publicly shame perpetrators, and reframe the meaning of their own previously experienced shame. The Confucian framework helps explain how collective agency can transform both power relations and the moral significance of shame.

The Confucian account also emphasizes the audience-sensitivity of shame. Because relational standing is constituted through relationships, the judgments of others—whether actual or internalized—serve as signals of relational disruption. These audiences may include elders, peers, the broader community, or one's own cultivated moral sensibility, each of which plays a role in eliciting and sustaining shame. As Mencius puts it, a morally upright person is one who, "looking up, is not disgraced before Heaven, and looking down, is not ashamed before human beings" (*Mencius* 7A20; Van Norden, 2008, p. 176).

Responsiveness to such judgments is essential for maintaining proper relational standing. A person who is indifferent to the perspectives of others cannot function as a relational being or navigate shifts in social relationships. Accordingly, both actual and imagined audiences are integral to how the Confucian self experiences and responds to shame.

At the same time, Confucians stress the importance of self-understanding for sustaining harmonious relationships (Van Norden, 2002). Without a sense of self-worth grounded in one's own values, an individual cannot form stable or genuine ties with others. To be a good parent, for example, one must take oneself to genuinely possess the relevant qualities, not merely appear virtuous in the eyes of others. For this reason, Confucianism rejects a sharp division between autonomous and heteronomous shame.[10] Internal and external perspectives are treated as mutually necessary for preserving relational integrity. This integration underlies the Confucian practice of "vigilance in solitude" (*shèn dú* 愼獨), which aims to sustain one's sense of shame even in the absence of an external audience.

One of the Confucian account's distinctive strengths is its ability to explain why shame can arise even when one does not endorse the relevant norms. Shame, on this view, does not require wholehearted acceptance of a particular rule or standard; it may instead arise from a perceived threat to one's relational identity or recognition. A child, for instance, may feel ashamed for embarrassing their parents even while rejecting the moral standard being invoked. This is not a form of moral inconsistency, but an expression of the child's sensitivity to relational embeddedness, where acknowledgment and harmony matter even in the absence of full normative agreement.

Moreover, such experiences of shame are not merely passive acknowledgments of authority. Within the Confucian framework, shame functions as a prompt for moral reflection. Feeling ashamed for embarrassing one's parents invites questions about the alignment between one's values and those of significant others, the sources of those values, and the impact of such divergences on the relationship. Most importantly, it directs attention to how harmony might be restored despite moral disagreement.

In the following section, I show how this Confucian understanding of shame more effectively addresses the challenges facing Maibom's appeasement-based interpretation, including aggressive responses to shame, second-hand shame, and the relationship between autonomous and heteronomous forms of shame.

7.3.2.2 *How Confucian Shame Responds to the Challenges Facing the Appeasement Account*

Maibom's appeasement-based interpretation offers an evolutionary explanation of shame as a submissive behavior that helps defuse social tension. However, it struggles to account for several well-documented features of shame, including (1) aggressive responses, (2) the phenomenon of second-hand shame, and (3) the apparent continuity between autonomous and heteronomous shame. The Confucian conception of shame—as an affective response to threats to the relational self—provides a more flexible and context-sensitive framework that better accommodates these complexities.

7.3.2.2.1 AGGRESSIVE REACTIONS TO SHAME: THREAT VS. FAILURE

Maibom's model predicts that shame typically motivates appeasement through withdrawal or submission. Yet empirical evidence suggests that shame can also trigger anger, defensiveness, or aggression—especially when individuals feel cornered or humiliated. The Confucian account explains this variability by distinguishing between two sources of shame: failure and threat.

Failure implies a settled normative judgment and often naturally encourages submission. Threat, by contrast, signals uncertainty and the possibility of action. On the Confucian interpretation, a person experiencing shame does not necessarily feel defeated by falling short of an ideal or a social standard. Instead, they feel endangered: shame registers as an alarm and generates urgency to protect one's self-worth.[11] If shame is understood as a signal that one's relational self is endangered—but not yet destroyed—then the response depends on how the individual assesses their capacity to manage or counteract the threat. Aggressive reactions can therefore be intelligible, especially when assertiveness seems like the most accessible way to address the danger.[12] Still, people will not respond uniformly.

The intensity of shame also varies with a person's psychological and social resources. Shame Resilience Theory (SRT), developed by Brené Brown (2006), suggests that shame resilience can be cultivated through acknowledging vulnerability, engaging in critical reflection on social and cultural norms, and—most importantly—forming mutually empathetic relationships. Empathetic connection reduces shame's distress by easing the alarm triggered by threats to the relational self. Such relationships function as a natural remedy: they reaffirm self-worth and signal that the social bond remains intact. This effect is less easily explained by appeasement theories, which have a less direct account of why empathy should diminish shame.

Early Confucians likewise maintain that developing complementary virtues—such as bravery, wisdom, and benevolence—can mitigate shame's distress by supplying internal strength and moral clarity. Cultivation can transform raw shame into the virtue of righteousness, avoiding shame's destructive effects. Mencius, for instance, suggests that benevolence enables one to reflect on mistakes and take responsibility rather than projecting blame outward:

> Benevolence is like archery. An archer corrects himself and only then shoots. If he shoots but does not hit the mark, he does not resent the one who defeats him but simply turns and seeks for it in himself.
>
> (*Mencius* 2A7; Van Norden, 2008, 48)

With benevolence as an ethical capacity, the virtuous person is better equipped to manage shame's defensive tendencies and to use its pain constructively.

This framework also helps explain why some people experience shame with exceptional intensity. Individuals with depression, for example, may find shame particularly intolerable. They perceive threats to self-worth and experience the alarm those threats trigger, but their condition undermines hope—the belief that damage to the relational self can be repaired. As a result, the alarm is harder to silence, and threats are more readily experienced as definitive failures. This misperception intensifies suffering, as shame becomes not merely a signal of danger but a confirmation of personal inadequacy.

The Confucian tradition thus does not deny shame's painfulness; it holds instead that shame need not be overwhelming and that its intensity varies across persons and contexts. From this perspective, aggression is not a contradiction of shame but one possible outcome—especially when confrontation appears to be the only or most effective means of restoring endangered relational standing. This interpretation accords with the Confucian principle of situational propriety (*li* 禮), which recognizes that appropriate responses to relational tension differ by context: sometimes humility and withdrawal are fitting, while at other times active defense or confrontation is required.

The Confucian account also clarifies why shame is morally ambivalent. As an affective alarm, shame can function virtuously by drawing attention to genuine threats to relational integrity and motivating repair, reflection, and moral growth. Yet when shame is excessive, misdirected, or socially weaponized, it can become vicious—manifesting as rumination, humiliation, domination, or self-contempt. In such cases, shame no longer signals a reparable threat but entrenches a sense of worthlessness or subordination that undermines moral agency. Confucian ethics does not valorize shame as such; it emphasizes cultivating complementary virtues—benevolence, wisdom, and ritual sensitivity (*li* 禮)—to regulate shame's intensity and orientation. Properly cultivated, shame supports moral development; left unregulated, it becomes morally corrosive. This dual potential explains both shame's ethical importance and its capacity to deform moral life.

7.3.2.2.2 SECOND-HAND SHAME AND THE RELATIONAL SELF

Second-hand shame poses a serious challenge to appeasement theories, which treat shame as primarily self-focused and arising from an individual's own social failings. By contrast, Confucian ethics begins with the relational self—a self constituted through social and moral relationships. From this perspective, it is natural that people may feel shame not only for their own actions but also for the actions of those with whom they are closely connected, such as family members, close friends, or national representatives.

On this view, moral identity is inherently relational and embedded within shared ethical life. When someone with whom one shares a significant relationship violates moral expectations and appears shameless, the integrity of the relationship—and the shared moral framework that supports it—is threatened. Shame, in such cases, is an affectively attuned response. As Mencius famously asserts, "People may not be shameless. The shame of being shameless is shameless indeed" (*Mencius* 7A6; Van Norden, 2008, p. 173). For a Confucian self situated in networks of mutual recognition and responsibility, disruptions of this kind are felt as threats to relational integrity. Second-hand shame is therefore not anomalous but a natural extension of relational self-awareness and moral concern.

Consider a father who feels ashamed when his son responds shamelessly to a moral failure. The son's behavior reflects on his own character, but it can also threaten the father's relational identity as a morally responsible parent—someone who has raised a child capable of recognizing and responding to ethical norms. Here, shame is not reducible to personal failure; it registers the vulnerability of moral relationships and the fragility of shared roles.

As Weiss notes, gender and racial dynamics can intensify second-hand shame. People across the globe often report shame when predominantly male political and civic leaders behave shamelessly (Weiss, 2018, pp. 537–538). This reaction can be especially pronounced among members of marginalized groups, who may experience second-hand shame in response to the dominant group's refusal to exhibit shame at all.

From the standpoint of appeasement theory, this phenomenon is difficult to explain. If shame is fundamentally a mechanism of submission designed to defuse tension with powerful agents, it is puzzling that marginalized individuals would feel shame on behalf of those who hold greater power.

The Confucian account, grounded in the relational self, provides a clearer explanation. Because the self is embedded in social and moral networks, shame can arise not only from one's own failures but also from failures within one's community or representational group, especially when those failures threaten shared social integrity. This framework also helps explain why marginalized groups disproportionately bear the burden of second-hand shame: their social position exposes them to more frequent threats to relational standing while limiting access to resources for resisting or repairing those threats. They may therefore be more sensitive to such disruptions and less likely to feel entitled to shamelessness, even when the shame-inducing conduct is not their own.

By contrast, members of dominant groups are often insulated from relational vulnerability. Their social power affords them both the privilege of disregarding shame and the capacity to deflect or suppress it. This asymmetry helps explain why dominant groups may appear indifferent to shameful conduct, while marginalized individuals continue to experience shame on behalf of a broader moral community.

7.3.2.2.3 BRIDGING THE GAP BETWEEN AUTONOMOUS AND HETERONOMOUS SHAME

Maibom acknowledges a conceptual divide between heteronomous shame (responsive to external judgment) and autonomous or moral shame (responsive to internal standards), treating the latter as a secondary development (Maibom, 2010, p. 588). As discussed in Section 7.2.2, however, this claim raises unresolved questions: in what sense is autonomous shame "secondary"? Is it less frequent, less authentic, or merely less aligned with shame's evolutionary origins?

Even if we grant an appeasement-based story about shame's origins, it would be premature to dismiss autonomous shame. It is a common form of experience and plays a significant role in moral transformation, motivating individuals to become better versions of themselves. Consider a top athlete who feels ashamed of failing to live up to a personal ideal—one no one else imposed. Such shame can motivate self-improvement and may be integral to the athlete's integrity, understood as commitment to excellence on self-defined terms.

The Confucian view helps dissolve a rigid divide between autonomous and heteronomous shame. On a relational conception of the self, both are manifestations of the same underlying capacity: sensitivity to threats to one's relational standing. What differs is not the structure of shame, but the standpoint from which the threat is registered.

In heteronomous shame, the threat is perceived through the responses of others who matter to one's relational identity. In autonomous shame, the threat is registered through one's cultivated moral perspective—an internalized but not isolating standpoint shaped by roles, exemplars, and practices of reflection. In both cases, shame functions as an affective alarm that signals a potential disruption of relational integrity and calls for a response.

This unified account explains why autonomous shame can carry genuine moral authority rather than merely echoing social pressure. It also clarifies why heteronomous shame may remain morally significant even when the relevant norms are not fully endorsed: the emotion responds less to the legitimacy of a standard as such than to the vulnerability of one's standing within a shared moral world. Autonomous and heteronomous shame are therefore best understood not as primary and secondary forms of shame, but as context-sensitive expressions of a single relational capacity.

7.4 Conclusion

This chapter has argued that a Confucian understanding of shame offers a compelling alternative to both agent-centered and appeasement-based evolutionary accounts. While Heidi Maibom's appeasement interpretation illuminates shame's biological and social origins, it struggles to accommodate key features of human shame—most notably aggressive responses, second-hand shame, and the moral centrality of reflective, self-evaluative shame. These difficulties point not to the irrelevance of evolutionary explanations, but to their incompleteness.

By grounding shame in the concept of the relational self, the Confucian framework reconceives shame as an affective alarm that signals threats to relational integrity rather than as a mere reaction to violated standards or a reflex of submission. This reconceptualization dissolves the rigid divide between autonomous and heteronomous shame, showing that both internal moral reflection and responsiveness to others' judgments express the same underlying concern: preserving one's standing within a shared moral world.

The Confucian account also situates shame within a broader story of cultural evolution. While human beings may inherit a biological disposition to experience shame in response to social threat, Confucian ethics exemplifies how cultures can reshape this disposition through ritual, moral education, and character cultivation. These practices refine what counts as a legitimate threat to self-worth, regulate shame's intensity, and guide agents toward responses that sustain moral relationships rather than merely defuse conflict. In this sense, Confucianism functions as a culturally evolved moral technology that transforms shame from a defensive emotion into a potential vehicle for virtue.

At the same time, Confucian thinkers are acutely aware of shame's ambivalence. Shame can support moral growth and relational repair, but it can also degenerate into vice—manifesting as humiliation, domination, rumination, or corrosive self-contempt. The Confucian emphasis on cultivation and contextual sensitivity reflects an effort to harness shame's moral potential while guarding against its destructive tendencies. Shame is thus neither unconditionally virtuous nor merely pathological; its value depends on how it is shaped, interpreted, and integrated into an ethical life.

Seen in this light, shame is best understood neither as a relic of our evolutionary past to be overcome nor as a simple mechanism of social control. It is a morally significant emotion whose meaning is continually negotiated through cultural practices. The Confucian tradition offers a powerful illustration of how an evolved emotional capacity can be ethically transformed—redirected toward the cultivation of virtue, the maintenance of relational harmony, and the flourishing of relational beings. In doing so, the Confucian perspective invites us to view shame not as a mere reaction to moral failure or a remnant of social control, but as a profound expression of care—care for oneself, for others, and for the relational bonds that sustain human flourishing.

Notes

1 The first, known as the traditional or agent-centered view, holds that these standards are personal ideals. On this view, shame is defined as the painful emotion we feel in response to our own failures to live up to our ideals or values (Thomason 2018, p. 19). John Rawls (2003), Gabriele Taylor (1985), and Deonna et al. (2012) are key proponents of this view. For Rawls, shame results from a blow to self-esteem when one fails to exercise certain excellences or virtues (Rawls, 2003, p. 388). Taylor defines shame as essentially a painful feeling when a person comes to see herself as inferior to what she had believed or hoped herself to be (Taylor, 1985, p. 64). This view emphasizes the autonomous dimensions of shame. The second approach, the naturalistic or group-centered view, regards the relevant standards as socially constructed norms (Thomason, 2018, p. 40). On this account, shame is the painful emotion we experience when we fail to live up to socially established expectations. Maibom, for instance, defines shame as an implicit acknowledgment of a demand to live in accordance with public norms and standards (Maibom, 2010, p. 587).

2 Fessler (1999), Gilbert (1989, 1992), Keltner and Harker (1998), and Weisfield (1997, 1999).

3 Although aggressive and defensive responses to shame may not be salient dispositions for everyone, some individuals do exhibit a strong tendency to respond to shame with aggressive behaviors and attitudes. My interpretation of the Confucian account of shame—grounded in the concept of the relational self and acknowledging the cultural–evolutionary origins of shame—can better capture this behavioral divergence than accounts that focus solely on the biological evolutionary similarities between non-human animals and human beings.

4 A more detailed comparison between Maibom's view and the Confucian perspective will be presented in Section 7.3.2.

5 All translations of *Mencius* are from Van Norden (2008).

6 The concept of appropriateness will be examined in the discussion that follows.

7 Maibom's view might be able to accommodate the perceived threat to one's standing if she acknowledges that shame can be an adaptation of cultural evolution rather than merely biological evolution, since the practical identity of relational beings is unique to human beings.

8 All translations of *Analects* are from the Chinese Text Project (available at: http://ctext.org/analects/xue-er).

9 Note that a subject does not need to feel disappointed in themselves when facing a threat to their relational standing. Such disappointment arises only when the subject perceives a lack of resources to address the threat.

10 I will return to this point in the next section.

11 One might object that mere threats to the relational self should be handled by emotions such as anger or anxiety, reserving shame for cases of actual relational damage. I do not deny that anger or anxiety are often co-present in

such situations. My claim, however, is that shame is best understood as an affective *alarm* whose function is to register the vulnerability of one's relational standing, rather than to track only completed damage. On the Confucian view, shame arises when one perceives that one's standing as a relational being is endangered, whether or not degradation has already occurred. This differs from appeasement accounts, which tend to treat shame primarily as a response to realized loss of status that motivates withdrawal or submission. By contrast, the alarm model explains why shame can arise prior to submission and can organize a range of downstream responses—including withdrawal, repair, or even confrontation—depending on how the agent assesses the threat.

12 According to Thomason, violence provoked by experiences of shame can be understood as a natural response to the sense of powerlessness that often accompanies such experiences. However, Thomason's characterization of shame as rooted in powerlessness differs from the Confucian account, which emphasizes shame as an affective alarm system. I do not further examine the similarities and differences between Thomason's and the Confucian perspectives here, as her theory does not appeal to either the evolutionary origins of shame or any objective moral standard. As she states, 'there is no value or object that shame tracks or responds to, and thus there is no independent standard for proper shame' (Thomason, 2018, p. 171). That said, I believe Thomason's account could be brought into closer alignment with the Confucian perspective if the tension between self-endorsed conceptions and social identities is reinterpreted as a form of moral alertness rather than as a sense of powerlessness.

References

Brown, B. (2006). Shame resilience theory: A grounded theory study on women and shame. *Families in Society*, *87*(1), 43–52.

Chinese Text Project. Available at: http://ctext.org/analects/xue-er

Deonna, J., Rodogno, R., & Teroni, F. (2012). *In defense of shame: The faces of an emotion*. Oxford University Press.

Fessler, D. H. T. (1999). Toward an understanding of the universality of second-order emotions. In A.L. Hinton (Ed.), *Biocultural approaches to the emotions* (pp. 75–116). Cambridge University Press.

Gilbert, P. (1989). *Human nature and suffering*. L. Erlbaum Associates.

Gilbert, P. (1992). *Depression: The evolution of powerlessness*. The Guilford Press.

Gilbert, P. (2001). Evolution and social anxiety: The role of attraction, social competition, and social hierarchies. *Psychiatric Clinics*, *24*(4), 723–751.

Gilbert, P., & McGuire, M. T. (1998). Shame, status, and social roles: Psychobiology and evolution. In P. Gilbert & B. Andrews (Eds.), *Shame: Interpersonal behavior, psychopathology, and culture* (pp. 99–125). Oxford University Press.

Keltner, D., & Harker, L. A. (1998). The forms and functions of the nonverbal signal of shame. In P. Gilbert & B. Andrews (Eds.), *Shame: Interpersonal behavior, psychopathology, and culture* (pp. 78–98). Oxford University Press.

Kurth, C. (forthcoming). Shames and selves: On the origins and cognitive foundations of a moral emotion. *British Journal for the Philosophy of Science*.

Kurth, C., & Nelson, A. (2022). Shame, selves, and morality. *Philosophical Psychology*, *35*(1), 22–140.

Maibom, H. (2010). The descent of shame. *Philosophy and Phenomenological Research*, *80*(3), 566–594.

Mason, M. (2009). On shamelessness. *Philosophical Papers*, *39*(3), 401–424.
Mencius (2008). *Mengzi: With selections from traditional commentaries, translated with introduction by Bryan W. Van Norden*, Hackett Publishing.
Rawls, J. (2003). *A theory of justice* (revised ed.). Harvard University Press.
Tangney, J., Wagner, P., Fletcher, G., & Gramzow, R. (1992). Shamed into anger? The relation of shame and guilt to anger and self-reported aggression. *Journal of Personality and Social Psychology*, *62*(4), 669–675.
Tangney, J. P., Stuewig, J., & Martinez, A. G. (2014). Two faces of shame: The roles of shame and guilt in predicting recidivism. *Psychological Science*, *25*(3), 799–805.
Taylor, G. (1985). *Pride, shame and guilt: Emotions of self-assessment*. Oxford University Press.
Thomason, K. K. (2018). *Naked: The dark side of shame and moral life*. Oxford University Press.
Thompson, K. O. (2017). Relational self in classical Confucianism: Lessons from Confucius' analects. *Philosophy East and West*, *67*(3), 887–907.
Van Norden, B. W. (2002). The emotion of shame and the virtue of righteousness in Mencius. *Dao*, *2*(1), 45–77.
Weisfeld, G. E. (1997). Discrete emotions theory with specific reference to pride and shame. In N. L. Segal, G. E. Weisfeld, & C. C. Weisfeld (Eds.), *Uniting psychology and biology: Integrative perspectives on human development* (pp. 419–443). American Psychological Association.
Weisfeld, G. E. (1999). Darwinian analysis of the emotion of pride/shame. In J. M.G. van der Dennen, D. Smillie, & D. R. Wilson (Eds.), *The Darwinian heritage and sociobiology* (pp. 319–333). Praeger Publishers/Greenwood Publishing Group.
Weiss, G. (2018). The shame of shamelessness. *Hypatia*, *33*(3), 537–552.
Wong, D. B. (2004). Relational and autonomous selves. *Journal of Chinese Philosophy*, *31*(4), 419–432.

8 Irrational Virtue, Unnatural Vice

Iskra Fileva

8.1 Introduction

Suppose a person became enraged with a stone he tripped on or perhaps, instinctively came to dislike – and strongly – a colleague for no reason other than that the colleague visually resembles his own abusive stepfather. Such tendencies seem irrational. Stones, not being agents with intentions, are not responsible for our tripping on them. Nor are agents with intentions responsible and blameworthy for things that have nothing to do with their intentions, for instance, for visually resembling another person, whatever the other's intentions and character and however bad. Yet there are cases in which sentiments of this sort – sentiments directed at agents who are not at fault or else at inanimate objects – and even action tendencies based on such sentiments appear good and perhaps required for virtuous character. For instance, we may expect that a horse rider whose horse, through no fault of the rider's own, hurts a passerby, would experience something akin to guilt even while judging that he is not at fault. We may expect him also to rush to apologize and perhaps to offer to compensate the injured person. If the rider has no such impulse, if *all* he experiences is a detached feeling of regret on par with that of a bystander to the scene, we may suspect a character flaw.

This type of case would be familiar to many readers under a label suggested by Bernard Williams: "agent-regret" (Williams, 1981, pp. 20–40). By "agent-regret," Williams means a kind of regret – typically coupled with a wish that one had acted otherwise – that a person may experience in the face of bad consequences brought about through her own causal involvement but without moral responsibility. Williams discusses a case very much like the horse rider's, but featuring a lorry driver who through no fault of his own runs over a child. We expect the driver to feel something different from and more personal than the regret a mere bystander may feel, and we expect him, perhaps, to do something to compensate, if only symbolically, the people affected (Williams, 1981).

DOI: 10.4324/9781003632641-11

Williams takes the expectation that the driver should have such sentiments at face value and uses it as evidence that however popular and plausible the idea that responsibility requires intention, we cannot, in fact, detach ourselves from the involuntary consequences of our actions. Writing some 200 years earlier, Adam Smith (1817) discusses such cases but proceeds more cautiously.[1] Like Williams, Smith says that the person whose body is causally involved in harm that befalls another but through no fault of the agent's own, for instance, a person whose horse has accidentally taken "fright" and hurt a passerby, would tend to feel responsible and act accordingly:

> He naturally runs up to the sufferer to express his concern for what has happened, and to make every acknowledgment in his power. If he has any sensibility, he necessarily desires to compensate the damage…
>
> (TMS II.iii.2.10)

But unlike Williams, Smith describes this as an "irregularity" in our sentiments. Such sentiments are "irregular," in Smith's view, because they go against a general principle we endorse in the abstract, namely, that:

> The only consequences for which he can be answerable, or by which he can deserve either approbation or disapprobation of any kind, are those which were in some way or other intended, or those which, at least, shew some agreeable or disagreeable quality in the intention of the heart, from which he acted. To the intention or affection of the heart, therefore, to the propriety or impropriety, to the beneficence or hurtfulness of the design, all praise or blame, all approbation or disapprobation of any kind, which can justly be bestowed upon any action, must ultimately belong. When this maxim is thus proposed, in abstract and general terms, there is nobody who does not agree to it. Its self-evident justice is acknowledged by all the world, and there is not a dissenting voice among all mankind…
>
> (TMS, II.iii.2.10)

Smith goes on to suggest that while we embrace the centrality of intention when we consider the matter abstractly, in particular cases, our sentiments deviate from this principle. He proceeds to discuss several other cases of irregular sentiments; importantly for present purposes, however, Smith suggests that we may experience and expect of others "irregular" sentiments directed at inanimate objects incapable of intentions. In a suggestive passage, he describes a sailor after a shipwreck who uses a plank of wood to get safely to shore. We expect the sailor to have warm feelings, akin to gratitude, toward the piece of wood. If, by contrast, he uses the plank to start a fire as soon as he gets to safety, we may suspect him of callousness. Relatedly, Smith observes that we have a desire to exact vengeance on a material object that has become the cause of a friend's death, and we think that a decent person should have such a desire despite knowing full well that inanimate objects, having no mental states, are not proper targets of gratitude or reactive attitudes in general.[2]

While Williams appeals to what he calls "agent-regret" cases (he does not discuss the inanimate object case) to attack a certain Kantian – as he labels it – picture of agency and responsibility, Smith takes such cases to present a puzzle in need of a solution. On the one hand, we believe that moral responsibility requires a guilty mind. He calls vengeance against an inanimate object an "absurd sort of vengeance" (TMS, II.iii.1.1). And about the rider who, through no fault of his own, hurts another, Smith (TMS, II.iii.2.10) asks why the rider should make an apology given that "he was equally innocent with any other by-stander"? On the other hand, Smith suggests that we not only expect people, as a descriptive matter, to have such sentiments: we endorse those sentiments and the action tendencies they give rise to, and we think that the *impartial spectator* would, at least to a large extent, do so as well.[3] He goes on to suggest that we would regard a complete lack of desire to make an apology or offer compensation on the blameless rider's part as "the highest brutality" (TMS, II.iii.2.10). Again, we would tend to judge harshly both ourselves and anyone else who does not have a negative reaction toward a material object that has become the cause of a friend's death, writing that we would think ourselves "guilty of a sort of inhumanity" if we did not vent (the admittedly absurd sort of) vengeance upon the offending material object (TMS, II.iii.1.1). And the sailor who as soon as he got to safety, used the plank that helped him get to shore to stoke his fire, would be "guilty of an unnatural action. We would have expected him to preserve the plank with care and affection" (TMS, II.iii.1.2).

What gives? What difference, if any, is there between purely irrational sentiments such as those toward the stone we tripped on or the colleague who resembles a person who hurt us and the sorts of "irregular sentiments" and actions motivated thereby that may be endorsed by an impartial spectator and may seem indispensable to a virtuous character? While Smith attempts to offer what he regards as a satisfactory answer, he regards the issue as a very difficult one.[4]

In this chapter, I wish to take a closer look at the intuition that moral sentiments and action tendencies not properly based on moral judgments of intentions may be necessary for virtue. I will focus on the two types of cases mentioned: moral sentiments that target involuntary aspects of actions and sentiments directed at inanimate objects, but I will aim to put forth a more general account that applies to a broader range of cases. I will suggest that Williams's discussion of the issue is unsatisfying. Smith's discussion is illuminating, but we need to go further than Smith still and think about our moral responses and human character in the context of our evolutionary history. This is what I propose to do here. In addressing my task, I proceed as follows: I first discuss the problem known in the contemporary literature as that of "agent-regret" (Section 8.2). I then turn to the issue of reactive attitudes toward inanimate objects (Section 8.3). Finally, I summarize my findings and raise some additional points (Section 8.4).

8.2 The Blameless Rider

Williams, as we said, suggests that we cannot detach ourselves from the involuntary consequences of our actions and that this is not what mature agents attempt to do. However, he offers no argument for this claim beyond the observation that we do not, as a matter of fact, tend to detach ourselves in cases such as the blameless rider or lorry driver and that we expect others not to try to detach themselves either. He considers a possible objection: such sentiments are simply irrational. In response, he says:

> [I]t would be a kind of insanity never to experience sentiments of this kind towards anyone, and it would be an insane concept of rationality which insisted that a rational person never would. To insist on such a conception of rationality, moreover, would, apart from other kinds of absurdity, suggest a large falsehood: that we might, if we conducted ourselves clear-headedly enough, entirely detach ourselves from the unintentional aspects of our actions ... and yet still retain our identity and character as agents.
>
> (Williams, 1981, p. 29)

It is not entirely clear why, in Williams's view, only "an insane concept of rationality" would lead us to label the tendencies in question "irrational." Charitably, the claim can be taken to be that any conception of rationality which had that implication would be too narrow. Would it?

One can argue that it would. Consider an analogy to prudential rationality: While, as a general matter, the prudentially rational person acts with forethought and does not rashly or negligently undertake actions that may hurt her in the long run, we expect her also, occasionally, to act spontaneously and on a whim, even at some cost to her best long-term interests. A conception of prudential rationality that left no room at all for whims and guilty pleasures would be, if not "insane," then at any rate, unhelpfully and unreasonably narrow. The deviations in question are arguably part of a well-lived human life and so prudentially rational broadly speaking.

Whatever contributes to a well-lived human life cannot, in the final analysis, be prudentially irrational.

But this analogy is limited: There is an important difference between the moral case and the prudential case. While it may not always be prudentially irrational, on a suitably nuanced conception of rationality, to act spontaneously, on a whim, and for no reason, even at a cost to our interests, morality requires that our actions and even, to the extent possible, our sentiments be attuned to moral reasons. And what moral reason is there for the blameless truck driver to see himself as being at fault?

Daniel Jacobson (2013) suggests, alternatively, that agents in such cases experience guilt that *is* irrational but that is nonetheless praiseworthy from a character point of view. The idea is that guilt of this sort is evidence of a serious concern for the well-being of others.

This is an important suggestion. There certainly *can* be cases in which actions that are not quite rational seem typical of a virtuous person. Consider, for instance, the case of a fictional mother described by George Orwell (1949/2020, p. 10) who wraps her child in her arms before a bomb explosion even though she knows that she cannot protect the child. The mother's behavior will accomplish nothing, but it is fitting from a character point of view, because it shows proper concern for her child's well-being, concern that outlasts the mother's ability to help.

But while there may be virtue in a goodwill effort doomed to failure – such as that of the mother – if a person truly experienced *guilt* in the blameless rider and lorry driver cases, that would be misguided. There are two possibilities here. First, the blameless driver may judge that he is at fault. In this case, the feeling of guilt would be based on an incorrect judgment. The reaction would resemble somewhat that of viewers of the movie *Lassie*, which features a dog that dies, who then sent flowers to honor Lassie, though the real dog playing Lassie was alive and well, and only the fictional Lassie had died (Schelling, 1984, p. 328). If, on the other hand, the driver judges that he is not at fault, then the feeling would not be based on an incorrect judgment, but it would lack the necessary ground, namely, a judgment to the effect that one is at fault. At most, a person may, unproblematically, experience some *guilt-adjacent* feeling, a feeling that lacks a crucial element, namely, the belief that one is actually at fault. But if one judges that one is not at fault, why must one apologize? To do so would be irrational in something like the way in which sending flowers for Lassie's death would be if you don't believe that any dog died except it would be worse in an important way: it would involve injustice. While a gesture of goodwill known in advance to be ineffective, such as that of Orwell's fictional mother, may be morally appropriate, an apology involves accepting a moral blemish, and accepting a blemish when one is not at fault involves perpetrating a wrong, albeit against oneself.

David Sussman (2018), in a discussion of agent-regret cases, makes a novel and original proposal that aims to have us rethink the principles which apply. Sussman suggests that the innocent pedestrian may see the driver as an innocent threat. But we have a right, Sussman argues, to protect ourselves against innocent threats. Of course, those who pose an innocent threat, in turn, have a right to defend themselves against the person they are a threat to. The pedestrian and the driver have been thrust into a state of nature:

> Just as the pedestrian may defend herself from this innocent threat posed by the truck driver, so too may that driver defend himself against these acts of justified self-defense directed against him. As the pedestrian raises her disintegrator to vaporize the approaching truck, so too may the truck driver draw his own disintegrator in an effort to preempt her attempts at what he knows to be completely justified self-defense.
>
> (Sussman, 2018, p. 803)

On Sussman's reckoning, the pedestrian, on the one hand, and the driver, on the other, must later reconcile and emerge out of the state of nature they have found themselves in due to unfortunate circumstances. The driver is not at fault, but neither is the pedestrian. The driver's apologies and offer to make amends are, on this view, part of a much needed attempt at reconciliation that may allow participants to return to normal moral life.

If this view succeeds, we will have a way to avoid both insisting, without an adequate supporting argument, that the reactions in question are not irrational, as Williams does, and arguing, as Jacobson does, that such actions are irrational but virtuous nonetheless.

It is, however, far from clear that the pedestrian has a right to harm the driver in this type of case. Suppose that the pedestrian slips through no fault of his own and falls onto the path of the truck. Suppose further that the pedestrian does happen to carry a disintegrator that can vaporize the truck and its driver. Does the pedestrian now have a right to vaporize the truck driver in order to avert the accident? That is highly doubtful, both legally and morally.[5]

Moreover, we can make the case against Sussman's interpretation even stronger by imagining that the truck driver, instead of accidentally running over a pedestrian, runs over an object of great sentimental value to another. Clearly, in that case, while there will be antagonism between the driver and the object owner, there is no "state of nature" that the two parties have been thrust into, and the owner of the valuable object has no right whatsoever to harm the driver in order to protect her prized possession. Even in that case, however, we would expect some sentiments misaligned with judgments of responsibility and intentions to arise in the breasts of the parties involved. If the driver is virtuous, he will have an urge to apologize after being made aware of the fact that his truck has smashed an heirloom of considerable sentimental value to another. In addition, Sussman's account, which appeals to innocent threats, does not help explain intuitions about the inanimate objects case.

Are we out of options? At this point, I wish to turn to Adam Smith's discussion, because I think it pays to revisit it. Smith, as we said, begins with the observation that when we examine the matter in the abstract, we conclude that the only thing relevant to moral judgments are agents' intentions, yet, in particular cases, our moral sentiments may be aroused by outcomes quite independently of mental states. Moreover, Smith points out that we rely on such sentiments when making moral judgments.

Smith gives three types of examples: chancy outcomes for which the agent is only partly responsible; chancy outcomes for which the agent is causally but not morally responsible; and attitudes such as gratitude and resentment toward inanimate objects that lack mental states altogether. The first two anticipate discussions of both moral luck and agent-regret. We allow chance – or as he calls it, "fortune" – to affect our moral practices. Anticipating discussions of moral luck, Smith suggests, for instance, that negligent behavior which results in serious harm is punished much more severely than negligent behavior which doesn't lead to a bad outcome despite the fact that the respective agents' intentions and mental states may have been identical. Similarly, we are prone to feel and express more gratitude toward a person who was successful in his attempt to help us, compared to a person who put in the same amount of effort but whose efforts remained in vain, despite the fact that success and failure are due to factors beyond the candidate benefactors' control.

The blameless rider case and the cases of gratitude toward a plank of wood or vengeance against a material object, for Smith, belong under this general umbrella: they are cases in which our sentiments are aroused by consequences independently of mental states. And as mentioned in the beginning, he suggests that the sentiments in question are irregular and sometimes absurd, but that their absence would be a sign of a serious character flaw (the "highest brutality"). Why?

Smith's answer has to do with consequences: nature was wise in implanting in us the "irregularity of sentiment" in question for were we to experience the same passionate fury against a person we merely suspect of harboring ill will toward us, thought and intentions would become objects of punishment even without any action, and every court would become "a real inquisition" (TMS, II.iii.3.2). On the other hand, good mental states that result in no action would amount to "indolent benevolence"

(TMS, II.iii.3.3), and we ought not content ourselves with them. This is what the impartial spectator grasps, and this is why she goes along with the irregularity of sentiment, at least in part. It may well be, Smith suggests, that God would judge human beings solely on the basis of thoughts and intentions, but human beings should not attempt to do the same *vis-à-vis* each other.

I discuss the problem of moral outcome luck elsewhere and will not do so here. I will mention only that I don't think a pure consequentialist argument succeeds.[6] But even if the consequentialist argument were to succeed in some moral outcome luck cases, it would not suffice in cases of agent regret and, even less so, reactive attitudes toward inanimate objects. (If anything, the person who has no qualms about burning the plank of wood would do better, from the point of view of consequences, compared to a sentimental person.)

Smith, however, makes some additional suggestions about the agent-regret and inanimate object cases and does not rely solely on the consequentialist argument. I wish to focus on the former case for the moment. Smith argues that the blameless rider in the agent-regret case enters the sentiments of his victim even while recognizing their irrationality. He apologizes and offers to compensate the other in order to appease "that animal resentment, which he is sensible will be apt to arise in the breast of the sufferer" (TMS, II.iii.2.10).

It is not immediately clear why we should try to appease "animal resentment" that we think likely to arise in the breast of another. "Animal resentment" may, after all, be due to a serious flaw in the other's values and character, as when blinded by tribalism, a person hates an outgroup member.

Smith would cede my point that sentiments of this sort must *sometimes* be overridden. For instance, while we are apt to experience resentment of just this immediate, unreflective sort toward a bearer of bad news, we should be careful, on his view, not to vent our spleen upon an innocent messenger in the way King Tigranes of Armenia is said to have done to the man who brought news of the approach of the Roman army (TMS, II.iii.2.6). But in other cases, such as that of the blameless rider, even the impartial spectator feels "some indulgence for what may be regarded as the unjust resentment of that other" (TMS, II.iii.2.10). Whence the indulgence?

I wish to begin untangling the threads by double-clicking on the word "animal" in "animal resentment." While appeals to evolutionary history must always be made with caution, I think that the following general story is highly plausible: In the overwhelming majority of cases in which our bodies are causally involved in harming another person or another's property, we, as agents, are responsible. The cases in which our bodies are causally involved in harm yet without negligence, recklessness, or ill will on our part are but a small subset of all cases. Our "irregular" moral sentiments are a kind of evolutionary "quick and dirty" heuristic that served ancestors without sophisticated moral abilities reasonably well. These sentiments are not properly attuned to the possibility of overshooting. They may, thus, give rise to the seeming that we are responsible even when we are not.[7]

Importantly, this seeming need not give rise to a belief. If it does give rise to a belief, that belief would be unjustified. We may end up in a Moore-paradoxical state or something like it since we have strong evidence that we are not at fault.[8] A person in this state would be thinking something like, "I am at fault, but my evidence suggests that I am not."[9]

The situation is analogous to other kinds of seemings based on immediate, automatic appraisals of the evidence that conflict with higher-order cognitive appraisal, for instance, the seeming that we are in danger on a glass bridge even though we know that the bridge is safe.[10] The seeming is tied up with our evolutionary history: being suspended many feet above ground and being vividly aware of this has, in our evolutionary past, always meant "danger" rather than, "you are on a safe glass bridge, and this is a tourist attraction." Not surprisingly, pre-verbal toddlers are scared of crawling on a platform that resembles a cliff but is padded with glass (Gibson & Walk, 1960, pp. 64–71). I wish to suggest that a parallel mechanism produces the seeming that we must have harmed the person who experienced harm related to our causal involvement even without any agential role on our part. We don't quite believe this seeming, and we would be irrational if we did, but the seeming is nonetheless compelling.[11]

No less importantly, as Smith points out, it appears to the harmed person that we harmed them, and we know that this is how things would appear. This is precisely why the other experiences "animal resentment" against us. We recognize that were we in the other's shoes, things would seem that way to us as well, which is why we want to appease the resentment despite the fact we know we are not at fault (and the other should, at least upon reflection, recognize this too).

Yet, clearly, we cannot simply endorse every evolution-based response of this sort. Our propensity to get angry with a rock we tripped on or with a person who merely visually resembles another ought to be counteracted. What if a person concluded that the same ought to be done here, that is, that the reaction more recently labeled "agent-regret" is simply misplaced and that no one should have it?

We ought to tread carefully. The answer is not straightforward. Two points are important to emphasize. First, if to begin with, a person did not have the seeming in question at all (and so did not need to counteract it), the likely explanation would be not that the person's sentiments are more attuned to reasons than the sentiments of the rest of us are, but rather that the person's automatic first-order dispositions are not very prosocial. Consider an analogy: A person who can lie on the train track and wait, without fear, for an incoming train to pass over him because he has measured the distance and knows that if he stays still, the train won't touch him is likely someone very low on fear, rather than someone much more rational than the rest of us (Fileva & Brakel, 2019). Similarly, a person who volunteers to perform an execution and does not have negative sentiments about it is someone whose motives we may suspect even if we fully believe that the execution is justified and ought to be performed. The problem is that in all these cases, the lack of relevant seeming may be a result of rational top-down information processing, but it may, alternatively, be a result of callousness or lack of prosociality at the level of dispositions. Since, in general, there are probably more people who lack first-order prosocial tendencies than there are people whose sentiments are highly responsive to reasons, we would be, at best, uncertain how to evaluate a person in the blameless driver case who did not have some seeming – though without a belief – that he is at fault.

Second, and relatedly, we expect a virtuous person to have an impulse to take on a greater moral burden than can be strictly required of her and to go along with that impulse, at least to some extent. This is true more generally. A virtuous person tends to focus on what she can do to improve the situation she finds herself in even if what she can do is not strictly required of her (Fileva, 2008). She shows a kind of generosity of spirit. Indeed, a virtuous person may tend to perceive supererogatory actions as required of her, as when it seems to her that she ought to help a person fleeing persecution – even at a risk to her own safety – although the action is altruistic and not strictly required.

On the other hand, however, how far a person in an agent-regret case may go in the direction of endorsing the impulse and acting on it depends on how virtuous the other agent in the situation is. If the other is virtuous, the other would resist the apology and refuse an offer of compensation. If, however, the other demands an apology or compensation, then the person who is not at fault may have to, courteously but firmly, explain that she is, after all, not to blame.

Indeed, we must recognize that our propensity to experience agent-regret, while possibly indispensable to virtue, may be highly burdensome and injurious. Consider, for instance, the legal case of *Jarrett v. Jones* (Lewis, 2008), in which a truck driver, through no fault of his own, collided with a car carrying a four-member family. The car's driver was negligent. The accident resulted in the death of a little girl in the car. The truck driver suffered only minor bodily injuries, but he experienced great emotional distress upon seeing the body of the dead girl, distress likely much more acute than that any bystander would have felt.[12] The obvious explanation of the driver's distress has to do with agent-regret: The truck driver felt implicated in the death of the little girl even though the accident was not his fault. The driver later sued the family of the girl and won in a 6-1 Missouri Supreme Court decision. Some may, upon reading about this case, think that having to compensate the truck driver imposed a great and unnecessary burden on a family that had already suffered the loss of a child. However, the fact that the family suffered a great loss does not mean that the driver did not unjustly suffer great emotional harm. The court in its ruling essentially recognized that agent-regret can lead to real emotional injury for which the agent may be completely blameless.

Where does all this leave us? The conclusion we must draw, in my view, is this: While a virtuous person in the blameless rider case would have an (evolved) impulse to apologize, a virtuous person in the pedestrian's place would (if not due to a first-order propensity, then as a result of reflection) resist the apology. If the injured person does not have this virtuous response, the blamelessly implicated person may, regrettably, have to remind the other that, objectively speaking, she is not at fault. Indeed, there may be cases in which it would be fitting for the party that experienced physical harm to compensate the other, as in the *Jarrett v. Jones* case.

So far, I discussed cases involving consequences brought about without a *guilty* mind. But Smith, as mentioned at the start, makes the more radical suggestion that sentiments directed at objects without minds at all may be required for virtue, and not because of those objects' intrinsic properties but because of the role they have played in bringing about good or ill. Is he right, and what are we to say about the relevance of those sentiments to character?

8.3 Unnatural Vice

There is an Ikea lamp commercial featuring the disposal of an old lamp. A woman is shown giving the old lamp a kiss before taking it out and leaving it outside, next to a bag of trash. The lamp looks lonely and sad. Clouds gather, and it begins to rain. Cars pass by, slinging mud at the lamp, and no one stops to take an interest in it. The camera zooms in on the window of the room where the lamp previously was. The viewer can see a new lamp in place of the old one, perched comfortably on the stand, inside a cozy room, providing foil for the old lamp now abandoned in the street after years of service. A man then appears on the screen and says in a Scandinavian accent, "Many of you feel bad for this lamp, because you are crazy [sic]. It has no feelings, and the new one is much better."

The man is right that many viewers are likely to feel bad for the old lamp (and this is the point of the commercial). But is this reaction irrational?

Adam Smith suggests that such a reaction is natural, where naturalness in the relevant sense is understood to be good. The plank of wood that has helped one get to safety or the material object that became the cause of a friend's death are, just like the lamp, inanimate objects, but on Smith's plausible proposal, a person who does not cherish the one or desire to exact revenge on the other would be guilty of "unnaturalness" and "inhumanity" that, in this context, would betray a flaw of character. Smith would not disagree that there is *something* irrational about these reactions – as we saw earlier, he calls vengeance against an inanimate object "absurd" – but he insists that our sentiments in such cases, while irregular, may be part and parcel of our humanity. The general idea seems to be that there is a certain way for humans and human sentiments to be, which is recognizably the proper human way.

Smith sometimes appeals tacitly to God here – though he calls God "author of nature" rather than our direct creator – in order to explain why our humanity would involve the sentiments in question, but we can tell the story in evolutionary terms. A human being is likely to have the sentiments Smith speaks of for evolutionary reasons: some inanimate objects have played the functional role of faithful and highly loyal friends that have helped us, without any ulterior motives, no less. An impulse to reciprocate and show gratitude and loyalty to agents is deeply ingrained in us, but it overshoots and persists even when the target that otherwise resembles a faithful friend is an inanimate object. From an evolutionary point of view, there are no strong reasons to prevent such false positives in this case. Contrast that with a case in which a tree branch shaped like a snake occasions a startle. The two cases admit of similar explanations: False positives in such cases would not be as costly as false negatives would be. So we have evolved to err on the side of caution. In the snake case, however, this tendency is only prudentially good. In the other case, it is good from the point of view of character: a complete absence of the impulse in question would be evidence that a normal human element is missing from the agent's constitution. What we have in the inanimate object case, then, much as in the agent-regret case, is a conflict between evolved dispositions and principled, reason-based moral judgments. Some evolved dispositions that conflict with judgments based on moral principles seem necessary for virtue. But why?

Smith's claim that their absence would be unnatural is suggestive, but we cannot stop here. Clearly, not every kind of unnaturalness detracts from virtuous character. It may seem unnatural to love one's enemy, but loving one's enemy hardly involves a sacrifice of one's humanity. It may seem unnatural to be sexually aroused at the sight of needles, but that's hardly inhuman either.

On the view I wish to propose, we must distinguish between three different senses of "unnatural." Unnaturalness in all three cases can be understood in opposition to biological evolution, but the moral profiles of the three varieties differ from each other. The first sense has to do with moral evolution: some tendencies, such as a disposition to forgive wrongdoing or wish one's enemy well, may seem wayward from a biological point of view, but they are a result of moral progress on the part of a person and, to some extent, on the part of her community. Those tendencies may improve character rather than undermining it. (On the flip side, morally bad tendencies may be perfectly natural in the relevant sense, for instance, children's tendencies to manipulate others for their own ends or to hurt each other in order to make themselves feel better.)

Other tendencies may be unnatural in ways that are orthogonal to morality and may constitute an aberration without moral implications, as when a person has highly unusual yet harmless sexual preferences.[13]

The third group involves dispositions such as the disposition to see nothing special in a plank of firewood that helps get one to safety, or no qualms about burning it once it is no longer needed for flotation. Such dispositions seem unnatural in a way which undermines full virtue.

For readers unconvinced that there is a species of unnaturalness that betrays a flaw of character, I wish to point out that there is a resemblance between the case at hand and a case involving dispositions that, I think, most readers would, without question, classify as problematic, for instance, a disposition to treat the dead body of a loved one without due care and respect. A dead body, too, is an inanimate object, but if someone had no qualms about cannibalizing it or throwing it in the trash, that would betray a serious character flaw in said person. It is deeply human to experience attachment to the body of a loved one even after death.

It is important to point out, however, that "unnatural tendencies" even of the third variety may have to be overridden too. While attempting to stifle or neutralize the initial impulse to experience gratitude toward a plank of wood that helped save our lives would be questionable, unwillingness to counter it when the circumstances call for that may be impermissibly irrational. If, for instance, the shipwreck survivor is so inordinately attached to the plank of wood that he finds himself unable to set fire to it even to save his life, something has gone wrong. Virtuous dispositions need not be fully and perfectly rational, and it may be that some attempts to make them more rational than they are would worsen our characters, but they ought to be *sufficiently* rational.

8.4 Conclusion

I began this chapter with a puzzle about moral sentiments disconnected from judgments that properly incorporate intentions and mental states. I focused on sentiments directed at consequences brought about without a guilty mind, specifically, what has come to be known as "agent-regret," and attitudes of gratitude, resentment, and related moral emotions directed at objects without minds. The puzzle was this: We do not think that people who are causally but not agentially involved in harming another are a fit object of moral attitudes such as resentment. Similarly, we do not think inanimate objects are ever an appropriate target of such attitudes. Indeed, such sentiments seem irrational. Yet we not only have such attitudes and expect that others will have them too: We are likely to suspect that a person who doesn't have them is exhibiting some kind of character deficiency. I asked why.

I considered several proposals for dealing with the agent-regret case and found them lacking, and I proposed an account that takes its cue from two sources: Adam Smith's discussion of "irregular sentiments" and our evolutionary history. On the view I put forth, sentiments of this sort result from evolved tendencies that we often do not have a reason to counteract and may have a reason to embrace, a reason that has to do with erring on the side of virtuous and prosocial intentions. Importantly, however, while acting on such impulses may not be perfectly rational, at least not in a narrow sense, it must be *sufficiently* rational. If and when evolved tendencies get in the way of either morality, as when we come to believe – and do not simply feel as though – we or someone else is at fault in agent-regret cases, or rationality – as when the sailor becomes so attached to the plank of wood that helped him get safely to shore, he cannot subsequently burn it to save his life – they ought to be counteracted.

It is worth mentioning here that there may be cases, though likely very rare, in which the initial impulse is so strong that not only overcoming it but attempting to overcome it proves impossible for a person even when acting contrary to the impulse would be best in her circumstances, while failure to do so would be costly. For instance, there may be cases in which cannibalizing the body of another, which is now but an inanimate object, is permissible and the most rational course of action, for instance, a Donner Party case. But a person who finds herself unable to do that or even to attempt to do it would not be showing herself to be impermissibly irrational, unlike the person who dies because he cannot bring himself to burn the dear to his heart plank of wood. Our evolved human nature may sometimes, not rationally but permissibly nonetheless, get the better of us.

There is a final question I would like to consider before closing this discussion. Suppose it became possible for us, through gene-editing or other technological advancements, to perfectly align our moral sentiments with moral judgments that properly reflect beliefs about intentions and mental states. We would then be creatures without what Adam Smith called "irregular" moral sentiments. For the purposes of the argument, we can stipulate that by opting to alter our genes or other aspects of our biological and psychological constitution, we would not cause any error in the opposite direction; that is, we would not be any less likely to have the appropriate sentiments *vis-à-vis* those who are apt targets thereof, rationally speaking. Should we do it?

I doubt that this would be a salutary development, but I cannot undertake to argue for this claim here. I would note two things. First, our evolved responses in general may provide a separate source of evidence and may, thus, be epistemically valuable. This is important because sometimes, our principled judgments may be wrong and evolved responses may provide a clue to this. The problem with re-engineering ourselves is that, in all likelihood, we can only ever make "irregular sentiments" regular by ensuring that they align with the judgments that we make rather than with the correct moral judgments.

Second, even setting the issue of the potential epistemic value of such sentiments aside, our connections with each other are richer and more meaningful on account of those sentiments. If we lose the tendency to show attachment to the material possessions of a lost loved one, for instance, or to have a negative reaction to the object that caused their death, something important would be lost. One can argue that biopsychological engineering changes that eliminate irregular sentiments would make us posthuman, and that this would be good. But while certain enhancements, such as those that involve cognition or health span, may make us posthuman without significant loss, alterations of the sort under discussion may make us posthuman only at the price of making us, in some very clear sense that Smith would have recognized, inhuman.

Notes

1 Williams does not reference Smith, but the cases are strikingly similar.
2 Blackburn (2015, p. 211) reminds us here of a practice that originated in Europe in the Middle Ages and persisted for several centuries of regarding objects that have caused great harm as "deodands" ("to be given to god") and of forfeiting them to the church or the state. Schliesser (2013) sees in Smith's discussion of inanimate objects an endorsement of superstition though he is sympathetic to the view he ascribes to Smith as he thinks that perhaps we do have a superstitious nature.
3 I must note here that there is some disagreement among interpreters regarding the proper interpretation of Smith's view. For instance, Simon Blackburn (2015), in an otherwise thoughtful and stimulating discussion, suggests that

ultimately, on Smith's view, the "candid and humane" part of us can keep character assessment separate from luck. I think that this is only partly right.

4 Blackburn (2015) is not the only author whose interpretation of Smith differs from mine. There is some disagreement among interpreters regarding the proper interpretation of Smith's view of irregular sentiments. I don't want to get sidetracked by discussions of Adam Smith here since my goal is not an interpretative one. I refer interested readers to Hankins (2016), which contains a good overview of prior interpretative attempts along with a new proposal.

5 Sussman draws on McMahon's work on just war theory. There is a separate discussion to be had about the rules that apply in a time of war, but what I will note here is that the cases are not analogous. This is true more generally. For instance, causing the death of innocent bystanders in a time of war is likely permissible in a much broader range of circumstances compared to ordinary life. The parties to a war are in something very much like a state of nature, but people in ordinary circumstances are not.

6 On the view I favor, in acting, we commit ourselves to being held responsible for multiple foreseeable outcomes, subject to the constraints of the existing moral and legal framework.

7 Jake Wojtowicz (2022) tells the story of a woman who accidentally killed a motorcyclist. Wojtowicz borrows the story from a *New Yorker* piece by Alice Gregory on accidental killers. Gregory reports that the woman, "Patricia," was upset when well-meaning friends suggested that the death was "just an accident." Wojtowicz (2022, p. 504) quotes the following passage from the interview with Patricia: "'No matter how much you want to dismiss it as an accident, I still feel responsible for it, and I am.' She cried, 'I hit him! Why does nobody understand this?'" Wojtowicz suggests that the well-meaning friends were failing to recognize that Patricia was an agent and that the accident happened as a result of something she did. I argue that if Patricia is not at fault, the case ought to be interpreted differently. What Patricia senses others do not fully understand is that she has a very compelling seeming that she is responsible and that this seeming cannot be easily dispelled top-down by focusing on the fact that the outcome was accidental. Her causal involvement can help explain why the seeming arises, but it need not lead either her or bystanders to interpret the case as one of agential involvement. There is, of course, a separate question about the factors which influence the genesis of a seeming of this type. How likely we are to have a strong seeming in such cases likely depends on the extent to which our causal involvement resembles agential involvement. For instance, I suspect that the bearer of bad news, to go back to another of Smith's examples, is not, at least not without powerful social conditioning, likely to feel responsible, because the news is what it is before he heads off to find the recipient; the messenger is transmitting a *fait accompli*.

8 There is a separate question here about whether a Moore-paradoxical state in this type of case would be rationally permissible. For arguments to the effect that in some such cases Moore-paradoxical states are permissible, see Fileva and Brakel (2019).

9 Wojtowicz's and Gregory's Patricia is in something like this ambivalent state, because she is quoted as saying, "Yes, it was an accident, and in a certain sense we were both to blame..." and then goes on to say, "...but, at the end of the day, I hit him, I took his life" (Wojtowicz, 2022, p. 504).

10 Cushman and Greene (2012) offer neuroscientific evidence that in cases of accidentally caused harm, two dissociable cognitive systems, one responding to

outcomes and the other to judgments of intention yield different verdicts. The focus of their discussion is on the judgments of observers, but there are clear resonances between their approach and mine.

11 Relatedly, authors such as Blackburn, Schliesser, and Hankins emphasize that on Smith's view, strictly speaking, we do not feel guilty when we cause harm accidentally; rather, we feel "piacular" – we feel as though atonement is required of us. These authors do not interpret the notion of piacularity in exactly the same way, and at any rate, I want to suggest that an evolutionary perspective can help shed light on this.

12 Hankins (2016, p. 740), in his discussion of Smith, writes that: "[W]hile it is easy to imagine someone being racked by an excessive sense of guilt, and her life suffering as a result, this is less easy to imagine in the case of ... agent's regret." But this may be a flaw in our imaginative dispositions, perhaps due to a tendency to focus on the person who has suffered physical harm, coupled with a belief that the agent who caused the harm is blameless. As a matter of fact, people can suffer great distress on account of causing harm accidentally. See, for instance, de Snoo-Trimp et al. (2025) for a discussion of this problem in medical contexts. I note also that a perceptive person may well imagine this easily. Alice Gregory (2017), in *The New Yorker* piece on accidental killers that Wojtowicz draws on, reports that a truck driver who came onto the scene of Patricia's accident began praying for Patricia: "He started praying, but he was praying for *me*. I heard him say, 'God, protect her. God, look out for her. God, give her strength."

13 Some cases, such as the incest case described by Haidt, may be in between. See Haidt (2001). A disposition to engage in incest in the Julie and Mark case described by him does not constitute a moral vice, but it does carry a sort of moral residue, and this residue is mostly negative (there may be openness to experience that's positive, but there is also a certain recklessness about damaging or changing the character of an important human relationship – that between siblings – that is negative).

References

Blackburn, S. (2015). Williams, Smith, and the peculiarity of piacularity. *Journal of the American Philosophical Association*, *1*, 217–232.

Cushman, F., & Greene, J. D. (2012). Finding faults: How moral dilemmas illuminate cognitive structure. *Social Neuroscience*, *7*(3), 269–279.

de Snoo-Trimp, J., van den Hoven, M., & Molewijk, B. (2025). Agent-regret in healthcare: Toward a more precise and empirical-based look into the dynamics of agent-regret experiences. *The American Journal of Bioethics*, *25*(2), 29–31.

Fileva, I. (2008). The neutrality of rightness and the indexicality of goodness: Beyond objectivity and back again. *Ratio*, *21*, 273–285.

Fileva, I., & Brakel, L. A. W. (2019). Just another paper on Moore's paradox, but we don't believe that. *Synthese*, *196*, 5153–5167.

Gibson, E. J., & Walk, R. D. (1960). The "visual cliff." *Scientific American*, *202*(4), 64–71.

Gregory, A. (2017, September 18). *The sorrow and the shame of the accidental killer*. The New Yorker.

Haidt, J. (2001). The emotional dog and its rational tail: A social intuitionist approach to moral judgment. *Psychological Review*, *108*(4), 814–834.

Hankins, K. (2016). Adam Smith's intriguing solution to the problem of moral luck. *Ethics*, *126*(3), 711–746.
Jacobson, D. (2013). Regret, agency, and error. In D. Shoemaker (Ed.), *Oxford Studies in agency and responsibility* (Vol. 1). Oxford University Press.
Lewis, M. E. (2008, July 29). *Jarrett v. Jones. Missouri Law Review*. http://missourilawreview.blogspot.com/2008/11/jarrett-v-jones.html
Orwell, G. (2020). *1984*. Amaryllis. (Original work published 1949)
Schelling, T. C. (1984). The mind as a consuming organ. In: *Choice and consequence* (pp. 328–348). Harvard University Press.
Schliesser, E. (2013). The piacular, or on seeing oneself as a cause in Adam Smith. In M. Lenz & A. Waldow (Eds.), *Contemporary perspectives on early modern philosophy: Nature and norms in thought* (pp. 159–177). Springer.
Smith, A. (1817). *The theory of moral sentiments*. Anthony Finley. (Original work published 1759)
Sussman, D. (2018). Is agent-regret rational? *Ethics*, *128*(4), 788–808. https://www.jstor.org/stable/26541889
Williams, B. (1981). *Moral luck*. Cambridge University Press.
Wojtowicz, J. (2022). Agent-regret, accidents, and respect. *Journal of Ethics*, *26*(3), 501–516

Part III

Justice and the Communal Architecture of Virtue

9 Evolution and Interpersonal Justice

Mark LeBar

In this chapter, I make a speculative case for a personal virtue of justice as the product of cultural evolution. It is necessarily speculative, in that much of it appeals to prehistorical interactions between people, so there is perforce no record. Still, it is worth considering, as, first, it seems to be empirically plausible, and, second, if I am right, this history informs the *content* of the norms for the virtue of justice in an important way.

In what follows, I do three things. First, I set out the conception of the virtue of justice for which I am arguing. Second, I sketch the evolutionary story which I take to underwrite it, as a matter of both intragroup and intergroup adaptation and selection. Finally, I offer a model of the kind of norm-evolution I have in mind from the law.

9.1 The Virtue of Justice

In this section, I introduce the conception of the virtue of justice we will be concerned with. This involves saying a bit about the very idea of a virtue as I understand it (since there are numerous conceptions of what virtue and virtues are), and then more about the specific character of the virtue of justice—what is distinctive about it as a virtue of character.

The framework for virtue I adopt is Aristotle's. In this framework, a virtue is an *excellence* of one of two sorts. Virtues of character are stable traits of character which have some crucial features, or *marks*, as I shall call them. There are also virtues of intellect, including the virtue of practical wisdom (*phronesis*), but our focus is on the specific virtue of character of justice.

What are the marks of virtues of character in this framework? What qualities do they share? For our purposes, they are three.

1 A virtue must be conducive to living well, to *eudaimonia* (as the Greeks referred to it).
2 A virtue must be bound up with other virtues in two distinctive ways. First, it must sustain the *reciprocity* Aristotle believes holds between

DOI: 10.4324/9781003632641-13

the several virtues of character. You must have them all to have any of them. Second, and relatedly, it must be interdependent with the intellectual virtue of practical wisdom (*phronesis*). All virtues of character require the direction of practical wisdom.

3 A virtue must be characterized by a *mean* (or intermediate) in passions and action. This mean is, in effect, the norm for the virtue: it is the standard for what counts as the excellence in question, as opposed to what falls short. The nature of this norm for justice will be our focus in the next two sections of this chapter.

What differentiates justice from other virtues of character? First, while he does not set out any explicit claim to this effect, Aristotle writes as though the several virtues of character each have a specific "concern" (Greek: *peri ho*). Thus, by way of example: "courage is concerned with confidence and fear";[1] "temperance must be concerned with bodily pleasures";[2] liberality is concerned with wealth.[3] Second, Aristotle seems to see virtues as generally associated with *passions* or emotions of particular sorts. These two qualities are what differentiate the virtues of character from one another, and my exposition of the virtue of justice consists in identifying these two elements.

First, the concern of justice. Aristotle's view is that that concern is the distribution of goods such as wealth, honors, and so on—what have come to be called "goods of fortune." For a variety of reasons, this is inadequate as an account of the aspects of our social lives which run both more deeply and more broadly than Aristotle's proposal can support. More deeply in the sense that the kind of equality Aristotle seeks for these goods is only one aspect of the more fundamental way in which we should see each other as equals, as I will suggest below. More broadly, intuitively strong cases of injustice, such as promise-breaking, do not seem to show up on that account at all.

Instead, I suggest that the concern of justice is the *authority to obligate* one another—an authority we should see ourselves as holding equally. The idea that this is an important form of equality is a much later idea—obtaining explicit expression only in the Enlightenment. I know of no clearer expression of it than Locke's, writing of what he takes our moral state to be outside of political community:

> A state also of equality, wherein all the power and jurisdiction is reciprocal, no one having more than another, there being nothing more evident than that creatures of the same species and rank, promiscuously born to all the same advantages of Nature, and the use of the same faculties, should also be equal one amongst another, without subordination or subjection. . .[4]

Though Locke expresses the idea clearly here, there are strong Kantian overtones in the idea, and indeed Kant explicitly endorses the same idea, calling it our "innate equality":

> *Freedom* (independence from being constrained by another's choice), insofar as it can coexist with the freedom of every other in accordance with a universal law, is the only original right belonging to every man by virtue of his humanity.—This principle of innate freedom already involves the following authorizations ...: innate *equality*, that is, independence from being bound by others to more than one can in turn bind them...[5]

And while Aristotle doesn't get anywhere near this point as a matter of the virtue of justice, he *does* take up an idea very similar when it comes to *political* authority. Here (and of course confining his attention to citizens, dismissing as irrelevant women, slaves, metics, and the like) he thinks those related by "civic friendship" ought to see each other as equals by way of authority: "That is why it is thought to be just that among equals everyone be ruled as well as rule, and therefore that all should have a turn."[6]

Now, what seeing each other as having equal authority comes to as a matter of practice is a longer story, and in fact the story we take up in the next section. But before moving on, I should observe that the counterpart passion to this equality of authority (in the way that, say, fear is the counterpart passion to courage) is, I propose, *respect*. This I intend in the roughly Kantian sense, of a *checking* or constraint. In Kant's account, respect for the moral law checks our inclinations and motivates us to act from duty.[7] Here, it is the moral authority of others—their standing or capacity to obligate us—that constrains us as we go about undertaking our projects.[8]

That is the nature of the virtue of justice as I propose we think of it. Now, Aristotle's account of virtues of character is well-known for its commitment to the "doctrine of the mean," as per the third mark above. Virtue "is a kind of mean, since it aims at what is intermediate," between excess and defect, he says.[9] He does not take this target to be a simple thing to hit; on the contrary, it is a challenge. Getting passions and actions right (the work of the virtues of character) requires feeling and acting "at the right times, with reference to the right objects, towards the right people, with the right aim, and in the right way..."[10] Moreover, this "intermediate" is nothing so simple as a midpoint on a scale or spectrum. Aristotle makes this point by contrasting the "intermediate in the object" and the "intermediate relatively to us," and insisting that it is the latter that matters for virtue.[11] His example is that 6 pounds of food may be intermediate between 10 pounds and 2, but be too much for us and too little for Milo the wrestler.[12] So the mean for Milo and the mean for us are not the same thing. That leaves us with the problem of determining the mean "relatively to us," and as I will now argue, however problematic that is for the other virtues of character (and it *is* problematic), it is even more so for the virtue of justice. Why?

Start with the mean for, say, courage. Identifying it requires the work of practical wisdom (*phronesis*). Aristotle says this mean is "determined by reason and in the way in which the man of practical reason would determine it."[13] There are two ways this notion can be interpreted; there are textual resources for both and they reflect deep disagreements in understanding the metaethical framework Aristotle is operating with. On the first, there is some normative standard awaiting recognition from the wise person. Call this the "norm-prior" picture.[14] On the second, the standard does not exist, and is not set, *until* the person of practical wisdom determines it via wise and virtuous judgment. Call this the "virtue-prior" view.[15] Neither picture is simple to articulate or defend, yet one or the other is necessary to make sense of what Aristotle must be thinking with the doctrine of the mean and thus his entire theory of virtue of character.

So this is a significant complication already, even for a virtue like courage. Here, I will not pause over this dilemma but will simply claim that the virtue-prior approach is a better fit with Aristotle's view overall and proceed on that basis. But that saddles us with the task of understanding just what it is that the wise and virtuous person is doing when determining the mean for acting courageously in a given situation. Whatever else that involves, that determination is specific to the agent doing the determining. Suppose that Al and Betty are both wise and virtuous and come across Gemma drowning in a lake. Al cannot swim and, being courageous, judges that jumping in to save her is *not* in order for him, as it very likely would result in both of them drowning. That would be *rash*, Aristotle says.[16] Betty, however, is an expert swimmer and trained lifeguard, and though doing so is not risk-free, she determines that courage requires her to jump in and rescue Gemma. The means "relative to" Al and Betty here differ.

However challenging that is, when we come to justice, things get much worse, at least given the conception I have set out. Why? Because the mean now must involve not just one person—the agent—but *two* people[17] involved in some sort of moral engagement or relation. Suppose now that Al and Betty are considering starting a business together, each contributing different amounts of capital, expertise, time, knowledge, and so on. How are they to split the profits in a way that is just? On the picture I have offered, doing so means Al must see Betty as obligating him in just the ways that he sees himself obligating her, and vice versa. Mean-seeking here requires respecting this mutual, equal, and reciprocal obligating capacity. So Al has the task of determining the mean relative to him, and Betty has the task of determining the mean relative to her, and that must be the same distribution! So justice—respect for equal obligating authority—calls for a coordinated determination of the mean (here, a plan to divide profits justly).

How are Al and Betty to undertake the task of determining this mean? In my view, that is where an evolutionary story kicks in, and we now turn to that.

9.2 The Import of History

The first thing to see, I believe, is that Al and Betty do not have a problem that is or can be resolved by reasoning in the abstract. Whatever else it is, the solution is a response to the many concrete particulars that characterize their situation now and as they anticipate the future. This is the opposite of abstract: it is empirically grounded in every way human decision-making can be. Equally important, Al and Betty do not have to take up this project in isolation. They live embedded in communities of various sorts and in history. They can look back on, and learn from, the cascade of the history of human interactions. Their judgments in determining this mean for the justice of their interaction draw on a long evolutionary history for such means, in a way that echoes John Stuart Mill's claims for how people would have come to identify utility-maximizing action. To the objection that people may not have time to calculate the effects of any given action on the general happiness, Mill replies: "there has been ample time, namely, the whole past duration of the human species. During all that time mankind have been learning by experience the tendencies of actions..."[18] Here the point is not the production of happiness or pleasure, but the justice of the mutual judgment as to how to interact.

The idea here is that the norms of justice emerge—and are emerging—through a process that is rightly seen as evolutionary, in a way I will explain. But the basic idea is simple: through a trial and error process, determinations of norms that do better at manifesting mutual and equal respect for the authority of one another will have selective and adaptive advantages, and because we have the capacity (and the incentives) to learn, they will be taken up with greater frequency than those that do not.[19] That is because just relations between people are the wellspring of the benefits of social life and cooperation. They are thus desirable both for their own sake and for the sake of the kinds of relationships with others they make possible. All of that is *not* to say that there is any guarantee that they emerge at all, let alone in any shortish period of time. Evolution has lots of dead ends, and we have ample proof of the robustness of profoundly unjust social norms (of which tribalism, slavery, and the subordination of women are prime candidates). But we do grasp and (at least many of us, much of the time) aspire to be guided by norms of justice, and we have good reason to. The question here is how those norms might emerge.

It is important to remember that in the first instance these are *particularistic* norms; that is, they are norms that apply to concrete particular situations, relations, and interactions. Norms in the form of rules can be derived from these particular judgments, but they are always general, in the sense that they must be specified in the use of judgment for particular concrete applications.

Also, it is not the survival of the process of selection that confers success on these norms. As indicated above, some deeply unjust norms have survived many generations of human evolution. The relevant canon of success here consists in manifesting mutual and equal respect for the obligating authority of one another. The process that develops these norms distributes the authority for doing so across all those interacting with others. The resulting evolutionary story is an account of how we—fallible and limited as we are—might come to arrive at recognizing and determining particular norms in a way that does exactly that, in a complex and constantly changing environment.

This evolutionary history may well have begun as it is supposed by Michael Tomasello. As Tomasello tells this story, early human societies built on proto-human societies by developing the capacity for collaboration in partnerships, as dyads.[20] The point here is not that this form of partnership or cooperation allowed for society; as Tomasello makes clear, there are several varieties of cooperation manifest in various species (human and non-human); his interest (and ours) is in what he refers to as a *moral* form of cooperation based on "fairness," evidently distinctive to us humans.[21]

These dyadic partnerships gave us occasion to develop a capacity to conceive of a joint "we" (the basis for conceptions of "we" for cooperative groups larger than dyads) and also generated norms of cooperation which emerged from that joint "we" perspective. If Al and Betty are involved together in an enterprise to meet their needs, their mutual reliance will begin to shape their thinking about one another and their expectations of one another. Thinking of a "we" induces Al to think of Betty, and Betty to think of Al, differently than they would otherwise. To borrow Rousseau's example,[22] if Al and Betty agree to hunt a stag together, and then Al wanders off to snare an easy rabbit instead, Betty will think about him differently than if he'd just headed off rabbit-hunting in the first place. Similarly, Al should (and likely will) think of Betty differently than he thinks of others whom he's deprived of deer meat. Al and Betty stand in a special relationship that is not shared (although, of course, it can be replicated) with others. It is a relation of mutual obligation to one another and (at least the potential of) respect for the other's authority to obligate in that way. And their success in collaboration can increase as they come to recognize this moral dimension of their relationship.

Tomasello's story is that this is how fairness morality came to be introduced among humans. Responsibility to one another and guilt for violating the "second personal" norms that are activated when individuals see one another in the relevant way were the first social moral attitudes and allowed our progenitors for the first time to have a sense of what they *ought* to do.[23] Individuals can and do participate in these dyadic relations with multiple other individuals, and these capacities are thus generalized into a sense of being born into a community bound by a shared moral culture. More general moral attitudes came in train, including moral responsibility to the community of others bound by the same moral norms. So our sensitivity to moral norms begins with those that bind us dyadically to particular others and expands to other potential partners and observers of our conduct. In this sense, Tomasello says, his account is grounded in a "natural second-personal morality."[24] We end up with communities bound by norms tying its members to one another not only materially (surviving together) but also morally.

Now, a pair or dyad, like Al and Betty, doesn't live pairwise but lives in a community. How does that matter? To see how that matters, we'll take up once again Tomasello's work on what some of the earliest communities of human beings might have been like, creating a context in which such norms might have begun to emerge.[25]

Let's suppose that we have a hypothetical community including not only Al and Betty but also additional members Gemma and Del. Of course, there will be others as well, but we can extrapolate from the dynamics that Gemma and Del introduce. The important thing is that we are expanding interactions beyond our original dyad, and we should understand these new actors as having roughly the same backgrounds, moral education, and the like as Al and Betty. What does this do to determine the mean for justice in interactions between them? There are three new mechanisms for the evolutionary story here.

The first mechanism by which the community can help do the work we need it to do, we might characterize as *learning*. Suppose that Al and Betty spend most of their time in each other's company, and they work together to meet their needs as cooperative partners. Most of the distinct, particular occasions on which their justice or lack of it comes into play will be with each other. Al and Betty may, in the first instance, learn from their own shared history. In effect: "Last time we agreed to do things this way, and that was a disaster. Let's try doing it differently."[26] To be clear once again, it is not necessary that Al and Betty consciously recognize these norms nor represent them explicitly; what matters is that they are sufficiently motivated by the success they realize in acting in conformity with them to be guided by them again.

Suppose the same is true of Gemma and Del. Given that they share roughly similar life conditions as do Al and Betty, it is not unreasonable to suppose that they may well face similar challenges, and similar occasions in which they will need to determine what is required to treat each other justly. All are imperfect in knowledge and character, and they make mistakes of various sorts. But the fact that Gemma and Del are grappling with the same sorts of challenges means that they may well stumble upon ways of meeting them—ways of negotiating how they will treat each other that each can see as satisfying the requirements of justice—that Al and Betty can see are more adequate in one or more dimensions than the arrangements they themselves have arrived at. We might think of these as something like small short-term "experiments in living" of the sort that John Stuart Mill is known for extolling.[27] The fact that they are living in community affords them the opportunity to learn from each other—both their successes and their mistakes. Obviously, up to some limit the more participants in the cooperative practices, the more experiments, and (at least in the optimal case) the more learning. In this way, the specification of the mean can proceed *socially* much more rapidly than it can in a perpetual dyad. But it doesn't take very many to get the process going. And, on the other hand, to arrive at the sorts of judgments as to complex concrete particular engagements that are part of ordinary life for us, some millennia removed from our hypothetical Al, Betty, Gemma, and Del, we have *many, many* people making judgments and assessing the results on *many, many* occasions over *many, many* years. That allows for the kind of richness and sophistication we are concerned with in norms of our day-to-day interactions.

Suppose, further, that Al (for example) does not exclusively cooperate with Betty. Sometimes he hunts with Gemma as well, and occasionally with Del. Gemma and Del are distinct individuals, from each other, and from Betty, as well as from Al, with their own outlooks and distinct perspectives on things, including the projects they undertake with others. Since the determinations they make as to what is just in particular cases are functions of their joint judgments in interaction, these may well—and typically *will*—differ from the judgments that Al might arrive at with Betty. So the range of possible patterns of just conduct is expanded in this way too, by the diversity of inputs from the range of diverse dyads arriving at particular judgments of what is just. More experiments, more trial and error. This process allows for a progressively improving specification of the mean in justice so long as fallible people are capable of learning.

It is worth once again emphasizing that it is not important (and in fact is unlikely) that these norms are explicit or that those who are guided by them do so consciously at this stage. Explicitness (as we shall see) is a much more difficult step, coming much later in human history—in fact, we are even now at the early stages of this undertaking. The point here is that the norms are doing the guiding *because* of their contribution to the successful lives of those guided by them.

A second mechanism builds on the first one. It is premised on our capacity for perspective-taking—for empathic identification, or "sympathy," as Adam Smith has it. Smith's focus is on our capacity for sympathy with others' passions or sentiments:

> ...we have no immediate experience of what other men feel, we can form no idea of the manner in which they are affected, but by conceiving what we ourselves should feel in the like situation.... [I]t is by the imagination only that we can form any conception of what are his sensations. ... By the imagination we place ourselves in his situation, we conceive ourselves enduring all the same torments, we enter as it were into his body, and become in some measure the same person with him, and thence form some idea of his sensations, and even feel something which, though weaker in degree, is not altogether unlike them.[28]

By imagination, in other words, we are able to *take up the perspective* of others and feel what they are feeling. Although Smith does not emphasize this point, this kind of identification extends to the judgments we make. We not only judge others' opinions and convictions by that same sympathetic mechanism but also form what we might think of as second-order judgments about the judgments others make, based on our own sympathetic identifications.[29]

Now, the way we do so requires that we enter into sympathetic identification with some considerable particularity:

> ... that there may be some correspondence of sentiments between the spectator and the person principally concerned, the spectator must, first of all, endeavour, as much as he can, to put himself in the situation of the other, and to bring home to himself every little circumstance of distress which can possibly occur to the sufferer. He must adopt the whole case of his companion with all its minutest incidents; and strive to render as perfect as possible, that imaginary change of situation upon which his sympathy is founded.[30]

In our case, we are not concerned with sympathy with suffering, but with taking the perspective of others, making judgments in particular situations about *what is just*—typically in dyadic relations, as to what is just in their treatment of one another. In the larger community around such dyads, others can take the perspective of the participants via sympathy. This mechanism, then, can contribute to the development of the norms of justice in two ways.

First, it represents yet another avenue for learning. Our capacity to enter into the perspective of others and see things as they see them is one important facilitator of this process.

Second, that same capacity can provide a backup for our failures to judge correctly in particular cases. This is where an important function of the broader community comes in. Others who properly invest themselves in the details of particular conditions of judgment can make their own judgments and offer correctives when the participants' judgments are errant, either to their own advantage or to their own disadvantage.[31] They can also add voice and strength—not just indignation, anger, and other "reactive attitudes," but potentially sanctions—to the outcry against injustices that threaten to go unchallenged. Thus, though the basic structure of judgments of justice is always dyadic, others can always through imagination enter into those dyadic structures, and the force and reach of the norms that are arrived at dyadically spread through a community and beyond.

Finally, community members can pose as *arbiters* for cases of disagreement between those aiming at justice, but not able to reach substantive agreement on what is just in those cases. The disputants can in such cases agree to a second-order norm that they be bound by the decision of an arbiter, chosen in some way that is satisfactory to both of them. The principle that we ought not be judges in our own case is well-known, and the role of arbitration in legal disputes is well-known as well. So what we are after here is a way of adjudicating in cases in which the two parties cannot achieve a meeting of the minds on how to govern their interaction. Their (equal) authority over themselves includes the authority to bind themselves by the judgment of an arbiter, and that function can only be served by a larger community than the original two.

In communities—groups larger than dyads—there is also the possibility of institutions of various sorts, engendering relations beyond those tracking virtuous interactions between dyads of individuals. Of course, the actions of individuals within the ambit of institutional roles are still the actions of individuals, and their institutional roles cannot justify the unjust treatment of others. If there are institutions and practices that require such treatment, then just people avoid them and aim to change or extirpate them.

That provides an account of how individuals within social groups may have learned to interact on terms that they can see as just, but that is not the full story for the resources Al and Betty have to draw on. There is, in addition, a story of *intergroup* norm evolution that augments it. Here I draw on the ideas of Friedrich Hayek. Hayek's single most significant contribution to social thought was his focus on the way that orders can emerge spontaneously, without intentional design.[32] That is, the order itself lacks intentional design; that is not to say that there is no intentionality in the elements of that design. Consider a natural language like English. The order in the syntax and semantics of English is not the product of any designer, but that is not to say that there is no intention or design in the fabrication of particular sentences or utterances by particular speakers; manifestly, there is.[33]

Hayek thought the rules of just conduct could be a spontaneous ordering of this sort: formed without any intention to *design* a system of rules, and in fact with the pertinent rules not even being explicitly recognized by those living by them. Nevertheless, by facilitating coordination and cooperation, such rules could and did govern a peaceful and productive society.[34]

On Hayek's view, the kinds of large and interactive market societies that have emerged in the past few centuries (what Hayek calls "the Open Society"), and have accounted for unprecedented growth in wealth and welfare, are the products of systems of norms that have themselves survived a process of natural selection within cultural evolution, working on the groups of people who are governed by those norms. An especially deep feature of human culture (and unique to humans among primates) is our orientation toward group membership, and toward in-group/out-group orientations.[35] The methods of coordination and cooperation that emerge in distinct groups play a special role in the cooperative viability of that group, and there is no reason to believe that the networks of norms that underlies those methods in different groups are identical.[36] Some systems of interpersonal norms are better than others at providing terms of peace and coordination and making cooperation possible. On Hayek's view, these norms typically are not aimed at advancing particular goals, in no small part because we do not generally share goals.[37] Instead, rules of just conduct regulate or constrain the means we may take to pursue our particular goals.[38] The mechanism Hayek has in mind to explain the emergence of these rules is a simple social evolutionary one: groups that are governed by more-effective sets of norms, or rules, out-compete and out-grow those governed by less-effective sets of norms.[39] Over countless generations, tradition and culture embody these norms and make a peaceful and productive social life possible.

The mechanisms that make this intergroup story plausible are not difficult to come by. The same kinds of norms of justice that make for cooperation and coordination within groups also make it possible across groups.[40] So when a mutation in culture emerges that allows for trade across group lines (that is, with out-group members), that mutation confers a selective advantage on the group in which those norms are realized. (This is due to the fact, known since Adam Smith, that the productive advantages of the division of labor increase with the size of the market.[41] The more extensive the network of trade, the more efficient the use of resources.) In addition to the ordinary selective advantages of such a development, since these norms are propagated culturally, they can be transmitted to the groups engaged in intergroup exchange, enhancing the fitness of the out-group as well.[42] So the process Hayek envisions certainly seems viable by the standards of contemporary thinking about selection across and within groups, with both biological and cultural evolutionary processes.

What features characterize this process? First, the process by which these rules of justice emerge is, once again, not a conscious one, and this is so in multiple ways.[43] To begin with, as indicated above, Hayek argues that we can be, and to a great extent throughout our history have not been, consciously guided by such rules.[44] In fact, we have, in general, not even known what they are.[45] For most of our history, most of the time, these rules have been simply patterns in our dispositions to act. They are, Hayek claims, in this respect much like the regularities of conduct we find in animals lower on the evolutionary scale.[46] The point is that we need to have no explicit representation of these patterns for them to do their work in making societies more (or less) successful; as long as they are manifest in our behavior, they can do their adaptive work.[47] Hayek distinguishes between law, as "enforced rules of conduct" which are "coeval with society" and "existed for ages before it occurred to man that he could make or alter it," and legislation, as the product of the exercise of the wills of legislators.[48] The first has emerged spontaneously, while the second involves an "intellectual task of great difficulty."[49] It is the first that counts as the "rules of justice" for present purposes.

Second (and relatedly), the process by which these rules have emerged is not rational, but causal: the familiar operation of natural selection, in this case selection of societies or groups based on the rules governing their interactions.[50] The rules of just conduct need not be objects of deliberation. Neither are they objects of rational choice. On Hayek's view, in fact, the order of influence goes the other way around: we have developed the thoughts we have because rule-governed social orders in which we have lived have inculcated them into us.[51] One way of understanding our collective moral deliberation and argument is as a vehicle for making explicit to ourselves the rules which have heretofore governed our interactions implicitly.

Finally, one key way these rules allow for selective advantage is that they enable the use of dispersed knowledge to make efficient use of scarce resources. Hayek speaks of this knowledge as being "of particular circumstances of time and place,"[52] but it may be helpful to think of it as knowledge of acceptable or desirable trade-offs in the use of resources, given their constantly varying costs and deployments. This is something we do increasingly well with the advances in abstraction possible to us as a concept-wielding species and is the vehicle by which scarce resources are allocated most effectively for the survival and thriving of individuals within the group and thus of the group itself. The effective rules of just conduct that have emerged over this process devolve decision-making and end-pursuit (and resource deployment) to individuals with the particular knowledge of their conditions, at the same time enabling them to make use of the much vaster knowledge of others, without conscious cognizance.

We have arrived at an account of how norms of just conduct might have emerged that provides an intelligible connection to their (putative) content. It (potentially) indicates why what we count as just, we count as just, in a way that does not involve recognizing truths from Plato's heaven. On this view, we are beneficiaries and trustees of a set of rules or norms for just conduct, which allows us to prosper via the kind of cooperation and exchange that seems to be characteristic of our kind. And as it turns out, we have seen a similar process of evolution of norms realized in the common law, which is useful as a model.

9.3 A Model for Norm-Evolution

A number of authors have made cases for the evolution of English customary and/or common law in ways that model what I have in mind for the evolution of norms of justice. To a significant extent, this is a matter of legal history. However, I also engage in some thematic reconstruction to bring out the commonalities between norm-determination in the case of the virtue of justice and in the legal case.

Begin with "customary law" and its roots in Dark Age history. Lon Fuller argues that we cannot understand "ordinary" law until we understand customary law.[53] As John Hasnas tells the story, this form of law begins with the withdrawal of the Romans and the need to fill a vacuum of dispute resolution to facilitate peaceful human interaction.[54] Attempts to reach peaceful solutions were negotiated before the public assembly: the moot. These settlements were case by case, but the moot had an "institutional memory" in virtue of which learning could occur, of settlements in the past which did not work (to be avoided!) and those which did (to be replicated). The patterns of these settlements forged expectations which became action-guiding for members of the community.[55]

The qualities of customary law important for my account carry over to what more properly can be called "common law," which is differentiated (for our purposes) by being judge-made, though still case law. In judge-made case law, the case-by-case, concrete particularity of customary law is maintained. Moreover, Hasnas emphasizes as a key difference from common law as it has come to be understood since (roughly) Blackstone that there was no notion of judges being bound by precedent—*stare decisis*—but rather a matter of judgment in accordance with fairness as understood by community standards expressed by juries.[56]

Still, even as judge-made common law became more prevalent, the kind of authority maintained by modern judges was distributed rather than concentrated. Competing court systems (polycentric adjudication) provided for competition among court systems, including the King's Bench, the Exchequer, the Court of Common Pleas, ecclesiastical courts, and many others competing for jurisdiction.[57] Of course, it would be hasty to suppose that the metric of success was justice or fairness, even by the lights of the communities in which these courts were functioning. It is certainly possible to have unjust peace—as a *modus vivendi* under an unjust lord, for example. And even with good motives, judges can make mistakes. But bad decisions are part of any human legal system. The point here is that (i) there was no claim to final, univocal obligating authority in the courts, and (ii) the structure allowed for robust learning as to what sorts of decisions were successful at maintaining peace and allowing for cooperation.[58] To be clear, peace and cooperation do not entail justice; various *modi vivendi* populate lots of human history with unjust stability for long stretches. Instead, the idea is that the process by which peace and cooperation emerged through the operation of the common law provides a model for seeing how the aim of living by norms of justice might emerge as a moral ideal. A number of features of just interaction, as discussed above, are modeled in this account of the emergence of common law. Each of these features—particular, case-based reasoning, distributed authority, implicit norm development, and learning—also characterizes how norms of justice emerge in moral life.

First, this law arises out of the determination of concrete, particular cases. "Customary and common law arises out of the settlement of actual cases," Hasnas claims. "Cases arise only when there is an injured party."[59] This is in contrast with statutory law, which is by its nature general and forward-looking. Of course, the normative implications of customary and common law are forward-looking in the sense that they bear on the expectations the members of the community develop of each other and of how others will react to what they do. But here, too, the story models the evolution and determination of norms of justice.

Second, as indicated above, there was no unilateral authority making the law, as there is in the case of statutory law.[60] Such authority as there was for customary law—the moot—was authoritative as to the resolution of particular cases, but as a manifestation and expression of the community, it would always have been in flux. Moreover, the parties to any given case would themselves likely have been part of the decision-making body for other cases. Similarly, the polycentric nature of common law distributed authority as well. This tracks the distributed structure of authority we have seen in the case of the virtue of justice.

Third, the legal norms in question may or may not have been explicitly contemplated; they can work effectively even as partially or entirely implicit, in a fashion such as we experience with formulating speech or arriving at a comfortable conversational distance.[61] To some significant extent, this is because the expectations case resolutions give rise to are to apply to unforeseen future conditions and circumstances.[62] But it is also because our grasps of the concepts used in normative judgment (including legal judgment) are never complete and instead are emerging through an evolutionary process of learning—of trial and error.[63] As we have seen, it is plausible to suppose that we can say the same of norms of justice.

Finally, the case-by-case determinations and distributed authority mean that bad decisions—inevitable from fallible and often not-very-virtuous people—are limited in their reach. Because those involved in the process are capable of learning, and seeing what sorts of outcomes follow what sorts of decisions, the norms of law are capable of steady and progressive improvement.

This learning is really central to both the common law and the virtue of justice. In both cases, norms (whether, on the one hand, for justice in character and conduct or, on the other, for common law) are *emergent* from bottom-up processes (Hayek's "spontaneous orders").[64] Hasnas explicitly understands the process of making common law as an "evolutionary process," as a matter of (corrigible) judge-made determinations of what the law requires in particular cases.[65] Similarly, in the account of just interaction on offer here, norms for just character and conduct evolve spontaneously out of the interactions of countless dyads of people aspiring (imperfectly) to be just, and to treat one another justly. These norms are absorbed into larger moral communities and succeed or fail as the patterns suggested by particular dyadic resolutions are recognized, interpreted, and transformed as they are applied to new and different cases with greater or lesser success.[66] The selective forces at work may be either conscious and explicit (for example, progressively improving judgments about what justice and fairness consist in) or unconscious and implicit (for example, the instrumental benefits of the social peace and extended cooperation afforded by successfully determining norms).

This baked-in imperfection of the process is central to Robert Brandom's deployment of the common law model, in his case for the grasp of concepts, whether empirical concepts (COPPER, for example) or legal ones. Brandom's Hegelian approach to the grasp of conceptual content takes the path of progressive, and fallible, evolution as modeled by case-by-case judge-made law. That grasp is always incomplete—a work in progress:

"The process has no endpoint; the returns are never all in."[67] The important thing for Brandom's take on common-law adjudication is that it represents a reconstruction

> by careful selection and characterization of precedents, so as to take the form of the gradual emergence into explicitness of a norm that becomes visible as having been all along implicit in the deliberations of prior judges.[68]

Brandom's take is the most idealized appropriation of the evolution of common law of those I have mentioned, and I cannot speak to its fidelity to history. But as a model for the evolution of the norms of justice as a virtue of individuals, it is spot on. It tracks a process that began with the earliest forms of human moral cognition, at least as Tomasello sees it, and continues to our attempts today to grasp what it means to see and treat one another justly.

Notes

1 Aristotle, 1984 *Nicomachean Ethics* III.9, 1117a29.
2 *NE* III.10, 1118a3.
3 *NE* IV.1, 1120a8.
4 Locke (1823, §4).
5 Kant (1797/1996, Ak. p. 237); emphasis in the original.
6 Aristotle, 1984 *Politics*. III.16, 1287a16; see also *Pol.* III.6, 1279a8, *Eudemian Ethics* VII.10, 1242b28.
7 Kant (1788/1956, Ak. p. 73).
8 More on respect as the passion relevant to justice in LeBar (2025, ch. 4).
9 *NE* II.6, 1106b28.
10 *NE* II.6, 1106b21–3.
11 *NE* II.6, 1106a29–b7. See also *NE* II.8, 1109a1xi*ff* for yet more complications of this story.
12 *NE* II.6, 1106a32*ff*.
13 *NE* II.6, 1107a1.
14 For discussion, see LeBar (2025, §2.2). This position is ably laid out in Curzer (2016).
15 This is the position I defend in LeBar (2013, ch. 7) and LeBar (2025, ch. 2). See also Berryman (2019).
16 *NE* II.7, 1107b2.
17 Perhaps more, but I will focus on the binary case for reasons that will become apparent.
18 Mill (1863/2001, 23).
19 Importantly, the learning need not be conscious, intentional, deliberate, or explicit. Like other creatures, we can learn without these qualities at work. More on this point below.
20 Tomasello (2016).
21 Tomasello (2016, p. 2). I thank an anonymous referee for pressing this point.
22 Rousseau (1755/1992, Part Two, p. 47).

23 Tomasello (2016, 74, 82). The notion of a "second-personal" morality draws on Stephen Darwall's framework in which our responsibility is to each other, second-personally, as opposed to a third-personal framework of norms or moral requirements (Darwall, 2006).
24 Tomasello (2016, p. 160).
25 There are close affinities between the account that follows here and John Hasnas' account of the emergence of "empirical natural rights" (Hasnas, 2024, ch. 6).
26 We can augment this basic story with an account of the transmission of moral norms from one generation to the next via the evolutionary development of cognitive resources allowing us to learn by "apprenticeship." Kim Sterelny argues that from a very early age we observe moral exemplars in paradigm instances of the exercise of moral judgment. We then exploit our capacity for pattern-recognition to apply this learning, and these norms, to new situations (Sterelny, 2012, ch. 7).
27 Mill (1859/1978, ch. 3).
28 Smith (1759/1976, I.i.1.2).
29 Smith (1759/1976, I.i.3.2, II.i.5.5).
30 Smith (1759/1976, I.i.4.6).
31 On the latter point, see Russell and LeBar (2022).
32 His most well-known application of the operation of spontaneous orders was to prices, where he argued that the system of prices that emerge through free buying and selling produces efficiencies in the use of scarce resources that cannot be achieved in any other way ("The Use of Knowledge in Society" in Hayek, 1945 and "The Meaning of Competition" in Hayek, 1948). Prices enable economic planning to be done by individual consumers and producers, coordinated by the information conveyed by those prices, without anybody's oversight, intention, or design of an overall plan.
33 I thank an anonymous referee for pressing this point.
34 Hayek (1982, ch. 2).
35 Tomasello (2014, p. 84).
36 Cf. Henrich (2016, 199).
37 Hayek (1982, p. 13).
38 Hayek (1982, vols. II, p. 3, III, p. 22).
39 Hayek (1982, vol. II, pp. 16, 21).
40 Hayek's account of group selection faded into the twilight as group selection itself was thought suspect among evolutionary biologists, but in recent years that mechanism has returned to credibility as a form of explanation of the evolution of culture and determinant of cultural effects on human biological evolution, at least in qualified forms. An excellent formulation of the qualifications can be found in Richerson and Boyd (2005, ch. 6). They argue that group selection makes sense as an evolutionary force when there is "a great deal of behavioral variation" among groups (204). That is precisely the sort of variation Hayek has in mind. See also David Sloan (Wilson 2002, ch. 1; 2005). Jonathan Haidt builds on Darwin's *Descent of Man*, Wilson, and others in making an extended case for group selection (as part of a multilevel selection process—a process going on both between groups and within groups) in Haidt (2012, ch. 9); see also Haidt and Kesebir (2010, pp. 817–818) and Tomasello (2014, p. 121). Wilson notes that the case for group selection can be made only when the among-group adaptive advantages of some sort of trait swamp any within-group adaptive disadvantages (Wilson, 2005). So there is a residual

worry about the advantages of predation free-riding on generally adaptive dispositions to respect the rules of justice. But our propensity to detect and punish free-riders is also well established. See, for example, Cosmides (1989) and Cosmides and Tooby (1995).

41 Smith (1776, I.I.III).

42 Haidt cites private correspondence by evolutionary psychologist Lesley Newson that makes precisely this point:

> I think it is important not to give readers the impression that groups competing necessarily meant groups being at war or fighting with one another. They were competing to be the most efficient at turning resources into offspring.
>
> (Haidt 2012, p. 253)

43 Cf. Henrich (2016, p. 199).

44 Hayek (1982, vol. I, pp. 74–75).

45 Hayek (1982, LLL, vol. I, p. 43). In an earlier work, Hayek says of the "moral rules of conduct":

> Next to language, they are perhaps the most important instance of an undesigned growth, of a set of rules which govern our lives but of which we can say neither way they are what they are nor what they do to us: we do not know what the consequences of observing them are for us as individuals and as a group.
>
> (Hayek 1960, p. 64)

46 Hayek (1982, vol. I, pp. 74–75).

47 As Wilson observes, "cultures can evolve to be smart in ways that are invisible to their own members" (Wilson, 2002, p. 32).

48 Hayek (1982, vol. I, pp. 72–73).

49 Hayek (1982, vol. II, p. 41).

50 Hayek (1982, vols. I, p. 44, II, p. 4, III, p. 154).

51 Hayek (1982, vol. III, p. 157).

52 Hayek (1982, vols. I, p. 14, II, p. 20; 1948, "The Use of Knowledge in Society").

53 Fuller (1969).

54 Hasnas (2005).

55 Hasnas (2005, p. 87); Fuller (1969, p. 3).

56 Hasnas (2005, pp. 92–94).

57 Zywicki (2003, p. 1584).

58 In later work, Hasnas maintains that these are the ends that justify an obligation to obey the law, when (and only when) law serves those purposes. See Hasnas (2024, ch. 7).

59 Hasnas (2013, p. 477); Fuller (1968, p. 95).

60 Hasnas (2005, p. 87).

61 Fuller (1969, p. 9). Kim Sterelny assesses the likeness of the cognitive resources necessary for this sort of moral learning to the cognitive resources for learning grammar (Sterelny, 2012, ch. 7).

62 Fuller (1969, p. 26).

63 Fuller (1969, p. 12); Fuller (1968, p. 107); Brandom (2019, p. 449).

64 Hayek (1982, vol. I, ch. 2).

65 Hasnas (2013, 465, p. 470). On this point, also see Hasnas (2005); Fuller (1969, pp. 9, 18).
66 Cf. Hayek (1982, vol. I, pp. 43–44).
67 Brandom (2019, p. 449); compare Fuller (1968, p. 107).
68 Brandom (2019, p. 737).

References

Aristotle. (1984). *The complete works of Aristotle* (J. Barnes, Ed., Vol. 2). Princeton University Press.

Berryman, S. (2019). *Aristotle on the sources of ethical life*. Oxford University Press.

Brandom, R. (2019). *A spirit of trust*. Harvard University Press.

Cosmides, L. (1989). The logic of social exchange. *Cognition*, *31*, 187–276.

Cosmides, L., & Tooby, J. (1995). Cognitive adaptations for social change. In J. H. Barkow, L. Cosmides, & J. Tooby (Eds.), *The adapted mind* (pp. 163–228). Oxford University Press.

Curzer, H. (2016). Rules lurking at the heart of Aristotle's virtue ethics. *Apeiron*, *49*, 57–92.

Darwall, S. (2006). *The second-person standpoint*. Harvard University Press.

Fuller, L. (1968). *Anatomy of the Law*. Frederick A. Praeger.

Fuller, L. (1969). Human interaction and the law. *American Journal of Jurisprudence*, *14*, 1–36.

Haidt, J. (2012). *The righteous mind*. Vintage Books.

Haidt, J., & Kesebir, S. (2010). Morality. In S. Fiske, D. Gilbert, & G. Lindzey (Eds.), *Handbook of social psychology* (pp. 817–818). Wiley & Sons.

Hasnas, J. (2005). Hayek, the common law, and fluid drive. *NYU Journal of Law and Liberty*, *1*, 79–110.

Hasnas, J. (2013). Is there a moral duty to obey the law? *Social Philosophy and Policy*, *30*, 450–79.

Hasnas, J. (2024). *Common law liberalism*. Oxford University Press.

Hayek, F. (1945). The use of knowledge in society. *The American Economic Review*, *35*, 519–530.

Hayek, F. (1948). *Individualism and economic order*. Chicago University Press.

Hayek, F. (1960). *The constitution of liberty*. University of Chicago Press.

Hayek, F. (1982). *Law, legislation, and liberty*. Routledge.

Henrich, J. (2016). *The secret of our success*. Princeton University Press.

Kant, I. (1788/1997). *Groundwork of the metaphysics of morals*. Cambridge University Press.

Kant, I. (1797/1996). *The metaphysics of morals*. Cambridge University Press.

LeBar, M. (2013). *The value of living well*. Oxford University Press.

LeBar, M. (2025). *Just people*. Oxford University Press.

Locke, J. (1823). Essay two. In *Two treatises of government*, from *The Works of John Locke* (Vol. 5). Thomas Tegg.

Mill, J. S. (1859/1978). *On liberty*. Hackett.

Mill, J. S. (1863/2001). *Utilitarianism*. Hackett.

Richerson, P. J., & Boyd, R. (2005). *Not by genes alone*. University of Chicago Press.

Rousseau, J.-J. (1755/1992). *Discourse on the origin of inequality*. Hackett.

Russell, D., & LeBar, M. (2022). Doing justice to oneself. In G. Pettigrove (Ed.), *Neglected virtues*. Routledge.
Smith, A. (1759/1976). *The theory of moral sentiments*. Liberty Fund.
Smith, A. (1776). *The wealth of nations*. Liberty Fund.
Sterelny, K. (2012). *The evolved apprentice*. MIT Press.
Tomasello, M. (2014). *A natural history of human thinking*. Harvard University Press.
Tomasello, M. (2016). *A natural history of human morality*. Harvard University Press.
Wilson, D. S. (2002). *Darwin's cathedral*. University of Chicago Press.
Wilson, D. S. (2005). The new fable of the bees: Multilevel selection, adaptive societies, and the concept of self-interest. In R. Koppl (Eds.), *Evolutionary psychology and economic theory* (pp. 201–220). Emerald Group Publishing.
Zywicki, T. (2003). The rise and fall of efficiency in the common law: A supply-side analysis. *Northwestern University Law Review*, *97*, 1584–1620.

10 Social and Individual Learning in the Epistemic State of Nature

Justin P. Bruner

10.1 Introduction

Miranda Fricker's account of epistemic injustice is anchored in a simple but powerful thought experiment (Fricker, 2007). Imagine an epistemic state of nature: a setting where no courts, universities, or professional codes of conduct regulate how knowledge is shared. In such a world, testimonial practices—who is believed, whose word is dismissed—are governed directly by the character of individual agents. Virtues like testimonial justice, humility, and open-mindedness help sustain the circulation of knowledge. They ensure that hearers give speakers a fair hearing, resisting the pull of identity prejudice or undue arrogance. Vices, by contrast, disrupt this fragile economy. Prejudice skews credibility assessments, arrogance leads agents to discount others' contributions, and closed-mindedness prevents them from updating their beliefs when they should. For Fricker, then, the distribution of virtue and vice across a community is the primary determinant of whether testimonial exchange in the state of nature succeeds or fails.

This emphasis is important: it highlights the moral precarity of epistemic life before the rise of institutions capable of correcting for individual bias. But it is also incomplete. Testimony, after all, is only valuable if there is something of value to testify about. A community can only share knowledge if its members have first managed to discover it. This requires agents not just to listen well but also to inquire well. And here too, virtue and vice make a difference. Discovery-side virtues—curiosity, conscientiousness, perseverance—generate new information, while laziness, arrogance, and wishful thinking frustrate inquiry. Without discovery virtues, testimonial practices are starved of content; without testimonial virtues, the fruits of inquiry fail to circulate effectively.

Philosophers of science such as Philip Kitcher (1990) and Michael Strevens (2003) have long emphasized the benefits of an epistemic division of labor[1]: communities do best when some agents pursue one line of inquiry while others follow different paths, thereby avoiding redundancy and ensuring

DOI: 10.4324/9781003632641-14

coverage. Their analyses show how social structures can harness individual efforts for collective epistemic gain. Yet the evolutionary dynamics of individual and social learning reveal a darker possibility. As Alan Rogers (1988) famously showed, natural selection does not necessarily drive populations toward divisions of labor that maximize group benefits.

Instead, it can stabilize a mix of strategies—some agents relying on costly individual learning, others on cheaper social learning—that leaves the community no better off than if everyone had simply gone it alone.[2] In this way, what looks like a promising balance of epistemic virtues can, under evolutionary pressure, collapse into what is now dubbed a "Rogers' paradox": a stable but collectively unproductive equilibrium.

Our goal in this chapter is to explore a virtue-based picture of the epistemic state of nature by integrating these insights. Using a simple model of individual and social learning, we show how testimonial and discovery virtues jointly shape epistemic outcomes, and how vices on either side can entrench inefficiencies or reproduce Rogers' paradox. In doing so, we highlight the conditions under which epistemic communities reap genuine benefits from division of labor, and the conditions under which evolutionary dynamics frustrate those gains.

The paper proceeds as follows. Section 10.2 offers a characterization of the epistemic state of nature. Section 10.3 introduces our model of individual and social learning, highlighting how virtues and vices map onto its dynamics. Section 10.4 concludes with a discussion of the broader implications, both for understanding the fragility of epistemic communities and for appreciating the role of virtues in sustaining them.

10.2 The Epistemic State of Nature: Nescient, Benighted, and Short-Sighted?

Fricker, as well as many others writing in her wake, takes a cue from Edward Craig's genealogical project in *Knowledge and the State of Nature*. Craig asks what the concept of knowledge is for, and he suggests we can answer this by imagining a "state of nature" where no concept of knowledge yet exists, and then asking what role such a concept would play if introduced (Craig, 1991). Bernard Williams develops a similar genealogical approach, tracing our epistemic practices to the practical needs of agents trying to cope with the world (Williams, 2010). Fricker and others have built on these insights to consider what testimonial practices would look like in such a setting, and what forms of virtue and vice would matter most.

The core problem of this epistemic state of nature is straightforward. Agents face an informational environment that they cannot fully navigate on their own. They must often rely on others to extend their epistemic reach. Yet reliance is risky: hearers must decide whether to trust speakers, and speakers must balance accuracy against the temptation to mislead. The absence of institutions—no courts, no universities, no norms of professional testimony—means that nothing beyond the character of individuals regulates these exchanges. Testimony is thus precarious. If hearers systematically misjudge credibility, or if speakers systematically misrepresent, the flow of knowledge breaks down.

One promising solution to this problem is rooted in epistemic virtue. Consider testimonial justice—the disposition to correct for prejudice in credibility judgments. Testimonial justice ensures that hearers give speakers a fair hearing, resisting distortions introduced by social identity or entrenched bias. More broadly, virtues such as intellectual humility, open-mindedness, and conscientiousness enable agents to listen responsibly, weigh testimony appropriately, and integrate it into their stock of beliefs. In the state of nature, these virtues are not optional; they are the only safeguards that prevent testimonial exchange from collapsing into suspicion or error.

The corresponding danger lies in epistemic vice. Prejudice is perhaps the most obvious: when hearers allow social identity to determine credibility, the testimonial economy is systematically corrupted. But arrogance, dogmatism, and laziness can be just as destructive. An arrogant hearer discounts the contributions of others; a dogmatic one refuses to revise beliefs in the face of contrary testimony; a lazy one fails to engage with testimony at all. In the state of nature, where no institutional correctives are available, the unchecked spread of such vices is disastrous. Communities dominated by vice cannot sustain reliable testimonial practices.

One might conclude, then, that testimonial justice and allied epistemic virtues are the primary regulators of epistemic life in the state of nature. They allow testimony to serve its purpose: extending the reach of individual agents by pooling information across the community. The distribution of virtue and vice, then, directly determines whether the state of nature is habitable as an epistemic environment. Without sufficient virtue, the circulation of knowledge collapses into prejudice, arrogance, and mistrust. With it, testimony becomes a functional substitute for the institutions and norms that will later stabilize epistemic life.

Some social epistemologists working in this area—such as Fricker—isolate a fundamental problem of the epistemic state of nature: how to fairly and accurately assess the credibility of others. This foregrounds concerns with testimony and the conditions under which hearers can identify sincere and competent speakers, and surely such considerations are an important part of the picture. Yet it is not the only problem we confront in the state of nature. For even if testimony flows without distortion, its value depends on there being something worth disseminating in the first place.

This points us to a second, equally pressing problem: *the problem of production*. Knowledge must be generated before it can be shared. Who, then, will do the legwork of gathering information? In the absence of institutions or formal incentives, there are reasons to worry. Agents may prefer to free-ride on the costly inquiries of others rather than undertake their own investigations. They may also be tempted to cut corners—producing information that is sloppy, incomplete, or unreliable. Relatedly, individuals may suffer from misplaced confidence in their own abilities: attempting inquiry when they ought to defer, or deferring when they ought to investigate. In short, the epistemic state of nature generates its own suite of discovery-side virtues and vices. Curiosity, diligence, and humility sustain inquiry; laziness, overconfidence, and carelessness undermine it.

For our purposes, the most striking analysis of this problem comes from the biological anthropologist Alan Rogers. Rogers considers a setting that resembles the epistemic state of nature in its most austere form. There are no institutions to distribute incentives, nor even informal coordination among agents. Individuals choose between costly individual learning (direct investigation of the world) and cheaper social learning (relying on others). At first glance, this seems to open the door for an efficient division of labor: some do the hard work of discovery, while others economize by learning socially. Yet Rogers demonstrates that evolutionary dynamics block such efficient schemes from emerging. Social learners proliferate, individual learners bear the costs, and the mix of strategies that stabilizes provides no additional benefit to the group—or even to the individuals themselves.

The result is puzzling. Evolutionary forces, which we might have expected to secure epistemic efficiency, instead entrench epistemic stagnation. Rogers' paradox underscores the fragility of discovery in the state of nature. Absent institutions, absent coordination, the distribution of virtues and vices shaped by natural selection will always leave our collective inquiry at the second-best. However fair our testimonial practices may be, without sufficient discovery-side virtue, the epistemic state of nature cannot sustain a productive flow of knowledge.

10.3 Modeling Inquiry

Social epistemologists such as Fricker capture one fundamental challenge of the epistemic state of nature: how to regulate the circulation of testimony in the absence of institutions. Yet, as Rogers' work reminds us, there is another challenge that is no less pressing: how to secure the production of knowledge in the first place. Both problems are, at their core, problems of coordination and conflicting interests. Who is to be believed, and who will do the costly work of inquiry? To investigate these questions systematically, we turn now to the tools of *game theory*. Game-theoretic models allow us to formalize the strategic pressures agents face, to make precise how virtues and vices shape outcomes, and to reveal the conditions under

Table 10.1 The learning game ($1 > c > 0$)

	Individual learning	*Social learning*
Individual learning	$1 - c$, $1 - c$	$1 - c$, 1
Social learning	1, $1 - c$	0, 0

which epistemic practices succeed or fail. Rogers' model of individual and social learning provides a natural starting point.

At its most basic, a game consists of three components: a set of agents, a set of actions available to those agents, and a specification of their preferences over possible outcomes. Agents choose actions, outcomes follow, and the point of game-theoretic analysis is to identify which patterns of behavior are stable given the incentives at play. A central notion here is that of a Nash equilibrium: a profile of actions in which no agent can improve their payoff by unilaterally changing their choice. In equilibrium, everyone is doing as well as they can given what others are doing (Table 10.1).

To fix ideas, consider the following "learning game." Each agent has two possible actions: Individual Learning (*IL*) and Social Learning (*SL*). Individual learning involves investigating the world directly. It is highly reliable but comes at a cost, *c*. Social learning, by contrast, is cheap but parasitic: it only works if one's partner has actually done the work of inquiry. We can summarize the payoffs as follows. If both players play *IL*, each receives $1 - c$: the full benefit of discovery minus its cost. If both engage in *SL*, neither acquires new information, and both receive 0. If one plays *IL* while the other plays *SL*, then the *IL* player again receives $1 - c$, while the *SL* player, free-riding on the inquiry of their partner, receives 1.

This simple model highlights an important trade-off: individual learning is costly but secure; social learning is cheap but only pays off if someone else bears the cost of discovery. Given this structure, the game admits two pure-strategy Nash equilibria. In both, one player chooses *IL* while the other chooses *SL*. The *IL* player is willing to continue bearing the cost, since deviating to *SL* would drive their payoff to 0. The *SL* player, meanwhile, has no reason to deviate, as *IL* would only reduce their payoff from 1 to $1 - c$. Thus, the game captures the core difficulty of epistemic life in the state of nature: agents have reason to economize by relying on others, but someone must nonetheless bear the cost of producing knowledge.

The static analysis of the learning game already points to a problem: while individual learning is reliable, social learning is always tempting when someone else does the work. But what happens when we view this through an evolutionary lens? Will natural selection push us toward a population of self-reliant inquirers, a population of free-riding social learners, or some mixture of the two? And what will this mean for the success of inquiry at the group level? These are the questions Alan Rogers tackled in his influential 1988 paper.

Start with a population composed entirely of social learners. Here, everyone defers to others, but no one has anything worth sharing. Inquiry collapses, and all secure the meager payoff of zero. Now imagine the appearance of a single asocial learner—a mutant who does the work of individual inquiry. This agent fares exceptionally well compared to the natives, provided the cost of learning c is less than one. The reason is simple: by investigating the world directly, the individual learner always acquires valuable information, while social learners spinning in an informational void get nothing. Unsurprisingly, we expect individual learning to spread in such an environment.

Yet there are limits to this expansion. Consider the opposite extreme: a population composed entirely of individual learners. Here, everyone bears the cost of discovery but reliably acquires information. Now introduce a lone mutant who learns socially. This agent thrives. Regardless of whom they consult, they encounter someone with valuable information and acquire it at a trivial cost. Social learning, once rare, has an evolutionary advantage in this setting.

What results, in the long run, is a stable mixture of the two types. Neither strategy drives out the other. Instead, natural selection equilibrates at a point where the expected payoff to social learners equals the expected payoff to individual learners. And here is where matters become puzzling. At equilibrium, everyone secures an expected payoff of $1 - c$. That is, the community as a whole does no better than if everyone had simply relied on individual learning. Social learning, despite its apparent promise, generates no group-level benefit.

This is Rogers' paradox. Evolutionary dynamics, far from driving the community to a division of labor that maximizes epistemic efficiency, stabilize at a point where too many agents free-ride.[3] The result is a glut of epistemic vice. If fewer individuals opted for social learning, the community's overall payoff would rise above $1 - c$. But natural selection keeps the balance tilted toward inefficiency. Discovery is chronically hampered, testimony is starved of content, and the potential gains of social learning are squandered.

Rogers' paradox shows how easily evolutionary pressures can lock us into inefficient equilibria.[4] The problem is not that social learning is inherently defective—it plainly economizes on the costs of discovery—but rather that, left unchecked, it proliferates to the point where discovery itself is under-supplied. From the perspective of virtue theory, we might say that the evolutionary dynamics generate a community with too much epistemic vice. Agents over-rely on others, shirk the burdens of inquiry, and free-ride on the diligence of a shrinking minority.

The lesson, then, is that successful inquiry in the epistemic state of nature cannot be secured by social learning alone. Testimonial virtues may ensure that information flows fairly, but without discovery virtues—traits like diligence, conscientiousness, and humility—there will be little of value to circulate. To avoid the inefficiencies revealed by Rogers' model, we need agents who resist the temptation to free-ride, who take on the costs of inquiry, and who calibrate their confidence so as not to overestimate their abilities. In the next section, we formalize this thought. By extending Rogers' simple learning game, we show how testimonial and discovery virtues together determine whether the epistemic state of nature yields genuine benefits—or whether vice and evolutionary pressure conspire to leave inquiry permanently hampered.

10.4 Results

The analysis so far has emphasized the importance of virtue in sustaining epistemic life in the state of nature. Testimonial virtues—justice, humility, open-mindedness—help regulate the circulation of information, while discovery virtues—curiosity, diligence, conscientiousness—sustain its production. Testimonial vices distort credibility judgments and block the flow of knowledge; discovery vices generate free-riding, overconfidence, and under-investment in inquiry.[5] Our model provides a way to make these dynamics precise.

Each strategy profile can be read as a distribution of epistemic character traits. Agents who reliably engage in individual learning exemplify discovery virtues, willingly incurring the costs of inquiry for the sake of reliable knowledge. Those who rely exclusively on social learning are closer to vice: they economize on inquiry, leaning on the diligence of others. Mixed strategies reveal a more complicated picture, with competence calibrated to confidence—a form of epistemic humility when properly tuned, or overconfidence when not.

With this in mind, the equilibria described below can be understood as evolutionary portraits of virtue and vice in the epistemic state of nature. Some equilibria correspond to communities that, through a balance of virtues, secure epistemic gains over asocial baselines. Others reproduce Rogers' paradox, entrenching vice and ensuring that the benefits of social learning are squandered. The key question is whether the evolutionary dynamics stabilize communities with enough discovery virtue to sustain inquiry, and enough testimonial virtue to ensure its fruits circulate.

We now formalize these dynamics in the simplest case, assuming just two types of inquirers: high-types, who engage in successful inquiry with probability 1, and low-types, who engage in successful inquiry with probability s.[6] In the first stage of the model, individual learners consult nature. All agents are then paired at random, and those who learn socially can consult with their partner. If a social learner is paired with an individual learner, then they learn from their counterpart (assuming their counterpart was successful in their investigation) at no cost to themselves. If instead they're paired with a social learner, neither partner learns about the world, and both secure a payoff of zero.

Strategies are slightly more complicated because a strategy must now specify how an individual behaves as both types. That is, an agent's strategy must recommend behavior when she is a high-type, as well as when she is a low-type. We represent strategies as a vector, where <*SL*, *IL*> instructs the agent in question to learn socially as a high-type and individually when a low-type.[7] Think of competence levels as rungs on a ladder; higher rungs correspond to greater success in inquiry, and an agent must know how to behave at each rung on the ladder.

Three different evolutionary equilibria are possible. We discuss each in turn.

10.4.1 All Play <IL, SL>

In this equilibrium, high-types shoulder the burden of discovery while low-types defer. From the perspective of virtue, this looks like a community with a reasonable division of epistemic labor. Discovery virtues are on display among those with the greatest competence, while those less competent agents come to learn socially rather than attempt unreliable inquiry. The balance of virtue here ensures that inquiry succeeds and the average payoff exceeds the asocial baseline predicted by Rogers.

For this arrangement to be stable, two things must be the case. First, it must be the case that high-types have no incentive to learn socially. That is, $1 - c$—the payoff high-types get from learning individually—must be greater than or equal to q, the probability of being paired with a high-type who learns individually. Second, low-types must have no incentive to learn individually. In other words, q must be greater than or equal to $s - c$. So, it must be the case that $s - c < q < 1 - c$.

Average payoff at this equilibrium is $(1 - c)q + q(1 - q)$. Because a condition of this equilibrium is that q is greater than or equal to $s - c$, the average payoff at this equilibrium is greater than or equal to the payoff agents secure should they all simply learn individually $((1 - c)q + (s - c)(1 - q))$.

10.4.2 Stable Mixture Involving <IL, SL> and <SL, SL>

Here, the population fragments. Some agents exhibit discovery virtue, learning individually when competent but deferring when not. Others succumb to vice, learning socially regardless of their ability. The result is a community where free-riding is endemic. Although the system stabilizes, the average payoff is no greater than if everyone had learned individually. Social learning, which could in principle provide collective gains, is squandered by an excess of laziness and misplaced deference.

When is this arrangement stable? Let's assume that p proportion of the population plays <*IL*, *SL*>. Thus, the payoff associated with social learning is pq. Thus, for high-types to be incentivized to sometimes learn individually, the payoff to individual learning $(1 - c)$ must be equal to the expected payoff from social learning, meaning $pq = 1 - c$, and thus p, the proportion of high-types learning individually, is $(1 - c)/q$. For this fraction to be in the unit interval, it must be the case that $q > 1 - c$. Recall that a necessary condition for the monomorphic <*IL*, *SL*> equilibrium (a stable scenario where all agents adopt the same strategy) is that q is *less* than $1 - c$. Thus, these two equilibria cannot both exist for the same set of parameter values.

Finally, for low-types to have an incentive to learn socially, it must be the case that pq is greater than $s - c$, meaning that p must be greater than $(s - c)/q$. This inequality always holds because, by assumption, $s - c$ is less than $1 - c$ and, as we established in the previous paragraph, $p = (1-c)/q$.

At this equilibrium, the expected payoff to individuals is $q(1 - c) + (1 - q)pq$, which simplifies to $1 - c$, the payoff a high-type gets when learning individually. Thus, at this equilibrium it is as though *all* individuals are high-types engaging in individual inquiry.

10.4.3 Stable Mixture Involving <IL, IL> and <IL, SL>

At this equilibrium, some always learn individually, while others restrict individual learning to when they are competent. On the surface, this might seem like a balanced allocation of epistemic labor. But the payoff structure reveals otherwise: the average outcome is identical to a world in which everyone simply learns individually. Social learning adds nothing. In virtue terms, discovery virtues are present but not calibrated. A lack of testimonial justice—or perhaps an excess of overconfidence—prevents the community from harnessing the benefits of division of labor.

Let r be the proportion of low-types who learn individually. The payoff to social learning is thus $q + (1 - q)sr$. For this arrangement to be stable, it must be the case that the payoff of learning as a low-type is equal to the expected payoff of social learning. That is, it must be the case that $s - c = q + (1 - q)rs$, meaning r is $(s - c - q)/[(1 - q)s]$. This fraction is always less than one, and is greater than zero when $s - c > q$. Note that this is in direct contradiction to a necessary condition of the other polymorphic equilibrium, as well as the monomorphic equilibrium (both assume $s - c < q$).

At this equilibrium, the average payoff to the individual is $q(1 - c) + (1 - q)(s - c)$, which is equal to the expected payoff should *all* individuals engage in individual learning, regardless of competence level. Thus, while some are learning socially at this equilibrium, there is no group-level nor individual-level benefit from social learning.

Taken together, these equilibria illustrate how the distribution of virtues and vices determines the quality of intellectual life in the epistemic state of nature. Communities with sufficient discovery virtue and testimonial humility can outperform purely asocial inquiry. But when vice takes the form of excessive free-riding or miscalibrated confidence, social learning collapses into Rogers' paradox: stable, but epistemically barren.

10.5 *N* Types of Inquirers

The extension to *N* types provides a more fine-grained picture of how epistemic virtues and vices distribute across a population. Namely, in this section we add an indefinite (but finite) number of competence levels. The general result—that the most competent types will always engage in some individual learning while the least competent defer socially—maps neatly onto the discovery virtues of diligence and humility. Those best equipped to generate knowledge take on the costs of inquiry, while those less competent recognize their limitations and defer. This division of labor reflects a virtuous calibration of confidence to competence.

Yet the below model also shows how vice can intrude. At equilibria where some types insist on learning socially regardless of their ability, the familiar pattern of free-riding emerges. Laziness, overconfidence, or misplaced deference can corrode the efficiency of the epistemic division of labor. Where too many low-competence agents rely on others, the system drifts back toward Rogers' paradox. Inquiry is under-supplied, and social learning adds no value.

Interestingly, the general N-types framework almost always avoids Rogers' paradox. With many gradations of competence, the distribution of inquiry across types tends to track the underlying variation in ability. Discovery virtues are concentrated among the higher types, testimonial humility among the lower, and together they sustain productive inquiry. Only in the special case where the very lowest type mixes between individual and social learning do we see a collapse back to the paradoxical outcome. In that scenario, discovery is once again hampered, not because of an absence of virtue at the top, but because of vice or miscalibration at the bottom.

The broader lesson is that evolutionary dynamics need not doom communities to epistemic stagnation. Where virtues are properly distributed—competence coupled with diligence, incompetence coupled with humility—the division of labor emerges organically and sustains inquiry. But when vices proliferate, especially among those least equipped to bear the costs of investigation, equilibrium outcomes slide back toward inefficiency. The epistemic state of nature, in other words, is habitable only to the extent that discovery and testimonial virtues outweigh the temptations of free-riding, arrogance, and laziness.

10.5.1 *Evolutionary Model*

Let's now consider this scenario with N different types. Let Q be the vector $(q_1, q_2,..., q_N)$, where $\sum_{i=1}^{N} q_i = 1$ (that is, the proportion of all competence levels sums to one) and q_i is the proportion of the population that is type i. Likewise, we introduce a competence vector, $S = (s_1, s_2,..., s_N)$, where the probability that type i engages in successful individual learning is s_i, and for all i and j, $s_i < s_j$ if $i < j$. That is, S arranges types from least to most competent. Finally, we assume that $s_N = 1$; that is, individual inquiry of the most competent type is always successful.

Like in the previous section, a strategy is a vector specifying how to learn for each type. As in the previous section, the model proceeds as follows: in each round, individuals have the opportunity to engage in individual learning and then are randomly paired with a counterpart.

Those individuals who did not learn individually then have the opportunity to learn socially from their counterparts. There are a plethora of possible equilibria. We do not specify them all, but instead outline general features common to all the equilibria and discuss those arrangements that sidestep Rogers' paradox.

First, note that, similar to what we saw in the previous section, it will always be the case that the most competent types will engage in (some amount of) individual learning. Moreover, if type i engages in individual learning at equilibrium, then type j will always learn individually at equilibrium so long as $j > i$ (that is, j is more competent than i).

Proving this is straightforward enough. First, note that at equilibrium the payoff to social learning is independent of the agent's type. That is, an individual always gets the same expected payoff to social learning regardless of their competence level. Of course, the *opportunity cost* of social learning varies from type to type (some are better than others at individual learning), but the expected payoff to social learning is constant across all types.

Let π_{SL} be the expected payoff to social learning at equilibrium. The exact value of this, of course, depends on the proportion of individuals who engage in individual learning and the competence of those who learn on their own. If type i engages in individual learning at equilibrium, then it must be the case that $s_i - c \geq \pi_{SL}$. That is, for the arrangement to be stable, type i must do at least as well learning individually as they do socially. Now consider type j, where $j > i$. The payoff to type j from individual learning is $s_j - c$, which by assumption is greater than the $s_i - c$, the payoff to individual learning for type i. Thus, if type i does at least as well as learning individually as learning socially, then a more competent type (type j) will prefer to learn individually.

Likewise, it is pretty straightforward to show that if type k learns socially at equilibrium, then type l will also learn socially at equilibrium so long as $k > l$. The payoff to those type k's who learn socially is simply π_{SL}, and thus for the stated arrangement to be stable, it must be the case that $s_k - c$ is less than or equal to π_{SL}. Because $s_l - c < s_k - c$, it will also be the case that $s_l - c < \pi_{SL}$, meaning at equilibrium type l will never want to learn individually.

In summary, if there are N types, then the stable equilibria will involve a continuous segment of "low-types" who learn socially and a continuous segment of "high-types" who learn individually. There may exist some intermediate type that mixes between the two learning styles, as we saw in the previous section.

Where does this leave Rogers' paradox? As mentioned in the previous subsection, Rogers' paradox is *almost always avoided.* There is only one arrangement where the average payoff at equilibrium is payoff equivalent to the scenario where everyone learns individually. This occurs when all types learn individually, with the exception of the lowest type.

Specifically, these least competent individuals mix between individual and social learning. Because they mix, it is the case that the average payoff for these lowest types is just $s_1 - c$, the payoff to individual learning. And because everyone else learns individually, at equilibrium, social learning provides no payoff bonus to the individual or the group.

10.6 Discussion

Our analysis has emphasized how evolutionary theory can illuminate not only the familiar terrain of moral virtue and vice but also epistemic character. Traits such as diligence, humility, and fairness can be understood as discovery- and testimonial-side virtues that stabilize productive inquiry, while laziness, arrogance, and prejudice function as epistemic vices that entrench inefficiency. The models considered here demonstrate that, under some circumstances, evolution by itself is sufficient to ensure a desirable epistemic outcome. When competence and confidence are well calibrated, the evolutionary dynamics channel discovery to those best equipped to bear its costs, while others defer appropriately. In such cases, the epistemic state of nature is sustainable without additional scaffolding. But as Rogers' paradox makes clear, evolution does not always deliver efficient equilibria. Left to their own devices, populations may stabilize at states in which too many agents shirk the burdens of inquiry, or in which testimony is starved of content. In these settings, vice proliferates at the expense of virtue, and epistemic practices stagnate. The lesson is that when evolutionary pressures falter, successful inquiry depends on mechanisms of coordination beyond the individual level. Communities must cultivate informal norms that reward discovery-side virtues, or construct institutions that structure credibility judgments and distribute the burdens of inquiry more fairly.

This points toward a natural extension of the present project: the study of epistemic institutions.[8] Institutions—be they scientific communities, universities, journals, or funding bodies—can correct for the failures of evolution in the epistemic state of nature. They can provide incentives for discovery, safeguard against prejudice in credibility judgments, and channel division of labor into productive forms. By combining evolutionary insights with institutional analysis, we can better understand both the promise and the fragility of our epistemic practices. If testimonial virtues in fact stabilize the circulation of knowledge, our results show that discovery virtues are equally indispensable for its production.

Notes

1 See also work by Kevin Zollman (2018) and Weisberg & Muldoon (2009).

2 There is now a somewhat sizable literature attempting to "dissolve" the paradox. For just one recent example, see Bruner (2025a).

3 To be precise, the group maximizing the proportion of asocial learners is $1 - c/2$. The proportion of individual learners at the evolutionary equilibrium is $1 - c$. Thus, evolution produces an excess of epistemic free-riders.

4 A thriving literature has emerged in the wake of Rogers' original paper. For instance, a number of proposed resolutions to Rogers' paradox have now been explored. Many of these modify the learning environment by adding new learning strategies or altering the structure of interactions. Some researchers introduce more sophisticated learners—such as critical or conditional social learners—that fare better than naïve social learners and can increase average fitness at equilibrium (Ehn & Laland, 2012; Enquist et al., 2007). Other proposals focus on population structure, showing how assortment, group structure, or correlated encounters can influence the relative success of individual and social learning (Bruner, 2025a; Kobayashi & Wakano, 2012; Rendell et al., 2010). The model developed by Kharratzadeh et al. (2017) is closest to the present analysis. They examine hybrid learners who rely on individual learning along some environmental dimensions and on social learning along others, leading to a division of epistemic labor. Their account, however, does not depend on any underlying differences in competence across agents. Rather, specialization is simply an efficient way to handle multidimensional environments. To our knowledge, no existing model considers the form of heterogeneity explored here, where agents differ in their intrinsic reliability at individual inquiry and where evolutionary dynamics determine how learning burdens are distributed across competence types.

5 Yet as a referee helpfully pointed out, not all cases involving an under-investment in inquiry are cases of epistemic vice. One might under-invest in certain kinds of inquiry one is especially poorly placed to conduct and instead evince discovery virtues in other areas of inquiry where one has expertise.

6 We assume $s > 0.5$ and also that both kinds of agents correctly know their type. This, of course, is a somewhat strong assumption which we would like to relax in follow-up work.

7 Individuals have a fixed type but nonetheless use a conditional strategy which instructs them how to behave in both scenarios. This conditional strategy is what evolution acts on.

8 Much of the work cited in the introduction on the distribution of cognitive labor can be seen as a study of the institution of science. More recent work by social epistemologists such as David Henderson and Peter Graham (Bruner, 2025b; Graham, 2020; Henderson & Graham, 2017) has brought into the discussion informal institutions (social norms and conventions).

References

Bruner, J. P. (2025a). A note on the evolution of social learning. *Biology & Philosophy*, *40*(3), 11.

Bruner, J. P. (2025b). Assertions: Deterrent or handicap? A reply to Graham (2020). *Episteme*, *22*(1), 346–356.

Craig, E. (1991). *Knowledge and the state of nature: An essay in conceptual synthesis*. Clarendon Press.

Ehn, M., & Laland, K. (2012). Adaptive strategies for cumulative cultural learning. *Journal of Theoretical Biology*, *301*, 103–111.

Enquist, M., Eriksson, K., & Ghirlanda, S. (2007). Critical social learning: A solution to Rogers's paradox of nonadaptive culture. *American Anthropologist*, *109*(4), 727–734.

Fricker, M. (2007). *Epistemic injustice: Power and the ethics of knowing*. Oxford University Press.

Graham, P. J. (2020). Assertions, handicaps, and social norms. *Episteme*, *17*(3), 349–363.

Henderson, D., & Graham, P. (2017). Epistemic norms and the "epistemic game" they regulate: The basic structured epistemic costs and benefits. *American Philosophical Quarterly*, *54*(4), 367–382.

Kharratzadeh, M., Montrey, M., Metz, A., & Shultz, T. R. (2017). Specialized hybrid learners resolve Rogers' paradox about the adaptive value of social learning. *Journal of Theoretical Biology*, *414*, 8–16.

Kitcher, P. (1990). The division of cognitive labor. *The Journal of Philosophy*, *87*(1), 5–22.

Kobayashi, Y., & Wakano, J. Y. (2012). Evolution of social versus individual learning in an infinite island model. *Evolution*, *66*(5), 1624–1635.

Rendell, L., Fogarty, L., & Laland, K. N. (2010). Rogers' paradox recast and resolved: Population structure and the evolution of social learning strategies. *Evolution*, *64*(2), 534–548.

Rogers, A. R. (1988). Does biology constrain culture? *American Anthropologist*, *90*(4), 819–831.

Strevens, M. (2003). The role of the priority rule in science. *The Journal of Philosophy*, *100*(2), 55–79.

Weisberg, M., & Muldoon, R. (2009). Epistemic landscapes and the division of cognitive labor. *Philosophy of Science*, *76*(2), 225–252.

Williams, B. (2010). *Truth and truthfulness: An essay in genealogy*. Princeton University Press.

Zollman, K. J. S. (2018). The credit economy and the economic rationality of science. *The Journal of Philosophy*, *115*(1), 5–33.

11 Virtues and the Moral Division of Labor*

Ron Mallon

A division of labor is a crucial component of humanity's astonishing ability to flourish in larger and larger groups. In a famous passage, Adam Smith writes of pin-making that:

> One man draws out the wire; another straights it; a third cuts it; a fourth points it; a fifth grinds it at the top for receiving the head; to make the head requires two or three distinct operations; to put it on is a peculiar business; to whiten the pins is another; it is even a trade by itself to put them into the paper; and the important business of making a pin is, in this manner, divided into about eighteen distinct operations. ... Ten persons ... could make among them upwards of forty-eight thousand pins in a day. Each person, therefore, making a tenth part of forty-eight thousand pins, might be considered as making four thousand eight hundred pins in a day. But if they had all wrought separately and independently, and without any of them having been educated to this peculiar business, they certainly could not each of them have made twenty, perhaps not one pin in a day.
>
> (2007, p. 15)

Such specialization involves the acquisition of specific skills and specific knowledge, and it can be shocking to reflect on the extent to which such knowledge is distributed in social networks in a nonoverlapping way. The anthropologist Edward Hutchins (1996) notes the way in which the knowledge to operate a large naval ship is distributed throughout the ship, with no one person understanding every aspect needed to control the ship.

* Acknowledgments: I am grateful to Maria Merritt, Shaun Nichols, Jessie Sun, Isaac Wiegman, and to audiences at the Moral Psychology Research Group Meeting at Cornell University in October 2022, at the University of Missouri, and at the 2024 Central APA meeting for critical discussion of this work.

DOI: 10.4324/9781003632641-15

And the philosopher Hilary Putnam (1975) has noted the way that even our ability to refer using natural kind terms depends upon a division of linguistic labor.

I take it that divisions of physical, epistemic, and linguistic labor are ubiquitous, resulting in specialization of skills and knowledge throughout complex social systems. Here, I explore the possibility of a division of *moral* labor that may (as in other domains) bring with it a division of (in this case moral) aptitudes, perceptions, knowledge, dispositions, and capacities.

"It takes all kinds to make a world" is a truism, and perhaps it is even true. It seems commonsensical that certain moral dispositions suit one better or worse to various professions so that, for instance, compassion may be a virtue for caregiving professions while a degree of cold impartiality may suit the head of a commercial firm or a judge charged with deciding on the basis of written law. Might such diverse moral dispositions fit together to form some sort of *moral ecology*?[1]

11.1 Empirical Questions

However compelling one might find the idea of a moral ecology at the outset, it is in substantial part a descriptive hypothesis with empirical commitments. Here are three questions it raises:

i Is there, in fact, variation within human communities in the traits and capacities that result in variation in our judgments and behaviors in moral domains?
ii Might such variation not be simply the result of noise, or of random deviation from a species-typical norm, but rather be sustained by law-like processes that support continued variation?
iii Might such diversity be in some ways beneficial to us as individuals or as members of a community?

While there are many ways of approaching questions of moral capacities, here I focus upon those capacities that underlie the sorts of traits that in ordinary English and philosophical tradition are called virtues and vices, traits like compassion, courage, diligence, cruelty, cowardice, and carelessness.

Beginning at least with Confucius and Aristotle, the tradition of virtue ethics has held that virtues are something like traits of a person that are manifested in responses to a situation, and this remains the dominant view.[2] In recent decades, a disparate scatter of philosophical accounts of virtues has been articulated, including views on which virtues are motivations, skills, perceptions, or emotionally constituted.[3]

For our purposes, what matters is that on most accounts, virtues are assumed to be (or be in part) relatively stable properties of individual persons that systematically exert influence upon their behavior in various virtue-relevant situations, and we understand such properties to be behavioral dispositions where this is understood to be broadly inclusive of attitudes, traits, dispositions, motivations, skills, perceptions, or emotional biases.

Considered in this way, the answer to the first question seems obvious. It seems a conspicuous and ordinary fact that there is variation in our moral judgments and behaviors within our communities (variously construed) and that the best explanation of this is that individuals have different moral capacities and dispositions that give rise to these judgments and behaviors. For our purposes, it is irrelevant whether these differences in disposition emerge in the course of development as experiences gradually mold our nervous systems, or whether they are the result of genetic variation or some impossibly complex entangling of both. The point is simply that differences in dispositions to judge and behave in morally relevant situations – different "moral personalities" – apparently exist.[4] Below, we review a sample of empirical evidence in support of this claim.

The second question asks about how to understand such variation. In both moral philosophy and empirical and experimental moral psychology, it is often assumed that human morality exhibits, in key aspects, something like species-typical universality.

In moral philosophy, for instance, we might consider judgments that are "platitudes" or "moral fixed points" as a way of understanding the moral domain (Bengson et al., 2024; Jackson, 1998); disagreement with such fixed judgments would be a sign of some mistake, inability, or conceptual confusion. Or we might consider the capacities and nature of the virtuous person (e.g., Hills, 2015; Jones, 1999).

In empirical and experimental moral psychology, despite rising attention to the importance of individual and cultural differences (e.g., Graham & Yudkin, 2022; Henrich et al. 2010; Saxe & Young, 2011), much work continues to seek "moral universals" in moral judgments or capacities. For example, we might look for typical moral judgments that most people think that it is permissible to switch a runaway trolley onto a side-track, killing one in order to save five (Petrinovich et al., 1993). In some cases, an explanation of divergence from the modal or majority view is often not considered at all. Others seek to understand standing features of mental architecture that underlie our moral judgments, behavior, and experience, for example, by positing a species-typical moral architecture (Gray & Pratt, 2025; Mikhail, 2011). In short, a great deal of work tends, at least implicitly, to treat deviation from an ideal type or from a central trend, as a kind of "noise" or as a kind of failure to fully realize the species-typical capacity.

But given what we know about human variation and biological variation more generally, it is plausible to think that moral dispositions *within* human communities may be variable or *polymorphic* (exhibiting more than one form) and that such polymorphisms are not simply an accident, but are driven by law-like processes that sustain and reproduce such variation. The possibility of human polymorphism, in the context of moral psychology (Dean, 2012; Mealey, 1995), personality psychology (Buss, 2009; Nettle, 2006), and more broadly, in the context of considering whether the mind has an evolved, species-typical structure (e.g., Kitcher, 1986), has been raised before. Here, I highlight and extend the discussion of this possibility.

Answers to our third question are more speculative still. If it is true that moral capacities are polymorphic in the sense mooted here, one explanation is that the particular capacities we have are in some way beneficial for ourselves within a community or to our group in its context. And one way such diversity could be beneficial is if it amounts to a division of moral labor, a division in which people are in some way adapted to different (perhaps mutually) beneficial roles in a larger society. My proposal is that we understand our moral dispositions as emerging as part of a moral ecology in which we are co-determined by those around us and by our environments and that doing so may have positive implications for how we view our moral opponents in cases of disagreement and conflict. I return to these conjectures below.

11.2 Illustrating a Moral Ecology

It may be helpful to have simple illustrations of how such a moral ecology might operate. Here are two.

11.2.1 Strict and Permissive Parents

Zado is a teenager with two loving parents. One of them, Strict, is a disciplinarian, a real hard ass. The other, Permissive, is very easygoing, a sympathetic softie. Suppose that:

- Sometimes Zado would do things that are dangerous or harmful, except for concern about the repercussions from Strict. In these cases, Strict's parenting keeps Zado safe.
- Other times, Zado avoids taking risks that carry just a small amount of danger but would lead to personal growth, also because Strict forbids them. In these cases, Strict's parenting inhibits Zado's personal growth.

- Sometimes, Zado does things that are dangerous or harmful, with knowledge that Permissive will take their side, and they won't get in much trouble. In these cases, Permissive's parenting led Zado to greater danger.
- Sometimes, Zado does things that carry just a small amount of danger but would lead to personal growth, with knowledge that Permissive will take their side, and they won't get in much trouble. In these cases, Permissive's parenting led Zado to greater personal growth.

This scenario exhibits one sort of moral ecology I have in mind; I imagine that Zado could benefit from having both a strict and a permissive parent, since, in some way, they balance out one another's more extreme tendencies, and each parent benefits since they value ZADO's flourishing. I suspect that pedagogical environments, in general, benefit from variations in personality of this sort since they allow the possibility of achieving a good "fit" among persons.

11.2.2 A Soldier, Entrepreneur, and Artist

Now consider specialization associated with occupational choice.

Three high school friends graduate and go their separate ways. Later, they catch up at their 30th High School Reunion.

- After graduation, Red became a soldier who served his country in the military. On two occasions, he was called upon to fight in military conflicts, and after retiring from the military, he joined his local police force. He is courageous and loyal, and he tends to see the world in simple terms. He enjoys the products and services entrepreneurs and artists provide.[5]
- Megan became an entrepreneur and businessperson who is good at creating businesses that efficiently provide goods to people while employing others. She is not especially courageous or loyal, but she is good at calculating risks and rewards, and she is good at empathizing with people, a skill that helps her understand her employees and customers. She pays taxes that support the military and police whose protection she enjoys. And she is inspired by and supports the arts.
- Robin grew into an eccentric individualist and an artist. They are terrible with money and also not especially courageous or loyal, but they create beautiful, imaginative artworks with a distinctive vision that inspires others, including Red and Megan.

These illustrations, like many toy models, are wild simplifications of contemporary situations. But the idea that distinctive psychological capacities and emotional skills are demanded by and adaptations to various social roles and occupations is familiar enough. For example, an emotional division of labor (where those emotions are themselves grounds of moral dispositions) has also long been part of the conception of and justification of gendered divisions of labor in the home and workplace. And it forms the core of the sociological theory that varied social positions and occupations require distinctive sorts of emotional labor (Hochschild, 1983).

The suggestion of a moral ecology we put forward here contrasts with a sort of Ant and Grasshopper fable that holds up a single moral personality as an ideal for all. In Plato's account of a division of labor in *The Republic*, for instance, he imagines distinctive classes of people, each with their own characteristic character traits (Shorey, 1961). The wisest and most virtuous become rulers, and others become soldiers, craftspersons, and workers. As I read him, Plato imagines moral varieties, but each located along a continuum of perfection (most closely realized by the philosopher rulers). In contrast, here I suggest that there could be multiple different sorts of virtuous personality and that individuals in a community, or even the community as a whole, benefit from this variety.

11.3 Empirical Evidence of Within-Group Morally Relevant Trait Diversity

Here, I review some suggestive evidence that there are varied moral dispositions within groups.

One sort of evidence can be found by looking at Moral Foundations Theory in social psychology.[6] Moral Foundations Theory holds that human judgment and behavior in moral domains are guided by a plurality of distinct underlying values including Care, Fairness, Loyalty, Authority, and Purity and that we can understand human moral commonality by appeal to these shared values, and human variation by appeal to different weightings or activations of these values between individuals and groups (Graham & Yudkin, 2022). So, for example, Hawkins et al. (2018) found that responses to an array of moral questions designed to elicit underlying beliefs and commitments (including commitment to various moral foundations) allowed them to cluster their US respondents into seven different political groups and that participants' commitments to moral foundations were correlated with political beliefs. For our purposes, the key point is that Moral Foundations Theory holds that underlying moral diversity in judgments and behaviors is a diversity of specific dispositions to evaluate that give rise to such judgments and behaviors.

Other lines of research also suggest moral diversity within populations. For example, it seems that disgust sensitivity is correlated with greater social conservativism, especially on "purity"-related issues like abortion or gay marriage (Inbar & Pizarro, 2022), suggesting that political judgments are influenced by emotional dispositions.

And other research suggests that those who tend to rely more on deliberation are more likely to support utilitarian judgments that seek to maximize the greatest good for the greatest number than those who rely on intuition (Bartels, 2008). Relatedly, some experimental philosophical work offers evidence that personality traits bias philosophical judgments, including specifically judgments in moral philosophy. For instance, Adam Feltz and Edward Cokely offer evidence that people high in the trait of emotional stability are more likely to attribute virtue to others and prefer actions done from virtue. They conjecture that this is because those high in emotional stability prefer less rule-based approaches to the ethical domain (Feltz & Cokely, 2016, p. 582).

This is just a smattering of evidence to corroborate the everyday sense that people have different moral dispositions within communities. What would it mean for these dispositions to be polymorphic?

11.4 Phenotypic Polymorphisms

So far, I've suggested that virtues and vices are appropriately understood as those behaviorally relevant dispositions that we aim to pick out with common lexical terms for virtues and vices and that such traits typically exhibit diversity within communities or populations.

Might such trait variation amount to a *phenotypic polymorphism?* In broad terms, "polymorphism" simply means many (or at least more than one) form. "Phenotypic" polymorphisms are distinguished from *genotypic* polymorphisms – variations in specific alleles between individuals or groups (that may occur within or between biological populations). Phenotypic polymorphisms, instead, are the developed traits of an organism that result from the interaction of genes and environment, and phenotypic polymorphisms can be produced by genetic variation, environmental variation, or both. Textbook examples in humans are the male and female sexes and various human blood types. Here, I use the term quite broadly to capture the broad class of systematic, recurring differences in a population's phenotypic traits.

Sometimes the term "polymorphism" is used very loosely to characterize nearly any phenotypic differences among individuals within a species. But I am arguing for the possibility of moral polymorphisms that meet two conditions:

P1. Polymorphisms exist within a community or population: Varied traits are co-present among individuals within the same group occupying the same habitat at the same time. So, differences within a population over evolutionary time, say, or differences that arise from geographic separation of conspecifics that form distinct populations are not phenotypic polymorphisms in this sense.

P2. Polymorphisms are nonaccidental: They exist because of various regular processes. In particular, polymorphisms may be sustained in a population because they adapt an organism or a group to one another or to an environment.[7]

P2 requires some unpacking. Here, I am especially interested in trait diversity that is stabilized over a relevant time scale by regular processes, including processes such as balancing selection or group selection governed by law-like generalizations, and I mean to include polymorphisms that stem from different sorts of causes (phylogenetic or ontogenetic), with attendant differences in time scales.

One classic question to consider when faced with variation in a population is why it exists at all. If a trait or behavior is beneficial to have or perform, why doesn't everyone have it or perform it? One answer to this sort of question is that the trait is a polymorphism which adapts its possessor to its circumstances, and different individuals face different circumstances.

In biology, some such processes are captured by the idea of *balancing selection*, which includes both *frequency-dependent selection* and *selection in a changing environment* (Buss, 2009).

Frequency-dependent selection occurs when there is selection for or against a phenotype in a population, conditional upon the relative frequency of other phenotypes in the population at that time. Such selection operates on individual organisms against the background distribution of phenotypes among other organisms in the community or population.

John Maynard Smith and George Price (1973) illustrated this while demonstrating the possibility of individual selection upon aggressive behavior in what is now often called a "Hawk-Dove" evolutionary game-theoretic model (Maynard Smith, 1982). Consider a population that competes for resources using one of two strategies: being a "Hawk" (always fighting) or being a "Dove" (sharing with other Doves but always retreating from a fight). Expressed in game-theoretic terms, a round of the payoffs might look like this (see Table 11.1).

Table 11.1 A round of Hawk & Dove

One Round of Hawks and Doves		*Player 2*	
		Hawk	*Dove*
Player 1	Hawk	–25, –25	100, 0
	Dove	0, 100	50, 50

Here, the idea is that Hawk-Hawk interactions, in which two Hawks fight one another over a resource, are typically so costly that the value of the reward is outweighed. However, a Hawk encountering a Dove will do well, taking away all the resources without a fight.

Now, if we imagine the game repeated over time in an evolutionary context, strategies that receive higher payoffs in one round will be more evolutionarily successful in the next round. In such a case, a population of mostly Doves (i.e., of players that exclusively play "Dove") will be overtaken by Hawks (i.e., those that exclusively play "Hawk") since Hawks always do better in direct competition with Doves. But the proliferation of Hawks will be checked since the more Hawks there are, the more likely they are to compete with other Hawks rather than Doves, leading to a declining – and ultimately negative – payoff.

Crucially, there is a point at which the population of Hawks and Doves becomes balanced so that a move in one way (more Hawks) reduces the payoff to Hawks, and a move in the other (more Doves) would increase the payoff to Hawks. The distribution of Hawk and Dove phenotypes is at an equilibrium point in the model, amounting to an "evolutionarily stable state" – stable in the sense that the system would tend to return to it if the system were perturbed (Maynard Smith, 1982).[8]

A different sort of balancing selection results from a changing environment. Consider a climate that oscillates between very wet years and very dry years. Suppose wet years favor a phenotype adapted for moisture, and dry years one favored for aridity, and that these two adaptations are incompatible. If the environment was always wet, the wet-adapted phenotype would be selected for and entirely displace the dry one, and mutatis mutandis for the dry-adapted. When the environment varies, however, a balance between two different phenotypes over long periods of time may emerge, as sometimes selection favors one, and other times favors the other. Returning to Hawks and Doves, if the environment oscillates so that sometimes the payoffs to Hawks are higher (perhaps because the available resources are easier for an individual to control and exploit) and other times lower (because the resources require cooperation to control and exploit), then we can see how changing environmental circumstances might also support sustained diversity of morally-relevant dispositions.

Both sorts of selection operate on individual organisms, where payoffs to an organism's fitness are mediated by the frequency of phenotypes in the population. Another possible source of polymorphism is *group* selection, the idea that a group might possess a trait that enables the group to survive or flourish relative to other groups. In the present case, the idea would be that a group ***P*** with a particular distribution of phenotypes (P_1, P_2, ..., P_n) might be more successful than a *monomorphic* group with only one phenotype (P_m), or than an alternative polymorphic group with a different distribution of phenotypes (P_1*, P_2*, ..., P_n*).

Why might this be? There could be a variety of causes. Ants, for example, are specialized in ways that allow a division of labor that benefits an entire colony. But while biological group selection is uncontroversial among social insects like ants that are highly interrelated, a possible role in human genetic evolution remains controversial because of greater competition among genetically distinct individuals within groups.[9] At the same time, much recent attention has been paid to the idea that group selection on cultural groups has shaped human morality, at least in our recent evolutionary past (e.g., Richerson & Boyd, 2005; Tomasello, 2016).

In short, while we have been borrowing ideas from evolutionary biology, similar ideas may play a role in the course of individual or group development on smaller time scales. If, in the course of development, an individual's learned dispositions are shaped by the frequency of other exhibited dispositions in their environment or if the manifestation of dispositions is shaped by oscillating environments or by group success, then some form of selection may shape the development or acquisition of adult phenotypes that are adapted to one's environment.

11.5 Moral Polymorphisms in Humans?

The idea that variation in the traits that give rise to moral judgments and behaviors in humans is polymorphic remains a working hypothesis, and in this section, I consider a couple of recent proposals of human moral polymorphisms and then consider what the suggestion means for moral psychology more generally.

11.5.1 Sociopathy/Psychopathy

Linda Mealey (1995) provocatively suggests that sociopathy (which she uses in a way that encompasses both the range of anti-social personality disorder and psychopathy) may be a polymorphism in human populations. Mealey's idea is that, given the prosocial and cooperative nature of human communities, there is a niche for those who would exploit trust and cooperation in order to cheat others. Just as Hawks invade a Dovish community, sociopaths may adapt to a cooperative and trusting community by exploiting it, extracting the benefits of prosociality without paying the costs. The ability of sociopaths to invade a community, however, is limited. Too many sociopaths result in increased vigilance against them as well as an increased chance of their encountering other sociopaths. So, as in the Hawk-Dove game, a certain but limited level of sociopathy could be, according to Mealey, an evolutionarily stable state.

Sociopaths are characterized by a variety of moral differences, including reduced or absent empathy or concern with harm to others, and this shows up in increased social conflict as well as legal troubles resulting from disregard for the law. But if Mealey's conjecture is correct, it represents a distinctive adaptation to group living.[10] Sociopathy, thus understood, would be a moral polymorphism sustained by frequency-dependent selection either on genes or upon ontogenetically developing personalities within a community, and in particular, it would be a polymorphism that underlies a kind of moral viciousness or lack of virtue since sociopaths plausibly possess a moral deficit or impairment of some kind – an inability to respond normally to morally relevant facts.

So, on the one hand, Mealey's case provides a plausible account of a moral polymorphism, but on the other, it reminds us that we cannot infer from the fact that a disposition may be adaptive and functional in some evolutionary sense to its being a good or right disposition to have in a moral sense.

11.5.2 Personality Traits

Looking elsewhere, it seems that the closest analogues to virtues and vices in contemporary psychology are personality traits, and this suggests that what we have learned about such traits may illuminate virtues and vices.[11] Whatever slippage there may be between the philosophical virtues and psychological character constructs, character traits as studied by personality psychologists look to be the best candidates for psychological dispositions that may underlie virtues and vices in the philosophical sense.

Might such personality traits themselves be polymorphisms? A number of authors have suggested this may be the case (Buss, 2009; Dean, 2012; Nettle, 2006; Price, 1969), noting that different personality traits and their opposites typically tend to reward behaviors in different situations and that varying social circumstances and changing environments suggest a role for balancing selection in sustaining personality diversity within a population.

For instance, Nettle (2006) argues that the trait of agreeableness (which is negatively correlated with anti-social personality disorder and positively correlated with an ability to empathize as well as avoidance of conflict and smooth interpersonal relationships) may carry downsides in contexts where it could lead one to trust others too much and be exploited. In short, the benefits (whether evolutionary or personal) of agreeableness (or its absence) likely depend upon specific features of the context, suggesting that it could be a polymorphic personality trait.

11.5.3 Toward a Population-Sensitive Moral Psychology

Psychologically, while the last three decades have brought a great deal more attention to the possibility of psychological variation among individuals and cultures in general, and in moral psychology in particular, little attention has been paid to the hypothesis of polymorphism of moral characters.

For instance, defenders of Moral Foundations Theory tend to regard evaluative differences as providing a common explanation of moral and political similarity and diversity both within and across various populations (Graham & Yudkin, 2022). In contrast, Tim Dean has noted that polymorphism proffers the possibility of a *different* explanation of variation within a group and variation among groups (Dean, 2012). So, if the hypothesis of phenotypic polymorphism is true, then psychology could benefit from attending to populations as wholes, and to features of their specific environments, in attempting to understand and explain the psychological features of individuals (Ereshefsky & Matthen, 2005).

One emerging research program that does fit nicely with the hypothesis of polymorphism focuses upon the "niche diversity hypothesis." This hypothesis:

> holds that within human populations, individuals occupy different niches: micropopulations within a larger population with different affordances and cost–benefit structures (e.g., organizations, occupations, social and cultural groups, coalitions, families). For example, in human societies, some niches may favor high levels of patience and industriousness with low levels of anxiety but not specific levels of imaginativeness. Other niches may favor high levels of sociability and risk tolerance without specifically favoring any level of honesty. And some may not incentivize any particular levels or combinations of traits.
>
> (Durkee et al., 2022, p. 2)

It is important to remember that people do not merely conform to niches, but also select which niches to occupy and shape niches to their own needs, so the niche diversity hypothesis, like the idea of a moral ecology, suggests an enormously complex causal network.

11.6 Intimations of a Moral Ecology

The possibility of polymorphisms encourages us to imagine that individual moral characters may exist within a larger, communal context. That virtues might be polymorphic remains an empirical hypothesis, and I do not take the discussion here to have proven that it is true. But it is a hypothesis whose truth bears upon both descriptive moral psychology and moral philosophy and so is worth exploring further.

At the outset, we suggested that within-group diversity of moral personalities might be in some way beneficial to individuals or group members. In employing evolutionary ideas, we have implied that the benefit is to their (individual or group) differential fitness in some biological or cultural selective process, and we have considered some possibilities of such adaptations.

In what remains, I want to suggest other sorts of advantages of centering the polymorphic diversity of people in our thinking of the moral domain.

11.6.1 Moral Disagreement and Moral Realism

While our focus here is on virtues rather than attitudes or beliefs, the most common discussion of moral diversity in ethics occurs in meta-ethical debates over moral realism. Many anti-realists have taken moral diversity in moral beliefs or judgments to be an argument against realism, and many realists have also taken it to pose a challenge since it undermines the sense that we perceive or converge upon agreement about a common moral reality.

Insofar as one takes moral judgments about cases to be key evidence for the truth of moral claims, variation in those judgments that are produced by differences of moral personality does seem to pose a prima facie challenge to the idea that, over time, we are gradually converging on the moral truth. There are, however, ample resources to sustain such moral realism in the face of such diversity. On the one hand, the moral realist can attempt to explain away apparent variation as the result of different background beliefs or information about the cases in question, or about an incomplete grasp of the relevant moral concepts. On the other, the moral realist can simply deny that variation in people's judgments in the moral domain bears on the question of whether the facts in that domain are the sorts of mind-independent, objective facts the realists suppose them to be (Doris & Plakias, 2008).

For present purposes, it is perfectly possible to agree with what I have said here about variation among moral personalities while maintaining that there are moral or ethical judgments that are genuinely true and false (whether these be judgments about general moral facts or about particular acts or specific persons). For similar reasons, there is nothing I have said that precludes the possibility of genuine moral disagreements in which one party is in the right and the other is in the wrong.

It could be, for instance, that in any specific case requiring parental guidance, Strict or Permissive has the true judgment about the best way of parenting Zado, whether it be whether a behavior should be allowed, how a transgression should be punished, or what a general rule ought to be. Similarly, it could be that, in any particular case, the soldier, the entrepreneur, or the artist has the correct view about the appropriate rate of taxation, the need for military action, or the morality of permitting abortion. Alternatively, it could be that only by compromising with others on a particular occasion do our agents arrive at the objectively right answer so that each, individually, is morally wrong, but the decision resulting from compromise is morally correct.

11.6.2 Incompatible Virtues

This raises the question: why not simply possess all the virtues required for wise and appropriate action in each situation? Aristotle notably maintained that to possess one virtue required possessing them all (*the unity of the virtues*), a thesis generally regarded as implausible nowadays.

In fact, some have wondered if co-possession of some virtues is even possible. Can a person who readily experiences righteous indignation at injustice also readily exhibit sympathy, compassion, and forgiveness toward those who commit it?[12] Can a person who is meticulous about following the rules also be carefree in harmlessly disobeying unimportant, counterproductive rules whose purpose is now past? Can the parent, coach, or dissertation advisor who is demanding the best performance from those they oversee also be accommodating, charitable, and supportive in the same instance?

In psychology, personality traits are theorized as located on polar continuums with opposite ends, and they are said to represent stable dispositions to respond to situations over time; this body of work thus suggests the nomological inconsistency of simultaneous possession of some contrasting personality traits. And in virtue ethics, a congruous suggestion has sometimes been made that "certain pairs of virtues can no more flourish in the same personality than different species of tree can flourish in the same soil" (Walker, 1989, p. 352; cf. Flanagan, 1991; Foot, 1978 for similar suggestions).

These considerations suggest:

> *The Nomological Incompatibility of (some) Virtues:* Possession of some virtuous traits is, at least for creatures like us, nomologically incompatible with possessing some others.

Why would simultaneous possession of traits or virtues be impossible? One possibility is that developing a trait or the virtues that it grounds requires possession of a certain developmental resource (like a gene or an environment) that is incompatible with the possession of the developmental resource required for a different virtuous trait. Another possibility is that the development of a virtue requires engagement with a certain developmental pathway inconsistent with the pathway required for an alternate trait or virtue. Such path dependence fits nicely with the Aristotelian idea that virtuous traits might be acquired via habituation or training of some sort (Annas, 2011; Barnes, 1984).

In short, the hypothesis of moral polymorphism implies not only that the unity of virtues is implausible but also that it is impossible, at least for creatures like us.

11.6.3 The Incompleteness of Atomistic Virtue

One thing the idea of virtues as polymorphisms sustained by law-like processes adds to the idea of their incompatibility is that dispositional differences emerge in a context of causal interdependence of people with morally relevant character traits.

> *The Causal Interdependence of Inconsistent Character Traits:* The ability of one person or group to possess some virtue or vice depends causally upon the possession of some alternate, nomologically incompatible virtue or vice by others within their larger community.

If the hypothesis of moral polymorphism is true, then part of the causal explanation for our having the moral dispositions we do is that there are other people or groups with different, perhaps incompatible dispositions with whom we are historically and causally intertwined.

I close with the suggestion that this polymorphic perspective thus gives us a different way of thinking of, and appreciating, our counterparties in a moral disagreement.

In a disagreement, we think that we are right, and our interlocutor is wrong. Suppose we are right. In one sort of case, we may recognize that, though we are right about the matter at hand, past cases have carried a mixed verdict. Each of us has sometimes been right and sometimes wrong regarding important moral matters. In such a case, we can value our interlocutor's contribution to our ongoing ethical dialogues rather than regarding them as a morally benighted soul that must be converted to our view, sidelined, or ignored.

In a second sort of case of disagreement, we are persistently right, and our interlocutor (possessed of a different moral personality) is persistently (and perhaps even always) wrong. However, even in this case, we might nonetheless appreciate that we are only able to occupy the position we occupy because they have occupied the position they occupy in our larger community.

In *On Liberty*, John Stuart Mill influentially argues that engaging a variety of differing opinions enhances our own epistemic standing by allowing us, via reflection, to have greater justification for holding our own view (Mill, 1975). That is no doubt true. But the present proposal is that we ought to appreciate the *causal* role that opposing moral personalities may play in our own moral dispositions, and in our having the views we do. The point is not simply that engaging with such views offers us evidence for our own, but rather that other moral personalities, other constellations of character, may be causal prerequisites for the existence and sustainability of our own.

This sort of causal interdependence is not an intuitive thought for many, though the Buddhist idea of interdependent arising (pratītyasamutpāda) or "interbeing" gets at the spirit: one on which we acknowledge a non-individualistic, relational conception of our causal origins and ongoing changes to our characters.[13]

The idea of a moral division of labor offers one conception of how this might be. Large groups tend to differentiate themselves into increasingly specialized roles, producing a division of labor that promotes excellence and efficiency in performing the roles. And the roles may depend generatively upon one another. It would be, for instance, impossible to devote one's life to philosophy, religious study, or the arts if others were not devoting their lives to the numerous other jobs that make such specializations possible. If we think of (at least some) moral virtues as grounded in dispositions that are specializations for niches, and we think of such roles as causally depending upon the existence of other specializations to other niches, some involving virtues with nomologically incompatible grounds, and others perhaps even vices, then we have a context in which our ability to have our own virtues and whatever dispositions – perceptual, theoretical, moral, or practical – they ground or entail nomologically depends upon those whom we are morally different than, but nonetheless inextricably causally entangled with. Insofar as we value our own virtues and dispositions, it gives us some reason to value the different others that are their causal prerequisites.

Notes

1 I borrow this term from Dean (2014).
2 See Slingerland (2003) and Barnes (1984) for early views.
3 See Hursthouse and Pettigrove (2023) for an overview of the philosophical landscape.
4 Some situationists deny that character traits (including those that underlie ordinary talk of virtues and vices) exist in the way folk psychology and much

philosophical work suggest (Doris, 2002; Harman, 1999; Ross and Nisbett, 1991). However, these situationists do not deny that there is interpersonal variation in the dispositions underlying our judgments and behaviors. Rather, they insist that such dispositions do not have broad, explanatory scope across virtue-relevant situations (cf. Doris, 2002, 2022).

5 The idea that a community benefits from varieties of moral personality, and especially that the peaceful benefit from those willing to serve in war is not a new one. Perhaps its most common expression nowadays comes via quotation of the conservative Richard Grenier who, paraphrasing George Orwell, wrote, "people sleep peacefully in their beds at night only because rough men stand ready to do violence on their behalf" (Grenier, 1993). Opinions about the value and role of the military vary widely, but what Grenier (and Orwell) were suggesting was that, at least in some cases, even peaceful persons benefit from some who would engage in violence on their behalf. A similar sentiment is expressed in the famous "You can't handle the truth!" speech by Jack Nicholson in *A Few Good Men* (Reiner, 1992). These statements imply a one-way causal dependence of civilians upon members of the military. In contrast, here the proposal is of a thoroughgoing causal interdependence of moral personalities.

6 For a recent overview and introduction see Graham and Yudkin (2022).

7 It is often required that that phenotypic polymorphism be of different *types* rather than just different degrees so that they are discontinuous and categorical (rather than graded). Here, I do not require this since satisfaction of this additional condition plays no role in my argument.

8 Maynard Smith's use of "Hawk" and "Dove" evoked the moral-political dimensions of proponents and opponents of military action, and it is easy enough to imagine the idea as not merely a metaphor, but representing a possible distribution of morally relevant personality traits held in balance within a population by frequency-dependent selection.

9 See Sober and Wilson (1998) for explanation and defense of the possibility of more general group selection.

10 Mealey believes there are two paths to sociopathy, the first being largely genetically driven, and the second being the result of a gene-environment interaction, wherein sociopathic tendencies are activated by challenging or traumatic life experiences.

11 Indeed, since Harman (1999) and Doris (2002), arguing over the connection between the psychological constructs of personality psychology and philosophical posits of virtue theory has spawned a vast literature.

12 I am grateful to Maria Waggoner for discussion and references on this point, in person and in her (2023, Chp. 3).

13 "Interbeing" is a way of putting the idea popularized by Thich Nhat Hanh and the Plum Village community. For a brief introduction, see Holst (2021).

References

Annas, J. (2011). *Intelligent virtue.* Oxford University Press.

Barnes, J.(Ed.) (1984). Nichomachean ethics. In: *Aristotle. The complete works: The rev. Oxford Translation.* Princeton University Press.

Bartels, D. M. (2008). Principled moral sentiment and the flexibility of moral judgment and decision making. *Cognition, 108*(2), 381–417. https://doi.org/10.1016/j.cognition.2008.03.001.

Bengson, J., Cuneo, T., & Shafer-Landau, R. (2024). *The moral universe*. Oxford University Press.
Buss, D. M. (2009. How can evolutionary psychology successfully explain personality and individual differences? *Perspectives on Psychological Science*, *4*(4), 359–366. https://doi.org/10.1111/j.1745-6924.2009.01138.x.
Dean, T. (2012). Evolution and moral diversity. *The Baltic International Yearbook of Cognition, Logic and Communication*, *7*, 1–16. https://doi.org/10.4148/biyclc.v7i0.1775
Dean, T. (2014). *Evolution and moral ecology*. [Doctoral Dissertation, University of New South Wales]. https://unsworks.unsw.edu.au/entities/publication/e07240ec-9dc0-43e4-a3e6-777af55c4097
Doris, J. M. (2002). *Lack of character: Personality and moral behavior*. Cambridge University Press. http://www.loc.gov/catdir/description/cam0210/2002071504.html
Doris, J. M. (2022). *Character trouble: Undisciplined essays on moral agency and personality*. Oxford University Press.
Doris, J. M., & Plakias, A. (2008). How to argue about disagreement: Evaluative diversity and moral realism. In W. Sinnott-Armstrong (Ed.), *Moral psychology, vol 2: The cognitive science of morality: Intuition and diversity*. MIT Press.
Durkee, P. K., Lukaszewski, A. W., von Rueden, C. R., Gurven, M. D., Buss, D. M., & Tucker-Drob, E. M. (2022). Niche diversity predicts personality structure across 115 nations. *Psychological Science*, *33*(2), 285–298. https://doi.org/10.1177/09567976211031571
Ereshefsky, M., & Matthen, M. (2005). Taxonomy, polymorphism, and history: An introduction to population structure theory. *Philosophy of Science*, *72*(1), 1–21. https://doi.org/10.1086/426848.
Feltz, A., & Cokely, E. (2012). The philosophical personality argument. *Philosophical Studies*, *161*(2), 227–246.
Feltz, A., & Cokely, E. T. (2016). Personality and philosophical bias. In: *A companion to experimental philosophy* (1st ed.). John Wiley & Sons, Ltd.
Flanagan, O. (1991). *Varieties of moral personality*. Harvard University Press.
Foot, P. (1978). *Virtues and vices: And other essays in moral philosophy*. Oxford University Press.
Graham, J., & Yudkin, D. A. (2022). Variations in moral concerns across political ideology: Moral foundations, hidden tribes, and righteous division. In M. Vargas & J. M. Doris (Eds.), *The Oxford handbook of moral psychology*. Oxford University Press. https://doi.org/10.1093/oxfordhb/9780198871712.013.38
Gray, K., & Pratt, S. (2025). Morality in our mind and across cultures and politics. *Annual Review of Psychology*, *76*(1), 663–691. https://doi.org/10.1146/annurev-psych-020924-124236
Grenier, R. (1993). Perils of Passive Sex. *The Washington Times*, April 6. Access World News.
Harman, G. (1999). Moral philosophy meets social psychology: Virtue ethics and the fundamental attribution error. *Proceedings of the Aristotelian Society*, *99*(1999), 315–331.
Hawkins, S., Yudkin, D. A., Juan-Torres, M., & Dixon, T. (2018). *Hidden tribes: A study of America's polarized landscape*. More in Common. https://hidden-tribes.us/media/qfpekz4g/hidden_tribes_report.pdf
Henrich, J., Heine, S. J., & Norenzayan, A. (2010). The weirdest people in the world. *Behavioral and Brain Sciences*, *33*, 61–135.

Hills, A. (2015). The intellectuals and the virtues. *Ethics*, *126*(1), 7–36.
Hochschild, A. R. (1983). *The managed heart: Commercialization of human feeling*. University of California Press.
Holst, M. A. (2021). 'To be is to inter-be': Thich Nhat Hanh on interdependent arising. *Journal of World Philosophies*, *6*(2), 17–30.
Hursthouse, R., & Pettigrove, G. (2023). Virtue ethics. In E. N. Zalta & U. Nodelman (Eds.), *The Stanford Encyclopedia of Philosophy*. Stanford University. https://plato.stanford.edu/archives/fall2023/entries/ethics-virtue/
Hutchins, E. (1996). *Cognition in the wild*. MIT Press.
Inbar, Y., & Pizarro, D. A. (2022). How disgust affects social judgments. *Advances in Experimental Social Psychology*, *65*, 109–166.
Jackson, F. (1998). *From metaphysics to ethics: A defense of conceptual analysis*. Oxford University Press.
Jones, K. (1999). Second-hand moral knowledge. *Journal of Philosophy*, *96*(2), 55.
Kitcher, P. (1986). The transformation of human sociobiology. *PSA: Proceedings of the Biennial Meeting of the Philosophy of Science Association*, 1986, 63–74. https://doi.org/10.1086/psaprocbienmeetp.1986.2.192790
Maynard Smith, J. (1982). *Evolution and the theory of games*. Cambridge University Press.
Maynard Smith, J., & Price, G. R. (1973). The logic of animal conflict. *Nature*, *246*(5427), 15–18. https://doi.org/10.1038/246015a0.
Mealey, L. (1995). The sociobiology of sociopathy: An integrated evolutionary model. *Behavioral and Brain Sciences*, *18*(3), 523–541. https://doi.org/10.1017/S0140525X00039595
Mikhail, J. M. (2011). *Elements of moral cognition: Rawls' linguistic analogy and the cognitive science of moral and legal judgment*. Cambridge University Press.
Mill, J. S. (1975). On the liberty of thought and discussion. In D. Spitz (Ed.), *On liberty*. W. W. Norton & Company. (Norton Critical Edition)
Nettle, D. (2006). The evolution of personality variation in humans and other animals. *American Psychologist*, *61*(6), 622–631.
Petrinovich, L., O'Neill, P., & Jorgensen, M. (1993). An empirical study of moral intuitions: Toward an evolutionary ethics. *Journal of Personality and Social Psychology*, *64*, 467–446.
Price, J. S. (1969). The ritualization of agonistic behaviour as a determinant of variation along the Neuroticism/Stability dimension of personality. *Proceedings of the Royal Society of Medicine*, *62*(11 Pt 1), 1107–1110. https://doi.org/10.1177/003591576906211P117
Putnam, H. (1975). The meaning of 'meaning.' In: *Mind, language and reality: Philosophical papers* (vol. 2). Cambridge University Press.
Reiner, R. (1992). *A few good men* [Theatrical]. Columbia Pictures.
Richerson, P. J., & Boyd, R. (2005). *Not by genes alone: How culture transformed human evolution*. University of Chicago Press.
Ross, L., & Nisbett, R. E. (1991). *The person and the situation*. Temple University Press.
Saxe, R., & Young, L. (2011). Moral universals and individual differences. *Emotions Review*, *3*(3), 323–324. 10.1177/1754073911402383
Shorey, P. (trans.). (1961). Plato. "The Republic." In E. Hamilton & H. Cairns (Eds.), *The collected dialogues of Plato including the letters*. Bollingen Series, LXXI. Princeton University Press.

Slingerland, E. (Ed.). (2003). *Confucius: The analects*. Hackett.
Smith, A. (2007). An Inquiry into the Nature and Causes of the Wealth of Nations. MetaLibri. https://www.ibiblio.org/ml/libri/s/SmithA_WealthNations_p.pdf
Sober, E., & Wilson, D. S. (1998). *Unto others: The evolution and psychology of unselfish behavior*. Harvard University Press.
Tomasello, M. (2016). *A natural history of human morality*. Harvard University Press. https://www.hup.harvard.edu/books/9780674986824.
Waggoner, M. (2023). *The vices of virtues: Making room for moral testimony in the life of the virtuous person* [Arts & Sciences Electronic Theses and Dissertations]. 2999. https://openscholarship.wustl.edu/art_sci_etds/2999. https://doi.org/10.7936/1ds7-w405
Walker, A. M. D. (1989). Virtue and character. *Philosophy*, *64*(249), 349–362.

12 Retributive Justice, Free Will, and Evolutionary Naturalism

John Lemos

In Book III of Aristotle's *Nicomachean Ethics*, he notes that many adults have character traits that causally necessitate them to act as they do. For instance, persons with vices are often led by their bad character to act badly. The coward will often be led to act cowardly, and the unjust person will be led to act unjustly. On Aristotle's view, many bad people are determined to act badly by their character, but despite this, he thinks they may rightly be blamed and, in some cases, punished for being as they are and acting as they do. He says this because he believes we are morally responsible for having the kind of character we have. That is, we are morally responsible for whether we are virtuous or vicious. And he believes this because he believes our character is a product of our choices, and he believes that in making choices what we choose is up to us. Aristotle believes that in making choices we settle for ourselves what we shall do.

However, growth in our scientific understanding of the world and how it operates creates a potential problem regarding how we ought to respond to those who act badly. With the growth in our scientific understanding of the world and the laws of nature and how they have application to the functioning of the human mind and its effects on behavior, we learn that perhaps all human behavior, including our choices, is determined (causally necessitated) by prior events going back in time. So, for instance, my choice to type this sentence now may be causally determined by neuro-physiological events, E1, occurring in my brain just antecedent to and during my typing of this sentence and those events may have been necessitated by events, E2, prior to that and those events may have been necessitated by earlier events, E3, and so on, such that my typing this sentence now is just the determined, necessary, consequence of factors beyond my control. Indeed, the world may be such that all events are causally determined in this way by prior states and events. Such a scientific understanding of the world and human nature may be seen as a threat to the rationality of the belief that persons deserve to suffer blame or punishment for their wrongdoing. For there is a widespread and intuitive sense

DOI: 10.4324/9781003632641-16

of justice that tells us persons should be blamed or punished for their wrongdoing only if they have freely willed control over their conduct, and such free will involves the ability of persons to settle for themselves how they act. But if all events, including all human actions and decisions, are the causally necessitated consequences of prior states and events, then it seems as though no one settles for themselves how they act; rather, prior states and events extending far back in time and external to the agent seem to settle this and not the agent themselves. In this way, our scientific understanding raises a puzzle as to how we ought to respond to wrongdoers. What is the just response to wrongdoers, given that their wrongdoing may not ultimately be up to them but to factors that extend outside of them, and which are beyond their control?

Evolutionary naturalism is the view that biological organisms, including human beings, are material beings that are the products of evolution by natural selection. God is not the cause of our existence nor the existence of the universe, and we have no nonphysical minds nor souls. For the reasons noted above, some evolutionary naturalists may be skeptical about the claim that retributive justice is a virtue. While Aristotle thought that it was a virtue, an evolutionary naturalist might think that it involves inappropriate blaming and punishing of persons who don't deserve such treatment. However, in this chapter I argue that retributive justice can be vindicated against such charges by arguing that evolutionary naturalism is consistent with the belief that retributive justice is a virtue and the corresponding practices of blame and punishment are fitting responses to wrongdoing.

I will look at three different views on free will that the evolutionary naturalist might adopt and note their implications for matters of retributive justice. I consider two views, compatibilism and libertarianism, which support beliefs in free will, desert, and justified blame and punishment, and one view, free will denialism, which rejects desert and justified blame and punishment. Most evolutionary naturalists who believe in free will, desert, and justified blame and punishment are compatibilists, but I argue that compatibilism is problematic. I contend that, despite arguments to the contrary, libertarianism can be made to fit with an evolutionary naturalist understanding of human nature. My primary aim in this chapter is to give some reasons to think that any evolutionary naturalist who seeks to defend belief in free will, desert, and justified blame and punishment should endorse libertarianism, not compatibilism. Ultimately, I maintain that while we cannot be sure that we have libertarian free will (LFW), we should live and act as though we have it, because it plays a vital role in supporting important values which would be undermined if we embrace free will denialism. As such, the chapter provides a way for the evolutionary naturalist to defend retributive justice as a virtue.

12.1 Three Approaches to the Problem

Philosophers typically take one of a few different approaches in dealing with issues of free will and responsibility and the related notion of just deserts. Many, perhaps most, philosophers provide a compatibilist response to this problem, arguing that even if all human behavior is causally necessitated by prior states and events which extend beyond the control of the agent, the agent – the person who acts – can still rightly be seen as responsible for their conduct as long as they act from the kinds of mental causes which constitute having a sufficient level of self-control that makes them responsible for their action and justifies blame or possibly punishment if they have acted wrongly. For instance, Harry Frankfurt (1971) has argued that even if determinism is true, human beings can still act with self-control and be responsible for their actions if they act on first-order desires that they want to act on. So, for instance, if I have the first-order desire to steal your tennis racquet and I act on this desire while wanting to act on this desire (a second-order desire), then I am acting with a kind of self-control which establishes my responsibility. Notice that such an act of stealing would be very different from stealing your racquet due to an overwhelming compulsive desire to steal that I would rather not act upon. Those suffering from kleptomania seem to suffer from such compulsive desires, and when they act on these desires while not wanting to act on them, we feel sorry for them and don't blame them because the mental causes of their actions are not expressive of what they really want for themselves. There are other compatibilist theories that offer different accounts of the relevant kinds of mental causes of actions which are consistent with determinism and are supposed to make sense of responsibility grounding freely willed control over action.[1] Most evolutionary naturalists who believe that human beings are typically responsible for their actions and who think some persons justly deserve blame and punishment for their actions are compatibilists about free will and responsibility.[2]

However, some philosophers believe that compatibilism is false. Some of those who think compatibilism is false believe that we have the kind of free will that is the ground of responsibility, and they believe that, since we have such free will, some of our actions must be causally undetermined, freely willed actions. These people hold the view known as "libertarianism." Libertarians don't necessarily believe that all of the actions for which we are responsible are causally undetermined, freely willed acts. Some, such as Robert Kane, believe that many of the freely willed acts for which we are responsible are causally determined actions, but they can rightly be counted as freely willed if they are causally necessitated by mental states which issue from a self we have formed through prior causally undetermined free actions. Kane calls these causally undetermined free actions which form our self (or character) "self-forming acts (SFAs)." On this view, if my motivational states now causally determine me to steal your tennis racquet, then I can rightly be said to do this of my own free will only if I have engaged in causally undetermined free acts in the past that have made me into the sort of person who is now causally necessitated to steal due to the bad character I have freely formed in the past. Libertarians don't believe that in a fully deterministic universe we could have the kind of free will that makes us responsible for our actions and makes us deserve blame or punishment for wrong actions. There is a small but growing number of philosophers who believe such a libertarian view of free will and responsibility can be made to fit with evolutionary naturalism.[3]

There are other philosophers who reject both compatibilism and libertarianism. According to these philosophers, we do not have the kind of free will that grounds moral responsibility and the deserved blame and punishment of wrongdoers. Such philosophers believe that there is no way to reconcile free will with determinism and they also believe that LFW is either logically incoherent or incompatible with the science of human nature. Some of these free will deniers are hard determinists. Hard determinists believe (a) all human behavior is causally determined and (b) determinism is incompatible with free will. Thus, they maintain there is no free will. But most free will deniers today acknowledge that there may be some indeterminism in human action and decision-making. Thus, most contemporary free will deniers simply claim that all our actions are either determined or undetermined and that whether actions are determined or undetermined, they are not freely willed. They argue that determined acts are not freely willed because they are not up to the agent. If acts are determined, then events beyond the control of the agent settle the matter of what actions a person commits. Further, it is also argued that indeterminism doesn't help matters, because either all undetermined actions are just random happenings beyond the control of the agent, making it a matter of luck what the agent does, or, even if this problem of luck can be solved in some way, it is just not scientifically plausible to believe that any such libertarian view can be true.[4]

With these points in mind, what view should the evolutionary naturalist take regarding the proper or just response to known wrongdoers? Should they think that blame and, in some cases, punishment are justified because wrongdoers deserve such treatment? Or should they encourage some different kind of response to wrongdoing which does not involve either blame or punishment?

It may be felt that these concerns about free will and responsibility are misplaced. It may be felt that blaming and punishing wrongdoers can be justified and morally appropriate, whether or not the persons receiving it deserve it. If one is a consequentialist about ethics, one might maintain that whether blame or punishment should be applied to wrongdoers has to do merely with whether its use contributes to better consequences on the whole. The question of the desert of those subjected to blame or punishment is irrelevant. But such an approach to justifying blame and punishment is notoriously problematic. In criticism of such consequentialist views of blame and punishment, it is often noted that there may be situations in which overall consequences may be improved by knowingly blaming or punishing the innocent, but doing this violates our deep-seated sense of justice.[5] A different non-desert-based justification for blame or punishment may note that the point of blame or punishment is to provide discomfort to known wrongdoers so as to reform them or morally educate them and to protect society and to put wrongdoers on a path toward a better way of life for themselves. Whether deserved or not, it may be felt that putting persons on the path to reform in this way may be ethically justified on such grounds. But this approach to blame and punishment suffers from the problem that blame and punishment are not the only viable means to achieving these ends.[6] And given that there may be gentler methods we could use to bring about these results that don't involve blame and punishment, then blaming and punishing for the sake of moral reform or moral education isn't justified. Some free will deniers, such as Derk Pereboom, Gregg Caruso, Bruce Waller, and others, argue that there are, indeed, more gentle ways to bring about the reform of wrongdoers, and they say that since wrongdoers don't deserve blame or punishment, it is unjust to subject them to this, and we should deal with them in more gentle ways.[7]

While the issues addressed in the preceding paragraph could be explored further, I would note that they do give us some reason to think that if blame or punishment are ever justified responses to wrongdoers, then human beings must have the kind of free will that is the basis of just deserts. So, again, I ask what should the evolutionary naturalist believe regarding the existence of such a desert-grounding free will?

12.2 Compatibilism Examined

As noted, most evolutionary naturalists are compatibilists about free will and responsibility. They argue that human beings are complex material organisms governed by the laws of nature and their psychology is largely shaped by the forces of natural selection. But they also assert that even if all our actions are the determined result of neurophysiological processes and events that are themselves the result of genetic and environmental factors we do not control, we can still have free will and be morally responsible for our actions. Such philosophers contend that, despite such determinism, in many cases blaming and punishing wrongdoers is justified.[8]

Daniel Dennett is one of the most famous evolutionary naturalists who defends the compatibilist approach. In various works, he has argued that what makes persons morally responsible for their actions has to do with their meeting minimal self-control conditions. One meets these conditions when one (a) has the ability to understand the circumstances one is in, (b) can comprehend the moral and legal rules of one's society and understands how they apply in his/her circumstances, and (c) has the capacity to guide his/her conduct in accordance with these rules. Most adult human beings meet these conditions for being morally responsible agents. Thus, according to Dennett, most adult human beings are morally responsible for their actions, and when they act in violation of moral or legal norms, it is just to regard them as responsible for their actions and blame them and, possibly, punish them if their wrongdoing is sufficiently serious. Those human beings with significantly stunted rationality or impulse control won't meet these conditions; they don't meet the "plateau of responsibility." Thus, when they engage in wrongdoing, we do not blame or punish them; instead, we feel sorry for them and try to put them in situations where they won't harm others in the future, and we try to help them overcome their deficits in rationality and/or impulse control.

Dennett's view is a version of compatibilism, because even if our actions are causally determined, we can meet these conditions for responsibility. For example, suppose there is someone in Bob's neighborhood who has a very nice motorboat, and it crosses Bob's mind that he would love to have that boat and he considers stealing it. But Bob also knows the law doesn't permit such theft and he would hate to be arrested and punished for doing so. Thus, contrary to his desire to steal the boat, Bob doesn't do so. All the thought processes leading Bob to not steal the boat may be causally determined, but in thinking through the matters in these ways and conforming his will to the law despite his having some desire to do otherwise, he exhibits a kind of self-control which is the mark of a morally responsible being. In comparison, if, through a different mode of reasoning, Fred concludes that stealing the boat would not be so bad and that he might profit significantly from stealing it and then he steals the boat, then Fred would be responsible for such wrongdoing and justly blamed for his wrongdoing if he is caught. And this is because he, too, meets the self-control conditions which constitute the plateau of responsibility. By way of contrast here, a person who suffers from kleptomania and who steals the boat because of an overwhelming urge to do so, and who neither believes it is ethical nor in his interest to do so, does not meet the responsibility plateau and should not be held responsible for his action.

Bruce Waller has rightly criticized such compatibilist views of responsibility, using an argument from unfairness. He notes that if the reasoning of Bob and Fred go in different directions, leading Bob to forgo stealing and leading Fred to steal, then as naturalists and on the presumption of determinism we should expect there to be some reason as to why Fred reasoned differently and made the one choice, while Bob made the other. Indeed, on the presumption of determinism, it is likely that the causes that led Fred to make the choice to steal and the causes that led Bob to make the choice not to steal had to do with factors beyond their control. Sure, both act under the control of their understanding of the situation and the moral rules and their desires. But still, there is the question of why Fred thinks stealing is not so bad and why he thinks he would profit significantly from stealing. Perhaps, there were circumstances in Fred's life growing up that have made him care less about the law and that have made him more selfish. And these differences in his past may have contributed to the ways his reason guides him now. And perhaps there were circumstances in Bob's life that led him to reason in a manner leading him to lawful conduct. Waller contends that on the presumption of evolutionary naturalism, we should expect there to be causally determinative reasons why Bob and Fred reason differently, and he thinks that if they were determined to reason differently, then, it is unfair to blame or punish Fred because it is really just a matter of constitutive luck that he reasoned and decided in the one way while Bob reasoned and decided in another (Waller, 2011, 2015).

Contra Waller, Dennett argues it is fair that people who meet the plateau of responsibility are held equally morally responsible for their actions, because (1) having such rules is essential to the maintenance of a civilized society and (2) we contract into living in such a civilized society due to the benefits it offers. Regarding point (2), he states:

> You don't have to play the moral responsibility game; you can be a hermit on an otherwise deserted island, fishing and foraging. But if you want to enjoy the benefits of living in a civilized society, you have to play the game.
>
> (Dennett, 2012)

However, Dennett's response is not adequate. First, it's not sufficiently obvious that to maintain a civilized society, we need to blame and punish wrongdoers who meet the plateau of responsibility. According to Gregg Caruso, Derk Pereboom, and Bruce Waller, there are effective nonpunitive ways to maintain a civilized society. And second, we only opt into societies with moral responsibility systems and harsh blaming and punitive criminal justice systems because it is the default approach to dealing with wrongdoing and crime. Sure, we want protection from wrongdoing and crime, and so we are willing to live in societies that play the responsibility game. After all, it beats living as a hermit trying to make a go of it in the wilderness. *But*, given that there may be effective ways of dealing with wrongdoers that don't involve blame and punishment, this hardly supports the practices of blame and punishment that are widely prevalent in societies today.[9]

Waller's argument from unfairness is well-founded. And while I have noted how this argument works against Dennett's particular version of compatibilism, I would point out that it works against other versions of compatibilism as well. While other compatibilist theories may establish the plateau – the criteria – of moral responsibility in their own different ways, the problem of unfairness can still be raised. For instance, Harry Frankfurt argues that even if determinism is true, a person is morally responsible for his conduct whenever his conduct reflects his deep self. That is, he believes persons are responsible for their conduct when they act on first-order desires, and they want to be moved to act (a second-order desire) by those first-order desires. Thus, if Bob meets these responsibility conditions in his refusal to steal the boat and Fred meets these responsibility conditions in his stealing the boat, then they are both equally responsible for their actions. But this, of course, suggests that it would be acceptable to blame and, maybe, punish Fred for his conduct and not Bob. However, Waller would note that on the assumption that determinism is true, there is still the question of why they reasoned and decided in the different ways they did. Why did Fred have the second-order desire to be moved to act by his desire to steal? Why did Bob have the second-order desire to be moved to act by his desire not to steal? From the deterministic perspective, it is reasonable to think that there would be various factors beyond the control of each of these men that would be determinative of their character and reasoning processes. Thus, it is unfair to blame Fred and not Bob, as the difference between them is, once again, just a matter of luck.

All of this suggests that compatibilism fails to establish that wrongdoers deserve to be subjected to blaming and shaming and harsh prison terms.[10] But what about libertarian views of free will and moral responsibility, might they offer a way of making sense of such responses to wrongdoing? And can such libertarian views of free will fit with evolutionary naturalism?

12.3 Libertarianism Examined

Most evolutionary naturalists harbor doubts about the libertarian understanding of free will and moral responsibility and not just because they see it as inconsistent with evolutionary naturalism. Many philosophers, whether they are evolutionary naturalists or not, believe that (1) libertarianism is conceptually incoherent. It is often objected that if an action is not causally determined by the character or motivations of the agent, then it is not something over which the agent has control. The action would be a random happening having no clear connection to the character or motivational structure of the agent. How can agents be morally responsible for what they do if their actions are not determined by their character or motivations? However, even if this problem can be solved, many naturalists also believe that (2) it is not scientifically plausible to think that human actions are ever causally undetermined in ways that would be compatible with moral responsibility. Many evolutionary naturalists believe that there is too much evidence suggesting that the neural processes involved in decision-making and action operate in causally determined ways, and even if there were some quantum-level indeterminism at work in neural processing that allowed for indeterminism in decision-making and action, this would not help establish free will and responsibility. Other evolutionary naturalists argue that (3) the kind of free will libertarians endorse is just not the kind of free will that we can reasonably expect to have been produced by natural selection. As such, some argue that LFW does not fit with the picture of human beings as products of evolution.

Despite these concerns, there are a number of contemporary evolutionary naturalists who believe that we may well have a LFW that is the ground of moral responsibility.[11] In what follows, I will provide the outline of a libertarian view which can, I think, answer these various concerns. In doing so, I will consider the view of Robert Kane. My own views on free will are Kanean in nature though I have argued that the view needs to be modified in certain respects to improve upon it. But here I will just present the Kanean view.[12]

According to Kane, to have free will and be morally responsible in a way that can make us deserve blame or punishment for wrongdoing, some of our actions must be causally undetermined, freely willed actions. He thinks that many of the freely willed actions we perform are causally determined by our mental states at the time of performing those actions. And those actions are freely willed only if they are the result of mental states that issue from a character or self the agent has formed through prior causally undetermined freely willed acts, what he calls "self-forming acts (SFAs)." He says that there are times in our lives when we are torn between two options, A and B, and we can only choose one or the other. It is in these situations that we make the free, causally undetermined choices that are our SFAs. These would be situations in which persons are strongly attracted to both options, and it is difficult to decide which action to perform. He says such situations call for deliberation, and he says that in the process of deliberating, it may well be that persons simultaneously make efforts of will to choose each of the options they are considering. So, for instance, as I am deliberating about whether to do A or to do B, I am making an effort of will, trying to choose A, and I am simultaneously making an effort of will to choose B. As a committed naturalist, Kane believes that these efforts of will would be physically realized in distinct neural processes in the brain. He rejects any appeals to Cartesian dualism or Kantian noumenal selves to make sense of libertarianism.

He argues as well that in making a choice between two options, it may well be that during deliberations chaotic states in the neural processing of the brain might make these processes sensitive to quantum-level indeterminacies present in those processes. This could allow an agent's choice to be causally undetermined. Suppose I choose A. The idea here is that my choice would be caused in an undetermined way by my effort of will to choose A. The quantum-level indeterminacy opens a window in which either one of my competing efforts of will may be made to win out over the other. So, when I chose A, it was nonetheless open to me to choose B, because the choice was undetermined. Furthermore, even though my choice is undetermined, I am responsible for my choice because it is a product of my effort of will; it was something I was trying to do. These points are key to Kane's reply to the charge noted above that any theory of LFW suffers from a problem of conceptual incoherence. Recall that on this charge there is no way to make sense of an agent having responsibility grounding control over a causally undetermined choice. But Kane wants to say this is not true, because even if a choice is undetermined, if it is a result of what he was trying to do, then the agent will still be responsible for it.

In support of this, he appeals to Austin-style examples from the philosopher J.L. Austin. He says, suppose a person is trying to make a putt, and it is causally undetermined whether she will have a nervous twitch sending her putt off course. If she makes the putt, it is still obvious that she is responsible for making the putt. So, too, if an assassin is trying to shoot and kill some innocent person, and it is causally undetermined whether she will undergo a nervous twitch sending her shot off course, and if she nonetheless succeeds in shooting and killing her target, then it is still obvious that she is morally responsible for killing that person. Thus, if I make a causally undetermined choice of A due to my effort to choose A, then I will be morally responsible for that choice as well. And since I was at the same time trying to choose B, if I had made that choice, then I would have been responsible for it as well. It is for these kinds of reasons that Kane thinks the charge of the conceptual incoherence of libertarianism can be answered.

It may be objected here that Kane's solution only works if we assume that the agent is already morally responsible for making her efforts of will. Notice the assassin is only responsible for killing her target, if she was morally responsible for making that effort. The problem here looks to be serious because his theory is that it is the undetermined SFAs that are to be the ground of all moral responsibility, but his account of the nature of these SFAs and our responsibility for them seems to presuppose moral responsibility for the making of the efforts of will involved in them. In reply to this, it can be said that if as an adult I am choosing between A and B and I make an undetermined choice of A as a result of my effort to choose A, then I can be morally responsible for that choice if I am morally responsible for the effort to choose A. Furthermore, as an adult I can be morally responsible for that effort if it issues from a character, a self, I have formed through prior SFAs. Now this, of course, leads to the problem of how moral responsibility is built up through prior SFAs. Most notably, how do we make sense of moral responsibility for the very first SFA of a human being? The efforts of will involved in one's first SFA cannot be a product of prior SFAs, if it is the first SFA. But it seems one must be morally responsible for the first SFA if one is to build up responsibility for later SFAs through prior SFAs. Here, it must be conceded that we bear very little responsibility for our first SFA and other SFAs of our early childhood. This is because they issue from a self that is too much a product of genetics and environment. However, even in a child's first SFA, she makes one of her efforts win out over a competing effort, and she could have done otherwise. So, even in the first SFA, the child bears some very minimal responsibility for what she does. Again, this responsibility is very minimal because the child has not yet established much, if any, responsibility for the self from which these efforts arise. Slowly over time as we mature and face many conflicts in our will, leading to difficult decisions requiring deliberation and choice, we build our character by making choices in these contexts where we could do otherwise due to the undetermined nature of these choices. In this way, from a self that was originally merely the product of genetic and environmental factors, we build step by step a self the nature of which we have shaped through free undetermined choices. It is in this way that over time we become more fully morally responsible for our character.

Kane's view on the coherence of LFW has been subjected to a lot of criticism, and he and I have both responded to many of these criticisms. I cannot answer all the criticisms here, as the criticisms are too many and varied, but I hope to have shown that it is far from obvious that there is no way to provide a coherent conception of LFW.[13]

However, there are evolutionary naturalists who may grant that, yes, Kane's theory can make coherent sense of LFW, but they still reject the view because they don't think there is any good reason to believe there is any indeterminacy in neural functioning that allows for the kind of undetermined decision-making that Kane's view requires. In reply to this concern, it should be noted that there is a growing literature involving works by both scientists and philosophers arguing that we simply don't know that all human behavior is causally determined and that there may, indeed, be causal indeterminism at work in neural and mental functioning that could allow for LFW. See, for instance, the works of neurobiologists Bjorn Brembs (2011) and Martin Heisenberg (2013), neurophysiologist and computational neuroscientist Christof Koch (2009), as well as neuroscientists Peter Tse (2013) and Kevin Mitchell (2023). All these scientists have argued that there is significant evidence that causal indeterminism is at work in the behavior of biological organisms and this might be the basis for LFW. Philosophers, like Mark Balaguer (2010), Robert Kane (1996, 2002, 2007a, 2007b, 2011, 2024), and Helen Steward (2012, 2016), appeal to the findings of these scientists and others in providing support for their distinctive versions of LFW.

The computational neuroscientist, Christof Koch (2009), claims:

> The universe has an irreducibly random character to it. If it is a clockwork, its cogs, springs, and levers are not Swiss-made; they do not follow a predetermined path. Physical indeterminism rules in the world of the very small as well as the very large.

In defending his thesis that the existence of LFW is an open empirical question, philosopher Mark Balaguer (2010) writes:

> Current neuroscientific theory treats a number of different neural processes probabilistically, and any decent textbook on neuroscience will point this out. For instance, synaptic transmission and spike firing are both treated probabilistically. One textbook (Dayan and Abbott 2001, 179, 9) puts these points as follows:
>
> I. …synaptic transmitter release is a stochastic process. Release of transmitter at a presynaptic terminal does not necessarily occur every time an action potential arrives and, conversely, spontaneous release can occur even in the absence of the depolarization due to an action potential.
>
> II. Because the sequence of action potentials generated by a given stimulus varies from trial to trial, neuronal responses are typically treated statistically or probabilistically. For example, they may be characterized by firing states, rather than specific spike sequences.
>
> (Balaguer 2010, p. 153)

Balaguer adds, "by treating these processes statistically, neuroscientists do not commit themselves to the thesis that, in the end, they are genuinely indeterministic. But the important point here is that they aren't committed to determinism either" (2010, p. 153). In further support of this, he quotes the following from Christof Koch:

> At this point we do not know to what extent the random, i.e., stochastic neuronal processes we observe are due to quantum fluctuations (a la Heisenberg) that are magnified by chemical and biological mechanisms or to what extent they just depend on classical physics (i.e., thermodynamics) and statistical fluctuations in the underlying molecules.
>
> (from Balaguer's correspondence with Koch, in Balaguer, 2010, p. 154)

Such considerations lead Balaguer and myself to conclude that the belief that neuroscience gives us reason to accept neural determinism is "straightforwardly false" (Balaguer, 2010, p. 154). Hence, Balaguer and I believe the existence of LFW is an open empirical question.

Philosopher Helen Steward (2012, 2016) argues that organismal agency itself requires behavior to be causally undetermined, and since organisms have agency and free will requires agency, free will must involve causally undetermined action. In support of her argument for LFW, she appeals to ideas about non-equilibrium thermodynamics developed by Ilya Prigogine. She argues that animals are non-equilibrium systems and that they may well be nondeterministic in their behavior. Further, she notes that animals must expend effort to secure energy for their self-maintenance, and she maintains that due to the physics of non-equilibrium thermodynamics, "There is no *a priori* reason to suppose that deterministic mechanisms must be the best evolutionary solution to the question of how such a system might go about the business of deciding where, when, and how best to expend these efforts" (Steward, 2016). The neuroscientists Kevin Mitchell and Henry Potter (2022) provide empirical support for such a view of agency, and Mitchell extends the argument in making a case for LFW in his recent book, *Free Agents: How Evolution Gave Us Free Will* (2023).

In describing and citing this literature, it is not my intent to argue that we know there are causally undetermined human choices and behavior. Rather, my point is simply that there *may* be. There is no consensus regarding this in the scientific community. And assuming there is a coherent conception of LFW, which I have argued there is, it may well be that we have such free will.

As noted, other evolutionary naturalists, such as Bruce Waller (2015), argue that the kind of LFW that Kane and I endorse cannot reasonably be thought to fit with an evolutionary understanding of human nature. Waller believes we lack the kind of free will that grounds just deserts responsibility. But he *does* believe that we have a kind of free will that is worth preserving and protecting. The kind of free will he believes we have does not require that we engage in causally undetermined actions. Rather, it merely involves the ability to choose from different alternative courses of action given the conditions we are facing. This ability doesn't require indeterminism; rather, it just involves the ability to recognize different possible courses of action and the ability to act in accordance with one's reason-guided decision. For instance, an animal, maybe a human being, goes to its preferred water source only to find predatory wolves there. The animal is aware of another water source farther away and in a different direction. Guided by the understanding that the wolves are dangerous and that there is an alternative water source of which it is aware, the animal goes to this other water source. In recognizing the alternative course of action and being able to choose it, the animal exhibits a kind of free will. But, according to Waller, this kind of free will does not require causally undetermined action, and it does not support the idea that bad actors deserve blame and punishment.

Waller argues that this kind of free will is adaptive, aiding those organisms that have it in the struggle to survive and reproduce. As such, it is a valuable kind of free will that we can reasonably believe we have, since, among other things, it fits with an evolutionary naturalistic understanding of human nature. He goes on to argue that LFW is not a kind of free will that we can reasonably expect to be produced through Darwinian processes of evolution. In making his case for this, he focuses on Kane's brand of libertarianism and makes several arguments as to why it does not fit with our evolved human nature. Contra Waller, I believe a Kanean view of LFW fits with an evolutionary understanding of human nature. In what follows, I consider each of Waller's arguments to the contrary and demonstrate why they fail.

First, he says that Kane's naturalistic conception of free will is complicated. Organisms that have it would need to have quantum-level indeterminacy at work in their neurons and there would have to be amplifying chaos in neural activity. He says this makes it implausible as a product of natural evolution, since an adaptive free will only requires the ability to pursue alternative courses of action given an understanding of one's situation and the alternatives, and this doesn't require causal indeterminism (Waller, 2015, pp. 152–153). Here, I would note it is true that to have the kind of free will that Waller thinks we have, one does not need to engage in causally undetermined choices that involve magnified effects of quantum events. However, it does not follow that we don't have LFW or that its existence is inconsistent with a naturalistic evolutionary understanding of human nature. The neurobiologist Bjorn Brembs (2011) argues that to avoid being subject to predation from other animals, unpredictable behavior is vital, and he says that one way to achieve this is through engaging in causally undetermined behaviors. While causally undetermined behavior is not necessary for unpredictable behavior, it is nonetheless one way to produce unpredictable behavior. If you couple this with the fact that there's no good reason to believe that all animal and human behavior is causally determined – something I have just argued for – then it becomes reasonable to think LFW may exist in us as an adaptive trait.

Second, Waller writes that Kane's libertarian conception of free will, "makes it a rare and wondrous thing ... But psychologically satisfactory free will is not something reserved for special occasions; rather, it is a common process of choice and control in a world of changing conditions. As a wide range of psychological studies have made clear … it is the daily mundane operation of free will – often on details that seem insignificant and are certainly more modest than Kane's 'self-forming acts'" (2015, pp. 152–153). His point is that, on Kane's view, we only exert our free will on those relatively rare occasions in which we are deeply conflicted about what to do. But Waller thinks that an adaptive free will would be something that we would use often, like when choosing what time to get up in the morning, what to wear, when to eat, how to get to the store, etc.

This argument is based on a misreading of Kane. It is not true that Kane believes our LFW would only be exerted in soul-searching, difficult choices. It is consistent with his view that causally undetermined free choices may occur at any time we are puzzled about what to do, and this can occur even in trivial daily decisions. Should I set the alarm for 6:30 am or 7 am? Should I wear the black tie or the maroon tie? Shall I take the highway to work or go on the back streets? It is true that in his writings Kane focuses much on more soul-searching decisions, involving big decisions that involve important moral values, but this is because these causally undetermined decisions are key to shaping our moral character and making us responsible for later determined decisions shaped by our character. But that does not mean that we do not make causally undetermined free choices in dealing with more mundane matters.

In his final evolution-based argument against Kanean libertarianism, Waller states:

> Kane makes the operation of free will fixed and constant, rather than variable; that is, on Kane's account the operation of free will always involves a fixed degree of dual process control balanced by a very modest measure of *randomness*. But animal free will, which operates in constant relation to a changing and challenging environment, requires different degrees of control and open alternatives under varying environmental conditions.
>
> (2015, p. 153) [My emphasis.]

Waller notes that, on Kane's view, freely willed choices occur in those situations when agents are torn between competing choice options, A and B, and while deliberating, they simultaneously make one effort of will to choose A and another effort of will to choose B. In addition, Waller seems to think the decisions agents make in such situations are then settled by a random quantum indeterminate event. With this understanding of Kane's view in mind, he wants to say that this would not be an adaptive kind of free will. Let's go back to my example of the animal that goes to its usual watering hole and finds wolves there that are perceived as a threat. The animal must now decide what to do – (a) try drinking there near the wolves or (b) go to another watering hole it knows of that is a little farther away. Waller is right to think here that it is adaptive for the organism to be able to assess its situation and alter its behavior in a manner that enhances its adaptive fitness. This is an adaptive free will and it is highly valuable. But Waller is suggesting that the animal's choice between these options is better off being the *determined* result of its thinking here, and the wise choice here is to go to another watering hole farther away. A choice here that is subject to random quantum swerves would not be adaptive. Animals making choices in accordance with such "neural coin-tosses" would not do well at surviving and reproducing. In this way, Waller believes Kanean LFW is not adaptive.

In reply, I would first note that Kane does not think that *all* freely willed decisions are causally undetermined. His view is that to have desert-grounding free will, *some* decisions need to be causally undetermined free decisions. Kane can acknowledge that the animal in this case is better off if it makes a causally determined choice to go to another watering hole farther away. Indeed, it is consistent with Kane's view that many important decisions we make are causally determined choices shaped by our knowledge of options and an understanding of what is best given current circumstances. But this does not mean that there would not be any value in making *some* causally undetermined free choices. As I noted earlier, Bjorn Brembs argues that unpredictable behavior makes it more challenging for an organism to be subjected to predation, and one way to get unpredictable behavior is through engaging in causally undetermined behaviors. Think of animal, A, being chased by a predator, P. It may be beneficial if some of A's movements are made quickly and at random so that P has trouble responding appropriately to A's movements. If animals of type A always moved in the same sorts of ways, then animals of type P might catch on to this and be all too successful at catching them. Behaviors that are causally undetermined and have a random element to them can also have other adaptive advantages. An organism may profit from randomly moving within its environment. By randomly choosing to explore different paths to walk and different areas to explore in the environment, an organism may learn about different places to go and what is to be found in different areas. Organisms that don't do this may not know their environment as well and may not know all of the options open to them. Thus, our animal that sees the wolves at its watering hole may know of the other watering hole only because it has learned of it through prior random probing of its environment. Such causally undetermined and random probing of the environment can be adaptive in this way.

12.4 Conclusion

In the preceding section of this chapter, I argue that Kane's conception of LFW is conceptually coherent and that there is no good reason to think that all animal, including human, behavior is causally determined and that there is no good evolutionary biological reason to think that we cannot have the kind of LFW that Kane describes. I have not argued that we *know* that such free will exists; rather, my point is that we may have it and that there is nothing in evolutionary naturalism that shows we do not have it. In an earlier section of the chapter, I also provided reasons to think that compatibilist conceptions of moral responsibility will suffer from the problem of unfairness. Admittedly, a full defense of some of the points in this chapter would need further argument. To give a full refutation of compatibilism probably requires a book-length discussion of the issues and a full defense of the coherence of LFW probably requires a book-length discussion as well.

However, assuming (1) compatibilism about moral responsibility is implausible and (2) we *may* have an LFW that grounds desert for wrongdoing and which justifies blame and punishment for wrongdoers – then what does this imply about the question with which this chapter began – the question of what evolutionary naturalists should think about matters of justice and wrongdoing? How should the evolutionary naturalist respond to wrongdoers? How should they understand the nature of justice in the context of responding to wrongdoers?

Well, my arguments here suggest that based on empirical evidence and philosophical analysis, we don't know that we have the desert-grounding free will that justifies blame and punishment for wrongdoers, but we don't know that we lack such free will either. Further, if I am right that compatibilism is deeply flawed, then the options before us are to embrace living and acting as if we have LFW even though we don't know we have it or to embrace free will denialism. Regarding the latter, it is my view that free will denialism is simply too problematic to rationally embrace. I have argued elsewhere that when it is denied that we have desert-grounding free will, it leaves us in a position where it is too difficult: (a) to defend human dignity in the context of criminal justice; (b) to make sense of moral obligation; (c) to make sense of justified pride in our accomplishments; (d) to make sense of rationally warranted forgiveness and gratitude; and (e) to make sense of the full value of our love and friendship relations.[14]

Due to limitations of space, I cannot fully support any one of these points here. However, to give readers some sense of my concern, I would note that, regarding dignity in the context of criminal justice, we have historically believed that punishing the innocent and giving excessive punishments to those who are guilty of crimes is unjust. And the injustice of this is best defended on the basis that those who are innocent *don't deserve* to be punished and those who are guilty *don't deserve* to be punished any more severely than is deserved given the nature of the crime they have committed and the level of their responsibility for the crime. Thus, if we deny that anyone has desert-grounding free will, then in the context of criminal justice, we undermine the best means for defending the protection of the innocent and the rights of those convicted of crimes from being subject to excessive punishment. Derk Pereboom (2001), Pereboom and Caruso (2018), and Gregg Caruso (2021), who deny that we have desert-grounding free will, say that we should reject punitive detention for those convicted of dangerous crimes and instead place dangerous criminals in comfortable quarantine and offer them therapy. And they contend that such measures are justified on the grounds of societal self-defense. However, if the justification of such quarantine is the protection of society and if no one deserves to be quarantined and if we could get better protection of society by lowering the standards of evidence for criminal quarantine, then it is unclear what the rational basis for not lowering the standards would be. If no one deserves to be quarantined and we can get better protection from crime by lowering standards of evidence for criminal convictions, then it seems we should do so. In addition, on the quarantine model of criminal justice, persons are to be held in criminal detention until they are deemed safe to release. But this allows that persons convicted of lesser but dangerous crimes may be held in criminal detention for very long periods of time despite the relatively minor nature of their offense. The problem here is that quarantining the innocent is unjust and a very long criminal quarantine for lesser offenses is also unjust. It is for these kinds of reasons that abandoning belief in desert-grounding free will is a threat to human dignity in the context of criminal justice. As noted, this is not the only problem that comes with the claim that we lack desert-grounding free will. I have merely developed this point here to give readers a better sense of what motivates at least some of my concerns regarding free will denialism.

In the end, it may be false that we have desert-grounding LFW, but given the problems with compatibilism and given that we may indeed have LFW, we should live and act as if we have it. This means that in response to wrongdoers we are rationally warranted to presume they deserve our blame and punishment. But given the uncertainties about the existence of such free will, we should be tempered in blaming and punishing them.

Notes

1 See, for instance, Dennett (1978, 1984, 2003), Fischer (1994, 2006, 2012), Fischer and Ravizza (1998), Flanagan (2002), McKenna (2004, 2012a, 2012b), and Watson (1975, 1987, 2004), among others.

2 For some of those who clearly self-identify as evolutionary naturalists and hold the compatibilist view, see Dennett (1978, 1984, and, esp., 2003), Flanagan (2002, Ch. 4), Radcliffe-Richards (2000, Ch. 6), and Ruse (1987, 2001, Ch. 12).
3 See, among others, Balaguer (1999, 2004, 2010, 2014), Kane (1996, 1999a, 1999b, 2002, 2007a, 2007b, 2011, 2019, 2024), Lemos (2011a, 2011b, 2014, 2015, 2017a, 2017b, 2018, 2020, 2022, 2023a, 2023b), Mitchell (2023), Potter and Mitchell (2022), Steward (2012, 2016).
4 For defenses of the view that we lack free will, whether determinism is true or not, see Caruso (2021), Caruso and Pereboom (2018), Levy (2011), Pereboom (2001, 2007a, b, 2014), Strawson (1994, 2000), and Waller (2011, 2015). Waller actually believes that we have a kind of free will, but he believes we don't have the kind that supports just deserts, and his argument strategy for this is that, whether determinism is true or not, we don't have desert-grounding free will.
5 For accounts of such criticisms of consequentialist justifications of punishment, see, Caruso (2021), Caruso and Pereboom (2018), McCloskey (1963, 1965), Pereboom (2001, 2014), and Smilansky (1990, 2000).
6 For a defense of the moral education theory of punishment, see Hampton (1984). For criticisms of it, see Caruso (2021), Caruso and Pereboom (2018), and Pereboom (2014).
7 See, for instance, Caruso (2021), Caruso and Pereboom (2018), Pereboom (2001, 2014), and Waller (2011, 2015).
8 For the works of some evolutionary naturalists who are compatibilists about free will, see note 2 above.
9 For development of these gentler ways of dealing with wrongdoers, see Caruso (2021), Caruso and Pereboom (2018), Pereboom (2001, 2014), and Waller (2011, 2015). Also, for an interesting Rawlsian defense of a less punitive approach to punishment, see Vilhauer (2013, 2019). These philosophers defend their views on the assumption that we lack desert-grounding free will. For critical replies to such desert-denying approaches to this issue, see Lemos (2016, 2018, 2019, 2023a).
10 Here I have tried to outline the basis for my own rejection of the compatibilist approach to issues of free will and responsibility. But the debate is ongoing. A full refutation of compatibilism would probably require a book-length treatise, as there are so many different versions of compatibilism. For instance, a full refutation would, among other things, have to confront arguments from Frankfurt (1969) and John Martin Fischer (1982) that moral responsibility does not require a capacity to do otherwise. Waller's argument from unfairness presumes that the ability to do otherwise is necessary for one to have desert-grounding free will and moral responsibility. Frankfurt and Fischer use famous thought experiments, what have come to be called "Frankfurt-style" cases, to argue that such an ability to do otherwise is not necessary to act with desert-grounding free will and moral responsibility. These kinds of thought experiments have given rise to a vast literature. Not all philosophers, including myself, find them compelling. To learn more about these matters, see Michael McKenna and David Widerker (2003), which provides a fine collection of essays devoted to the critical analysis of the Frankfurt-style cases. See also Justin Capes (2023), in which he provides a book-length analysis of such cases and argues that they don't refute the alternative possibilities condition for free will and moral responsibility. In addition, some philosophers, such as Michael McKenna (2004), argue that a proper understanding of the nature of blame suggests that the capacity to

do otherwise is not essential for justified blame. For a reply to McKenna, see Lemos (2023a, Ch. 1).

11 For the works of some evolutionary naturalists who are libertarians about free will, see note 3 above.

12 For development and defense of my modified Kanean view of libertarian free will, see Lemos (2018, 2021, 2023a, 2023b). The key difference between my view and Kane's is our different accounts of the nature of self-forming acts. I replace his theory of efforts of will with a theory of agential evaluative weightings of options considered in deliberation.

13 Kane has explained and responded to criticisms of his views in various publications. For instance, see Kane (1999a, 2023b, 2002, 2011, 2014, 2019, 2024) and Lemos (2011a, 2013, 2014, 2015, 2017a, 2018, 2020, 2022, 2023a).

14 For development of these points, see Lemos (2016, 2018, 2019, 2023a).

References

Balaguer, M. (1999). Libertarianism as a scientifically respectable view. *Philosophical Studies*, *93*, 189–211.

Balaguer, M. (2004). A coherent, naturalistic, and plausible formulation of libertarian free will. *Nous*, *38*, 379–406.

Balaguer, M. (2010). *Free will as an open scientific problem*. MIT Press.

Balaguer, M. (2014). Replies to McKenna, Pereboom, and Kane. *Philosophical Studies*, *169*, 71–92.

Brembs, B. (2011). Towards a scientific concept of free will as a biological trait. *Proceedings of the Royal Society: Biological Sciences*, *278*, 930–939. doi: 10.1098/rspb.2010.2325

Capes, J. (2023). *Moral responsibility and the flicker of freedom*. Oxford University Press.

Caruso, G. (2021). *Rejecting retributivism: Free will, punishment, and criminal justice*. Cambridge University Press.

Dennett, D. (1978). On giving libertarians what they say they want. In *Brainstorms: Philosophical essays on mind and psychology* (pp. 286–299). MIT Press.

Dennett, D. (1984). I could not have done otherwise – So what? *The Journal of Philosophy*, *81*, 553–567.

Dennett, D. (2003). *Freedom evolves*. Viking Press.

Dennett, D. (2012). *Review of Against Morality Responsibility*. https://Naturalism.org

Fischer, J. M. (1982). Responsibility and control. *The Journal of Philosophy*, *89*, 24–40.

Fischer, J. M. (1994). *The metaphysics of free will*. Blackwell.

Fischer, J. M. (2006). *My way: Essays on moral responsibility*. Oxford University Press.

Fischer, J. M. (2012). *Deep control: Essays on free will and value*. Oxford University Press.

Fischer, J. M., & Ravizza, D. (1998). *Responsibility and control: A theory of moral responsibility*. Cambridge University Press.

Flanagan, O. (2002). *The problem of the soul: Two visions of the mind and how to reconcile them*. Basic Books.

Frankfurt, H. (1969). Alternate possibilities and moral responsibility. *The Journal of Philosophy*, *66*, 829–839.

Frankfurt, H. (1971). Freedom of the will and the concept of a person. *The Journal of Philosophy*, *68*, 5–20.

Hampton, J. (1984). The moral education theory of punishment. *Philosophy and Public Affairs*, *13*, 208–238.

Heisenberg, M. (2013). The origin of freedom in animal behavior. In A. Suarez & P. Adams (Eds.), *Is science compatible with free will?* (pp. 95–103). Springer.

Kane, R. (1996). *The significance of free will*. Oxford University Press.

Kane, R. (1999a). Responsibility, luck, and chance: Reflections on free will and indeterminism. *Journal of Philosophy*, *96*, 217–140.

Kane, R. (1999b). On free will, responsibility, and determinism: Response to Clarke, Haji, and Mele. *Philosophical Explorations*, *2*, 105–121.

Kane, R. (2002). Some neglected pathways in the free will labyrinth. In R. Kane (Ed.), *The Oxford handbook of free will* (pp. 406–437). Oxford University Press.

Kane, R. (2007a). Libertarianism. In J. M. Fischer, R. Kane, D. Pereboom, & M. Vargas (Eds.), *Four views on free will* (pp. 5–43). Blackwell.

Kane, R. (2007b). Response to Fischer, Pereboom, and Vargas. In J. M. Fischer, R. Kane, D. Pereboom, & M. Vargas (Eds.), *Four views on free will* (pp. 166–183). Blackwell.

Kane, R. (2011). Rethinking free will: New perspectives on an ancient problem. In R. Kane (Ed.), *The Oxford handbook of free will* (pp. 381–404). Oxford University Press.

Kane, R. (2014). New arguments in debates on libertarian free will: Responses to contributors. In D. Palmer (Ed.), *Libertarian free will: Contemporary debates*. Oxford University Press.

Kane, R. (2019). The complex tapestry of free ill: Striving will, indeterminism, and volitional streams. *Synthese*, *196*, 145–160.

Kane, R. (2024). *The complex tapestry of free will: A philosophical odyssey*. Oxford University Press.

Koch, C. (2009). Free will, physics, biology, and the brain. In N. Murphy, G. F. R. Ellis, & T. O'Connor (Eds.), *Downward causation and the neurobiology of free will* (pp. 31–52). Springer.

Lemos, J. (2011a). Kane's libertarian theory and luck: A reply to Griffith. *Philosophia*, *39*, 357–367.

Lemos, J. (2011b). Wanting, willing, and trying and Kane's theory of free will. *Dialectica*, *65*, 3–48.

Lemos, J. (2013). Hard-heartedness and libertarianism. *Philo*, *16*, 180–195.

Lemos, J. (2014). Libertarianism and free determined decisions. *Metaphilosophy*, *45*, 675–688.

Lemos, J. (2015). Self-forming acts and the grounds of responsibility. *Philosophia*, *43*, 135–146.

Lemos, J. (2016). Moral concerns about responsibility denial and the quarantine of violent criminals. *Law and Philosophy*, *35*, 461–483.

Lemos, J. (2017a). Hard-heartedness and libertarianism again: A rejoinder to double. *Journal of Philosophical Research*, 42, 319–323.

Lemos, J. (2017b). Thoughts on Dennett and Harris. *Rivista Internazionale di Filosofia e Psicología*, *8*, 231–246.

Lemos, J. (2018). *A pragmatic approach to libertarian free will*. Routledge Press.

Lemos, J. (2019). A moral/pragmatic defense of just deserts responsibility. *Journal of Information Ethics*, *28*, 73–94.

Lemos, J. (2020). Kane, Pereboom, and event-causal libertarianism. *Philosophia*, *48*, 607–623.

Lemos, J. (2021). The indeterministic weightings model of libertarian free will. *The Journal of Philosophical Theological Research*, *23*, 137–156.
Lemos, J. (2022). Kane and the physical indeterminism luck objection: A reply to Moore. *Philosophia*, *50*, 2597–2615.
Lemos, J. (2023a). *Free will's value: Criminal justice, pride, and love*. Routledge.
Lemos, J. (2023b). Haji and the indeterministic weightings model of libertarian free will. *The Journal of Philosophical Theological Research*, *25*, 101–118.
Levy, N. (2011). *Hard luck: How luck undermines free will and moral responsibility*. Oxford University Press.
McCloskey, H. J. (1963). A note on utilitarian punishment. *Mind*, *72*, 599.
McCloskey, H. J. (1965). A non-utilitarian approach to punishment. *Inquiry*, *8*, 249–263.
McKenna, M. (2004). Responsibility and globally manipulated agents. *Philosophical Topics*, *32*, 169–192.
McKenna, M. (2012a). Moral responsibility, manipulation arguments, and history: Assessing the resilience of nonhistorical compatibilism. *Journal of Ethics*, *16*, 145–174.
McKenna, M. (2012b). *Conversation and responsibility*. Oxford University Press.
McKenna, M., & Widerker, D. (2003). *Moral responsibility and alternative possibilities: Essays on the importance of alternative possibilities*. Routledge.
Mitchell, K. (2023). *Free agents: How evolution gave us free will*. Princeton University Press.
Pereboom, D. (2001). *Living without free will*. Cambridge University Press.
Pereboom, D. (2007a). Hard Incompatibilism. In J. M. Fischer, R. Kane, D. Pereboom, & M. Vargas (Eds.), *Four views on free will* (pp. 85–125). Blackwell.
Pereboom, D. (2007b). Response to Kane, Fischer, and Vargas. In J. M. Fischer, R. Kane, D. Pereboom, & M. Vargas (Eds.), *Four views on free will* (pp.191–203). Blackwell.
Pereboom, D. (2014). *Free will, agency, and meaning in life*. Oxford University Press.
Pereboom, D., & Caruso, G. (2018). Hard-incompatibilist existentialism: Neuroscience, punishment, and meaning in life. In G. Caruso & O. Flanagan (Eds.), *Neuroexistentialism: Meaning, morals, and purpose in the age of neuroscience* (pp. 193–222). Oxford University Press.
Potter, R., & Mitchell, K. (2022). Naturalising agent causation. *Entropy* 24. doi: 10.3390/e2400472
Radcliffe-Richards, J. (2000). *Human nature after Darwin*. Routledge.
Ruse, M. (1987). Darwin and determinism. *Zygon*, *22*, 419–442.
Ruse, M. (2001). *Can a Darwinian be a Christian?: The relationship between science and religion*. Cambridge University Press.
Smilansky, S. (1990). Utilitarinaism and the 'punishment' of the innocent: The general problem. *Analysis*, *50*, 256–261.
Smilansky, S. (2000). *Free will and illusion*. Clarendon Press.
Steward, H. (2012). *A metaphysics for freedom*. Oxford University Press.
Steward, H. (2016). Libertarianism as a naturalistic position. In K. Timpe & D. Speak (Eds.), *Free will and theism: Connections, contingencies, and concerns*. Oxford University Press.
Strawson, G. (1994). The impossibility of moral responsibility. *Philosophical Studies*, *75*, 5–24.
Strawson, G. (2000). The unhelpfulness of indeterminism. *Philosophy and Phenomenological Research*, *60*, 149–155.

Tse, P. (2013). *The neural basis of free will: Criterial causation*. MIT Press.
Vilhauer, B. (2013). Persons, punishment, and free will skepticism. *Philosophical Studies*, *162*, 143–163.
Vilhauer, B. (2019). Deontology and deterrence for free will deniers. In E. Shaw, D. Pereboom, & G. Caruso, (Eds.), *Free will skepticism in law and society* (pp. 116–138). Cambridge University Press.
Waller, B. (2011). *Against moral responsibility*. MIT Press.
Waller, B. (2012). *Reply to Dennett*. Naturalism.org.
Waller, B. (2015). *The stubborn system of moral responsibility*. MIT Press.
Watson, G. (1975). Free agency. *Journal of Philosophy*, *72*, 205–220.
Watson, G. (1987). Responsibility and the limits of evil. In F. Schoeman (Ed.), *Responsibility, character, and the emotions: New essays in moral psychology* (pp. 256–286). Cambridge University Press.
Watson, G. (2004). *Agency and answerability*. Oxford University Press.

Part IV

Applied Ethics

Evolving Characters in the Digital Age

13 Can We Learn to Be Virtuous from the Vicious?

On Moral Learning and the Pedagogical Value of Antihero Protagonists

Maria Waggoner and Daniel Kelly

13.1 Introduction

In efforts to pay for his cancer treatment and financially provide for his family, high school chemistry teacher Walter White takes up cooking meth. Walter teams up with his former student, Jesse, a novice distributor, and forms a father-son-like relationship with him. Despite his good intentions, Walter "slowly becom[es] irredeemably bad" (Vaage, 2015, p. 113). He poisons Jesse's girlfriend Jane's kid brother to frame an enemy drug lord. He stands passively by as Jane dies so she won't turn him in to law enforcement. Right before Jesse is to be killed, Walter gleefully reveals that he let Jane die, for the simple sadistic pleasure of it.[1]

Many of us find stories like *Breaking Bad*, which are centered around an antihero, to be strangely compelling. Despite their vicious character, we often find ourselves sympathizing, even identifying with the problematic protagonist.[2] This is both puzzling and worrisome given the role narratives tend to play in moral learning. There is good reason to think that stories are an important medium for all kinds of social learning and that evolution capitalized on this (Boyd, 2009; Sugiyama, 2017). Furthermore, it seems obvious that exposure to stories of morally good protagonists can support the acquisition of moral norms and other useful moral knowledge. Narratives featuring moral exemplars provide models who exhibit praiseworthy behaviors and, perhaps by activating feelings like admiration, motivate us to emulate them. This picture raises a series of questions, then, about stories with antihero protagonists. Why do we find them so compelling? What benefits, if any, are provided by stories that lavish so much attention on such figures? What benefits, if any, might these stories provide that are *distinct* from encounters with real-life bad actors?

In this chapter, we offer answers to these questions. We begin in Section 13.2 by sharpening them, raising a challenge that has been made within the virtue ethics tradition: If we often learn from whichever models we are exposed to, then those models should be positive ones, otherwise our

DOI: 10.4324/9781003632641-18

moral character will remain undeveloped, or worse, will break bad. In Section 13.3, we explain that while some stories featuring an antihero protagonist might proliferate simply because they are *selfish memes*, others can make genuine contributions to a specific kind of moral learning. Stories are equipment for living (Camp, 2024), and this genre of story, we argue, provides specific kinds of moral goods, which we call moral inoculation. In Section 13.4, we focus on a problem, generated by our evolved cognitive biases, that arises for moral learning from norm violators in our everyday lives. In Section 13.5, we argue that stories with antihero protagonists work around this problem and that the way they solve it also provides materials to respond to the challenge from the virtue ethics tradition we started with.

13.2 Moral Learning from a Virtue Ethical Perspective

Many strands of virtue ethical theory advise that when we (i.e., the non-virtuous, the morally mediocre) are trying to figure out the right thing to do, we would do well to pay attention to moral exemplars. Knowledge of virtuous people is thought to be action guiding – we should do what a virtuous person would do in similar circumstances (Hursthouse, 1999, p. 28). But the ideal of the virtuous person is also thought to provide a broader kind of guidance, offering a model of a good life.

In addition to providing epistemic guidance for identifying the right action or good life, moral exemplars are thought to serve an important motivational purpose. When we witness moral excellence, it often gives rise to elevation and admiration (Algoe & Haidt, 2009), emotions that can motivate emulation (Zagzebski, 2017). In this sense, moral learning, understood capaciously to include moving toward virtue, isn't merely epistemic. It can equip us with the ability to identify morally appropriate actions, but also with the affective and motivational responses needed to do the right thing and refrain from doing the wrong one. Both these epistemic and motivational goods acquired through moral learning will be germane to our argument.

Our focus will be on negative exemplars, but there are also challenges to accounting for the role of positive moral exemplars in moral education. First, to be effective, any moral exemplar needs to be *relatable*. We – the not-yet-virtuous – often fail to desire what is morally excellent. We crave social status, take pleasure in cruel jokes, and drink to excess. We might find ourselves viewing the virtuous person, or the moral saint, as alien and alienating: "dull-witted or humorless or bland" (Wolf, 1982, p. 422). Empirical studies have likewise found that attainable moral exemplars are more relatable than extraordinary exemplars, and so more effective in creating prosocial motivations in those who witness them (Han et al., 2017,

2022). It seems we need to be able to see enough of ourselves in them to be moved to become like them. If some threshold of identification isn't reached, we don't feel called to try and bridge the gap between us.

Second, even if a less-than-virtuous person can find the motivation, emulation might not be prudent. While a moral saint needn't have any qualms about accepting a dinner invitation with an attractive colleague while their partner is out of town, a less temperate person should "avoid [this] dinner like the plague" (Doris, 1998, p. 516). While "[the virtuous person] can attend the dinner without risk, an ordinary person cannot" (Doris, 1998, p. 518). Thus, for us, the morally mediocre, "doing what the virtuous person would do" can get us into trouble. Given our susceptibility to situational influences, the "virtuous [may be able to] tread without fear, but the rest of us cannot" (Doris, 1998, p. 519).

While much ink has been spilled over how positive moral exemplars contribute to moral learning, the role of negative exemplars and vicious persons has received less attention. Zagzebski (2017) briefly considers the idea that one could learn from negative exemplars via cautionary tales like "Pinocchio" and "The Boy Who Cried Wolf." She dismisses this possibility by invoking empirical findings from a study involving young children, concluding that the "evidence [indicates] that children emulate a [positive] exemplar but are not likely to learn from a negative example" (p. 131).[3]

While interesting, we think this point is orthogonal to our question. However an audience relates to, say, Pinocchio, it is significantly different from how it relates to a genuine antihero like *Breaking Bad*'s Walter White. Still, virtue ethicists may have other arguments against sympathetically engaging with a character like Walter White. They might claim, for example, that since antiheros are negative models, they should not be presented as objects of sympathy. Rather, they should be portrayed in ways that elicit negative emotions like disgust or contempt. Stories featuring antiheros do occasionally invite sympathy for their protagonists, though. Eaton (2010) worries about exactly this: "[Films and stories] make us vividly aware of a character's moral failings and simultaneously render her attractive, likeable, sympathetic ... [They] train our desires and sympathies on morally unworthy characters" (p. 519).[4]

Despite the antihero protagonist's manifest character flaws, the audience is lured into their perspective, and thus to empathize, even sympathize with them. Thus, the possibility that these stories exert a subtle but corrosive effect on the audience's moral character looms; they may be slowly being seduced away from virtue and down the pathway to a vicious life. Worries like this have a long history, going back at least as far as Plato (*Republic III*). As Jessica Moss explains, Plato criticized the "works of Homer based on their content – they represent people behaving immoderately – and psychological effect: as audience we weep and wail and behave as immoderately as the characters, and this undermines the order of our souls" (2007, p. 415).

So while there are problems that not-yet-virtuous people encounter in looking to positive moral exemplars, the problems associated with taking the perspective of and sympathizing with antihero protagonists seem more pressing, especially if doing so is likely to put one's moral character in jeopardy. In what follows, though, we push back on this line of thought. We propose a way in which moral learners can allow themselves to be drawn in by stories featuring antihero protagonists while also leveraging their sympathetic engagement for genuine moral learning, using the knowledge they acquire to steer away from becoming vicious versions of themselves.[5]

13.3 The View from Cultural Evolutionary Theory

Cultural evolutionists have developed a wide range of conceptual tools to help understand our species and the effects of its profound immersion in culture (Henrich, 2015; Morin, 2016; Richerson & Boyd, 2005; Sperber, 1996). Our reliance on culture – understood as any information or technology that is socially learned, rather than innately specified or individually discovered or invented – has deeply shaped human minds (Henrich et al., 2023; Kelly & De Block, 2022), and cultural evolutionists have put forward a range of hypotheses about how our minds interact with the cultural items they evolved to handle. This perspective yields several suggestions about why stories with antihero protagonists can be so curiously, perhaps alarmingly, compelling. We distinguish two kinds of answers.

Call the first *selfish meme* answers.[6] According to these, antihero stories provide little to no benefits to those who tell them, to those who consume them, or to those groups through which they circulate. Indeed, they may be harmful to humans, cultural parasites that "selfishly" exploit their hosts to get themselves replicated. Thus, the source of their cultural evolutionary success lies not in any moral lesson they convey nor in any other kind of benefit they bestow on their users. Rather, that success is mainly rooted in their ability to get themselves told, learned, and retold. And this is driven by the distinctive and powerful ways they push our psychological buttons.[7]

Selfish meme answers can be developed using the idea of a *cultural attractor* (Buskell, 2017):

> a term coined by Sperber (1996) and which scholars currently use to describe cultural possibilities favored during cultural transmission (Scott-Phillips et al., 2018). Sperber (2012) presented the happy ending of Little Red Riding Hood as an example of a cultural attractor; even if a person told the story with its happy ending cut off, others would be predisposed to reconstruct it during encoding and retelling, ensuring its reappearance and subsequent stability. Both

> psychological factors and aspects of the physical environment help determine which cultural traits are attractors (Miton et al., 2020; Sperber & Hirschfeld, 2004).
>
> (Singh, 2025, p. 3)

Many cultural attractors are generated by the interaction between minds and the cultural items they handle. On the one hand are particular facets of the psychological machinery that make up individual minds: perceptual apparatus associated with each sensory modality; cognitive mechanisms and heuristics underlying capacities for attention, parasite avoidance, folk biology, mate selection, cheater detection, imitation, mentalizing, reputation and status, social norms, moral judgment, etc.; as well as a whole range of social and epistemic emotions like fear, anger, shame, guilt, sympathy, awe, curiosity, etc. On the other are a range of cultural items – visual patterns, melodies, ideas, beliefs, norms, stories, and any other bits of information, skills, or technologies – that are passed by social learning between people and that can spread throughout a population.

For selfish meme answers, cultural items that fit better with some facet of our minds – that contain elements that are more salient, more compatible with intuitive concepts, more familiar, more memorable, more emotionally resonant, more motivating – will have an advantage over those that do not. By engaging any one of these facets of individual human minds, such cultural items will be more likely to proliferate. Those that fit with and engage *many* such facets will, as a result, enjoy an even larger advantage. Short catchy major key melodies are more likely to become ubiquitous earworms than long complex dissonant tone poems (see, e.g., Hoeschele & Fitch, 2022; Savage, 2019); dining norms that prohibit viscerally disgusting behaviors are more likely than arbitrary rules to remain in subsequent editions of etiquette manuals (Nichols, 2002); conceptions of supernatural entities are likely to spread and be more common if they are just the right amount of surprising, violating some but not too many of the intuitive expectations associated with our folk concepts (Atran, 2002; Boyer, 2001); electronic gaming machines and the social media interfaces and algorithmically guided endless scrolls that are based on them are so good at capturing and holding our attention that they have fundamentally reorganized many aspects of society (Schüll, 2012, also see Kelly, 2017).

Reframing our question: what is it about stories that feature an antihero protagonist that makes them "cultural possibilities favored during cultural transmission"? Versions of a selfish meme answer hold that there is a cultural attractor for antihero stories, but it is mostly, perhaps entirely, driven by these types of interactions. Stories centered on such characters are alluring and resilient, but just because they are psychological candy (Gottschall, 2012). They provide little moral nourishment, no ethically important skills or information. Neither do they earn their keep by contributing to evolutionary fitness, to either individuals or groups. Indeed, as virtue theory suggests, they may even be harmful, detrimental to their users, and a vehicle for the spread of vice. On this view, these stories spread because they contain a particularly potent blend of elements that fit well with different psychological mechanisms found in human minds, making them captivating even if they are not good for the humans themselves.[8]

A second kind of answer rejects the idea that antihero stories are purely or even mostly candy, selfishly replicating themselves while contributing little else.[9] Call these *moral inoculation* answers. These countenance the fact that, of course, tales of charismatic bad actors engage us psychologically – perhaps antihero protagonists are humanized, and these stories depict them as morally ambiguous in ways that especially engage our curiosity and mentalizing abilities (Wylie & Gantman, 2023). But, they claim, these stories provide crucial kinds of moral knowledge as well.

Adopting an evolutionary perspective on a topic makes it natural to think in terms of selection. Some traits will be selected *for*, because they provided a comparative advantage to those who possessed those traits versus those who didn't. Successful cultural traits would be expected to provide some benefit to those individuals who adopt them, or to generate group level benefits that accrue to an entire community, and that give them an advantage over other groups they are in competition with, often by improving coordination and cooperation (Richerson & Henrich, 2012; Richerson et al., 2016; Smaldino, 2014), key elements of morality (Curry et al., 2019). Those who defend moral inoculation answers accept that this is true of antihero stories as well.

They also accept that there is something special about *stories*, that the narrative format has special advantages when it comes to conveying of cultural information. Social learning, including moral learning, often happens via observation and imitation, sometimes scaffolded by explicit instruction given in the context of apprentice/mentor relationships (Sterelny, 2012). But people are limited in the situations they can directly observe or participate in firsthand, and crucial adaptive information, including moral knowledge, is often learned in other ways. Stories provide an especially effective kind of vehicle for the transmission of information of all sorts, but especially information about social dynamics. Stories are typically about characters doing things together, confronting challenges, navigating social conflict, making decisions, taking actions, and reckoning with the consequences. Stories allow people to encounter a much-expanded range of social situations than they would be able to experience or observe themselves. Stories can depict especially challenging situations and people and provide protective and preemptive knowledge that may be crucial to audiences if they are ever confronted with those types of situations and people "in the wild." For example, stories with antihero protagonists can serve as cautionary tales, providing knowledge that will help people resist the temptation of embracing a life of vice themselves. These stories can do this by laying out the disastrous outcomes to which a vicious character often leads. They can also prepare people for encounters in the real world with vicious others who are immoral but charismatic and seductive. Stories with antihero protagonists give their audience some low-stakes, imaginative experience with bad actors, thereby transmitting to them the beginnings of skills and virtuous traits they can use to identify, resist, and combat the allure of those bad actors when they are encountered in real life. In what follows, we will develop our own version of this kind of moral inoculation answer.

13.4 How Do We Learn Morality from Vicious People?

We make our case in stages. First, we note that even if it is correct that people can learn moral lessons about how to be virtuous from vicious people, there is still a question as to why, evolutionarily speaking, *stories* may be a particularly effective way to do it. One might be skeptical: we encounter enough vicious people in our everyday lives, what additional moral benefit could stories about them provide? To respond, we take a closer look at the moral benefit one can gain from encounters with actual bad actors. Such encounters, we argue, are likely to be influenced by several evolved heuristics and biases that can lead to in-group favoritism, out-group prejudice, and the fundamental attribution error (FAE). These heuristics and biases make it *difficult* to adequately extract moral knowledge from such real-life encounters. Yet, as we go on to argue in the next section, these difficulties can be circumvented by narratives, including antihero stories.

Several key features distinguish "antihero protagonists" in stories from "real-life vicious people." The protagonist of an antihero story can, of course, be a fictional character (Hannibal Lecter, Tony Soprano, Cersei Lannister), but can also be based on an actual person (Mussolini, Charles Manson, Anna Sorokin). For us, though, the idea of "real-life vicious people" is meant to refer to people known personally and interacted with in one's own life. The antihero protagonists, on the other hand, are characters in narratives, whether fictional or nonfictional, that moral learners come to know only through stories.

What kind of moral benefits can be acquired via interactions with bad actors? Many, perhaps, but the idea of a *feared self* is key to the answer we'll explore here. It's a platitude that positive role models can provide us with an ideal to aspire to, helping us to see a version of our best self that can function as a regulative ideal. Vicious people also can provide us with an ideal, only a negative one, helping us see versions of our possible but feared selves. Real-life vicious people can serve as negative exemplars, giving us a picture of who we might, but do *not* want, to become: "there but for the grace of God go I." They can provide us not just with an image of who we don't want to be, but can also instill strong motivations to avoid adopting their wicked ways (Athanassoulis, 2024; Waggoner, 2021). Just as other people can provide us with an image of our ideal or aspirational self, giving rise to admiration which motivates us to seek after that which we value, others can also serve as examples of our worst selves, giving rise to the powerful motivator of fear, used here to stay on the straight and narrow (Markus & Nurius, 1986).

But to reap these moral benefits from real-life bad actors, we need to think of them in the right way. We must, of course, understand their actions or way of life *as* bad, but perhaps more importantly, we also need to see it as a genuine possibility that *we, ourselves,* could end up like them. Real-life interactions, however, raise challenges for clearly seeing vicious people in both these ways at the same time because of the heuristics and social biases they activate.

Consider the FAE (Ross, 1977). To commit this error is to overestimate the influence of internal dispositional or personality factors in the production of other people's immoral behavior, while underestimating the influence of external situational factors. When judging oneself, alternatively, one is more apt to see immoral behavior as being strongly influenced by external (and often exculpating) situational factors, while underestimating the role of internal psychological factors like personality traits or other mental states. Those making the FAE judge others too harshly while being too lenient with themselves. The grace of this leniency, however, can also be extended to *some* others. When evaluating the moral slipups of those close to us (friends, family) or in-group members, we tend to be similarly forgiving, seeing greater importance in situation factors and glossing over character (Hewstone, 1990; Jones & Nisbett, 1971; Pettigrew, 1979). This extension of grace may be driven by another heuristic, one that generates in-group bias: when we witness an in-group member violate a moral norm, we might very well find ourselves pointing to situational factors that excuse them. We keep such violators in our in-group and write off their actions as not that bad and not their fault (Valdesolo & DeSteno, 2007).

However, even in-group members can do something so egregious as to lose the benefit of the doubt. If we take them to have done something seriously wrong, it can activate different evolved social heuristics that push our evaluation in another direction. For flagrant transgressions, our minds are outfitted with biases capable of "othering" the transgressor, creating psychological distance between them and us. Rather than excusing their behavior, we see the transgressor as entirely at fault, and see in their wrongdoing evidence of a flawed, perhaps morally tainted character. This, in turn, can be accompanied by strong motivations to punish and ostracize, which can push us to socially distance ourselves and relegate the wrongdoer to the fringes of our in-group. This "othering" can draw on a range of emotions, including, in the most extreme cases, disgust (Kelly, 2011). Disgust is an emotion that is "a response to witnessing violations of norms … expressions of disgust are a form of social rejection of the individual at whom they're directed" (Plakias, 2018, p. 5462). Disgust can bias social judgments, contributing to the exclusion of "individuals from in-groups through stigmatization and ostracism" (Rottman et al., 2018).

Given the presence of such biases, we will be disposed to other vicious people, relegating them to the social periphery and downgrading their status. This gives rise to the second part of the problem for moral learning: it will be difficult to see genuinely possible versions of ourselves in such marginalized and othered others. Rather than identifying and sympathizing with extreme wrongdoers, the activation of these evolved heuristics and biases makes us prone to dismiss the transgressor as *irrelevant* to our own self, actual, ideal, feared, or otherwise. The perceived risk of becoming bad like them is dramatically diminished; it becomes harder to admit that "there but for the grace of God go I" applies to her as well.[10]

In short, evolved features of our minds raise difficulties for learning moral lessons from the vicious people we encounter in our own everyday lives. Either we relate to the wrongdoer, see some of ourselves in them, and take them to be in our in-group, and thus become inclined to excuse their wrongdoing or downplay the transgressions. Or, if we see the wrongdoer as doing something grave and inexcusable, we become inclined to other them, to distance ourselves psychologically, marginalize them socially, and thus downplay any potentially edifying similarities between them and ourselves.

For us to engage in effective moral learning from vicious people, we need to be able to both (1) relate to them and identify with at least some of their choices but also (2) remain clear-eyed about the wrongness of their wrongdoing. We must be able to sympathize with them and to see a bit of ourselves in their viciousness. They must be sufficiently "close" to us, while also being removed enough that we can take their moral failings seriously. Antihero stories, we argue, are distinctively configured to meet these criteria.

13.5 How Antihero Stories Deliver Their Moral Benefits

The creator of any story selects the order in which they dispense information to their audience. A common (though not the only) way they get the audience on the side of antihero protagonists is by displaying their good moral features first, thus harnessing primacy effects and anchoring biases (Murdock, 1968; Tversky & Kahneman, 1974). When an antihero protagonist is initially portrayed as someone with "values or qualities roughly congruent with those we possess, or those that we wish to possess … we [will come to] identify with the character" (Smith, 2022, p. 2). Indeed, Kjeldgaard-Christiansen (2017) sees this strategy in *Breaking Bad:*

"Audiences take Walter's perspective and sympathize with him. ... We witness Walter's good qualities before experiencing his dark side ... [making use of the] bias to judge others based on first impressions..." (p. 111).

But antihero stories use a particular blend of other tricks to keep the audience engaged with their protagonist. They enable us to sympathetically identify with the character – which Smith calls *allegiance* – but they also allow for a more detached, merely epistemic identification, which he calls *alignment.* Alignment gives us a kind of third-person, God's eye view or "access to the actions, thoughts, and feelings of characters" (Smith, 2022, p. 6). These two different ways of identifying with a character need not overlap: one can have epistemic access to the ins and outs of a character's inner life without necessarily being sympathetic to them or what they are doing.

In many antihero stories, we are given to consistently experience alignment with the protagonist. But when it comes to allegiance, our identification is inconstant and often cyclical: we can lose it when they do something particularly evil, and narrative time and energy are required to rebuild it: "after a violent or immoral act, the viewer needs ... to 're-establish' their sympathy for the character" (García, 2016, p. 64). In our real lives, there is less room for this possibility: a person who "commit[s] crimes, abuse[s] and deceive[s is] the type of person that would repel us in real life" (García, 2016, p. 53), leaving less opportunity for regaining allegiance. We add to García's point the materials for an explanation: some of the main hurdles to regaining allegiance will be the evolved heuristics and biases described above. In short, allegiance, genuinely sympathetic identification with the antihero protagonist, can wax and wane over the course of a story, at times disappearing altogether. Even then, however, the audience can continue to be aligned with them via their deep access to the character's psychology. These fluctuations can happen at the same time as other compelling elements of the narrative retain the audience's attention, keeping their attention trained on the antihero and their machinations.[11]

Once again, our interactions with antihero protagonists differ significantly from our interactions with the bad actors we encounter in real life. These differences, in turn, influence how each can contribute to the acquisition of a particular kind of moral knowledge, namely, the development of an effective conception of our feared selves. A good antihero story can create and *sustain* in the audience prolonged identification via allegiance with the protagonist. As one's sympathy grows, one is better able to see parts of oneself, actual and possible, in the character. As the narrative manipulates that allegiance, often inviting and repelling it in cycles, audience members can both (1) condemn the antihero protagonist's bad actions as bad, rather than excusing them or simply writing them off as the result of situational factors, while also (2) not losing all identification with the antihero protagonist, and thus not falling into the trap of marginalizing and othering them. This

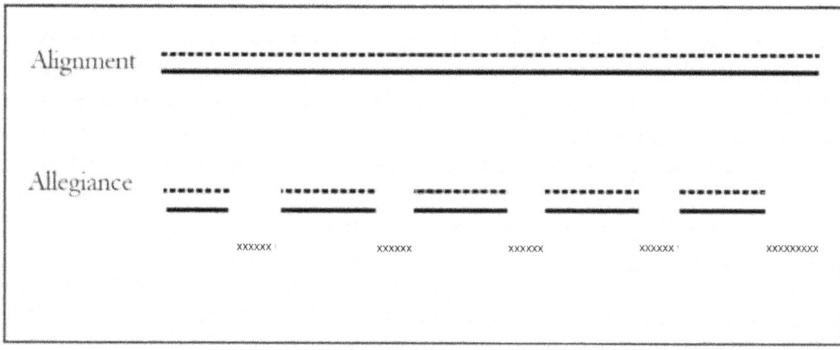

Figure 13.1 Anti-hero protagonist

is made possible by an initial allegiance that is minimized or lost when the antihero acts immorally, but then recovered again. An overarching narrative thread allows the audience's identification with the antihero protagonist to vacillate in its strength, but also to condemn their bad actions wherever they are in the cycle. This allows the audience to see their feared self in the antihero protagonist, ideally engaging the motivation to avoid becoming that version of themselves.

Consider Figure 13.1, which depicts how the audience can align with the antihero protagonist for the entirety of the story, if they have access to their thoughts, feelings, and actions. This is represented by the parallel dashed (audience) and solid (antihero protagonist) lines. At the same time, the audience's allegiance to the antihero protagonist can wax and wane, even eventually disappearing altogether. When allegiance is secured, the parallel dashed and solid lines are both present. But when the antihero protagonist commits some particularly egregious atrocity, the audience loses their allegiance. This is represented by the gaps in the dashed line (audience) and dips in the protagonist's line, depicted by a short line of x's. At these points in the narrative, the audience can more easily condemn the antihero protagonist's actions as bad, rather than excuse them.

But in the midst of the story, the audience typically doesn't completely abandon the antihero protagonist; allegiance can be regained. And it is this continual regaining of allegiance that allows us to see the antihero as someone close (enough) to ourselves. Where the dashed and solid lines are parallel, it is relatively easy to see ourselves in them. While we don't identify with the person who does the wrongdoing at the time of their wrongdoing, we are still consuming a story about how *someone like us* could also make horrible decisions, do things we clearly see as horrible. Additionally, we in the audience also have epistemic access to the antihero protagonist's psychological states and can identify via alignment. This enables us to discern patterns, to see how such perspectives can lead to such evil actions. In other words, alignment enables the audience to acquire a kind of explanatory template for how *Breaking Bad's* antihero protagonist, Walter White, gets from (i) wanting to build a nest egg for his wife and disabled child, in lieu of a Stage III cancer diagnosis, to (ii) becoming a murdering drug lord. The audience's cyclical alignment with Walter allows the audience to see themselves in Walter enough of the time that a line of thought like this becomes worth taking seriously: "If I see myself in Walter at t_1, t_3, t_5, and t_7, *and* I see how his relatable psychological states at those times lead to his irredeemable action at t_8, then why couldn't I, too, end up doing something like Walter? I'd better be careful!" The antihero protagonist has achieved the status of a genuinely possible self, but a cautionary, unwanted one. The fear generated by such a negative ideal can motivate us to beware similar situations and decisions, to recognize certain life patterns and turn away from certain paths, and, ideally, to turn toward, and remain committed to, more virtuous ones.

Life itself is not a story. Without narrative packaging, real life has less mechanisms that can sustain sympathy and identification with the bad actors we encounter, and we lack the access to their inner lives that would explain all their flaws and transgressions. If we are too close to the bad actor, one set of evolved biases can lead us to refrain from condemning them as bad, or even keep us from *seeing* them as bad. In terms of our graphic representation, we'd fail to register the "dips" at all. Alternatively, if their flaws and transgressions are too flagrant, a different set of evolved biases can push us to another extreme, marginalizing and othering the bad actor so that it is difficult for us to see much of ourselves in them at all. Once we've witnessed them at their worst, it is an uphill battle for them to regain our allegiance. Either way, we stop seeing them as a feared self and gain no moral benefit from them. These two alternatives can be seen in Figures 13.2 and 13.3.

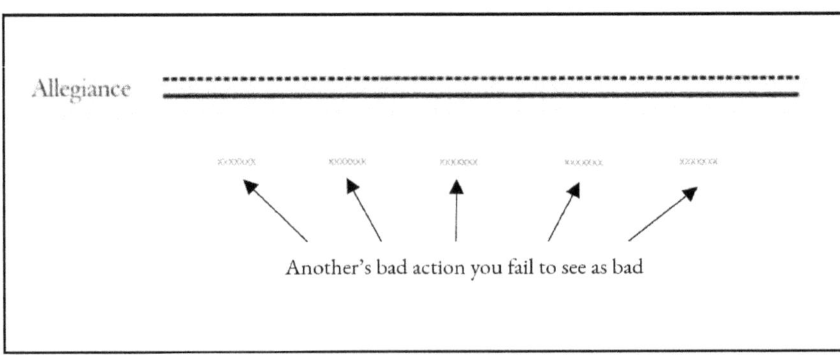

Figure 13.2 Real life close other

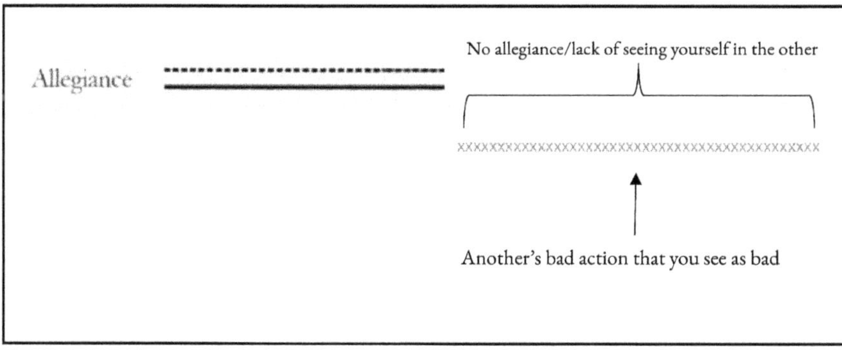

Figure 13.3 Real life out-group othered

We do not think it is impossible for bad actors in our actual lives to contribute to the construction of our feared selves. Rather, we think it *likely* that certain biases will be disruptive to the process. Stories of antihero protagonists, conversely, are particularly well suited to strike fear in our hearts – fear that takes the form of a clear picture of the kind of terrible person we could, if we are not careful, become.

13.6 Conclusion

Obviously, shows like *The Sopranos*, *Game of Thrones*, *Dexter*, *Mad Men*, and *Breaking Bad* have been extremely popular. We, like many of our friends and colleagues, have been successfully entertained. But we have also found ourselves puzzled, and a little alarmed, at how captivated we have been by the antiheroes in these stories, and wondered if we were getting edified or tarnished by our investment and the time we spent with their protagonists. We have argued here, however, that such tales proliferate for good reasons, and reasons that are good not just for the stories themselves, but for us. They are vectors of a distinctive kind of moral knowledge, inoculating us against moral bads in ways that our real-life experiences might not be able to. So go ahead and watch another episode of *Breaking Bad*. It just might help you become a better person.

Notes

1 In an interview with Stephen Colbert, the creator of *Breaking Bad*, Vince Gilligan, describes how Walt's confession to Jesse was done "with apparent glee" and out of "pure sadism" (Poniewozik, 2013).
2 Throughout this chapter, we focus on the character type of the antihero protagonist, rather than that of a standard villain. An antihero protagonist is presented as at least somewhat morally ambiguous, rather than straightforwardly

evil, and often manifests good intentions despite having obvious flaws and making bad decisions.

3 In this study, children were given classic moral stories that either depicted the protagonist as experiencing negative consequences of dishonesty ("Pinocchio" and "The Boy Who Cried Wolf") or positive consequences of honesty ("George Washington and the Cherry Tree"). While the former showed no effect on reductions in the children's lying, the latter did. Zagzebski seems to draw conclusions about how adults (don't) learn from negative exemplars from findings about children, ages 3–7. But it's far from clear that this is a valid move. This is because research shows that "[Y]oung people have some difficulties with considering the negative side of some behaviours, such as risky behaviours..." (Zhu & Tse, 2016, p. 145). These researchers likewise found that younger-mid adolescents (ages 12–17) were much less likely to have, and make use of, a 'feared self', compared to older adolescents (18+). Thus, while children may not be able to learn much from negative exemplars, given developmental changes regarding the registration of risks, we shouldn't think this likewise applies to adults.

4 For more on this issue from the lens of aesthetics and art criticism, see the exchange between Eaton (2012) and Carroll (2013).

5 Some have recently argued that a society can benefit, or develop societal-level virtues, from individual vicious people (see Astola, Bland & Alfano, 2024, for example). Our proposal is different from this insofar as we are considering how an individual person can benefit from, or develop *individual-level virtues*, from individual vicious people. Both of these stories might be relevant from an evolutionary perspective and need not be exclusive of the other. It should be noted that we are *not* claiming that a virtuous person can sympathize with the vicious person to help them be virtuous; see Bailey (2021) for discussion. Nor are we claiming that the not-yet-virtuous – the morally mediocre – always draw the right lessons from encounters with bad actors, even in narratives. Many a young lad has come away from *The Wolf of Wall Street* aspiring to be Jordan Belfort; Michael Lewis frequently notes his surprise at how many people who go into finance claim to have been inspired by his book *Liar's Poker*. We are here claiming, and trying to map out, how individuals *can* make use of vicious exemplars, especially antihero protagonists, to move away from vice and toward virtue.

6 A nod to Dawkins (1976); also see Dennett (1995) for discussion.

7 For an example of this kind of explanation for the success of certain urban legends over others, see Heath et al. (2001).

8 We acknowledge that many stories of villains might simply be entertaining without providing any moral benefits. But we think this isn't true for *all* stories of bad actors, especially those that are construed as cautionary tales. At least some will be well-positioned to play an important role in moral learning. However, it should also be noted that whether a story is effective in increasing moral learning related to moral improvement in the audience is at least somewhat dependent on the receptivity of the audience. Not all people will be influenced by stories in the same way.

9 We have presented selfish meme and moral inoculation as families of answers, instances of which might be worked out in more detail. We think of the terms as marking poles on a spectrum rather than discrete categories.

10 One might think this suggestion is empirically questionable. For instance, what has been called *The Black Sheep Effect* seems to indicate the opposite:

One is *harsher* on transgressors within one's in-group when compared to those in the out-group. This is because transgressors within one's in-group are a bigger threat to group integrity and harmony (Marques et al., 1998). But the story is a bit more complicated than this. Wang et al. (2016) found that the Black Sheep Effect holds only when the transgressor's intentions are clearly perceived; when the intention is perceived as ambiguous, in-group favoritism and leniency occur, rather than the Black Sheep Effect. Oftentimes, especially when it comes to close others, we are less likely to perceive another's intentions as straightforwardly hostile and instead see them with a certain amount of ambiguity. The in-group-out-group bias may, itself, impact how ambiguous vs. straightforwardly hostile we take the intentions of others to be.

11 Even in narratives there are ways to go too far, of course: "the possibility [also] exists that they can make you lose all of that sympathy gradually" (García, 2016, p. 60), as "the sympathetic allegiance that characterizes series starring antiheroes has its limits" (García, 2016, p. 85).

References

Algoe, S. B., & Haidt, J. (2009). Witnessing excellence in action: The 'Other-Praising' emotions of elevation, gratitude, and admiration. *The Journal of Positive Psychology*, *4*, 105–127.

Astola, M., Bland, S., & Alfano, M. (2024). Mandevillian vices. *Synthese*, *204*(1), 29.

Athanassoulis, N. (2024). The phronimos as a moral exemplar: Two internal objections and a proposed solution. *The Journal of Value Inquiry*, *58*(1), 131–150.

Atran, S. (2002). *In gods we trust: The evolutionary landscape of religion*. Oxford University Press. doi:10.1093/acprof:oso/9780195178036.001.0001

Bailey, O. (2021). Empathy with vicious perspectives? A puzzle about the moral limits of empathetic imagination. *Synthese*, *199*(3–4), 9621–9647.

Boyd, B. (2009). *On the origin of stories: Evolution, cognition, and fiction*. Belnap Press.

Boyer, P. (2001). *Religion explained: The evolutionary origins of religious thought*. Basic Books.

Buskell, A. (2017). What are cultural attractors? *Biology & Philosophy*, *32*(3), 377–394.

Camp, E. (2024). Stories and selves: A twisted love story about the meaning of life. *Royal Institute of Philosophy Supplement*, *95*, 157–179.

Carroll, N. (2013). Rough heroes: A response to A.W. Eaton. *The Journal of Aesthetics and Art Criticism*, *71*(4), 371–376.

Curry, O. S., Mullins, D. A., & Whitehouse, H. (2019). Is it good to cooperate? Testing the theory of morality-as-cooperation in 60 societies. *Current Anthropology*, *60*(1), 47–69.

Dawkins, R. (1976). *The selfish Gene*. Oxford University Press.

Dennett, D. C. (1995). *Darwin's dangerous idea*. Simon & Schuster.

Doris, J. (1998). Persons, situations, and virtue ethics. *Noûs*, *32*(4), 504–30.

Eaton, A. W. (2010). Rough heroes of the new Hollywood. *Revue Internationale de Philosophie*, *64*(4), 511–524.

Eaton, A. W. (2012). Robust immoralism. *The Journal of Aesthetics and Art Criticism*, *70*(3), 281–292.

García, A. N. (2016). Moral emotions, antiheroes and the limits of allegiance. In A. N. García (Ed.), *Emotions in contemporary TV series* (pp. 52–70). Palgrave Macmillan.
Gottschall, J. (2012). *The storytelling animal: How stories made us human*. Houghton Mifflin Harcourt.
Han, H., Kim, J., Jeong, C., & Cohen, G. L. (2017). Attainable and relevant moral exemplars are more effective than extraordinary exemplars in promoting voluntary service engagement. *Frontiers in Psychology*, *8*, 283.
Han, H., Workman, C. I., May, J., Scholtens, P., Dawson, K. J., Glenn, A. L., & Meindl, P. (2022). Which moral exemplars inspire prosociality? *Philosophical Psychology*, *35*(7), 943–970.
Heath, C., Bell, C., & Sternberg, E. (2001). Emotional selection in memes: The case of urban legends. *Journal of Personality & Social Psychology*, *81*, 1028–1041.
Henrich, J. (2015). *The secret of our success: How culture is driving human evolution, domesticating our species, and making us smarter*. Princeton University Press.
Henrich, J., Blasi, D., Curtain, C., Davis, H., Hong, Z., Kelly, D. and Kroupin, I. (2023). A cultural species and its cognitive phenoytpes: Implications for philosophy. *Review of Philosophy and Psychology*, *14*, 349–386.
Hewstone, M. (1990). The 'Ultimate attribution Error'? A review of the literature on intergroup causal attribution. *European Journal of Social Psychology*, *20*(4), 311–335.
Hoeschele, M., & Fitch, W. (2022). Cultural evolution: Conserved patterns of melodic evolution across musical cultures. *Current Biology*, *32*(6), R265–R267. https://doi.org/10.1016/j.cub.2022.01.080
Hursthouse, R. (1999). *On virtue ethics*. Oxford University Press.
Jones, E. E., & Nisbett, R. E. (1971). *The actor and the observer: Divergent perceptions of the causes of behavior*. General Learning Press.
Kelly, D. (2011). *Yuck! The nature and moral significance of disgust*. The MIT Press.
Kelly, D. (2017). Moral cheesecake, evolved psychology, and the debunking impulse. In R. Joyce (Ed.), *The Routledge handbook of evolution and philosophy* (pp. 342–358). Routledge Press.
Kelly, D., & De Block, A. (2022). Culture and cognitive science. In Edward N. Zalta (Ed.), *The Stanford encyclopedia of philosophy*. https://plato.stanford.edu/archives/sum2022/entries/culture-cogsci/
Kjeldgaard-Christiansen, J. (2017). The bad breaks of Walter White: An evolutionary approach to the fictional antihero. *Evolutionary Studies in Imaginative Culture*, *1*(1), 103–20.
Markus, H., & Nurius, P. (1986). Possible selves. *American Psychologist*, *41*(9), 954–969.
Marques, J. M., Yzerbyt, V. T., & Leyens, J. P. (1998). The 'Black sheep Effect': Extremity of judgments towards in-group members as a function of group identification. *European Journal of Social Psychology*, *18*, 1–16.
Miton, H., Wolf, T., Vesper, C., Knoblich, G., & Sperber, D. (2020). Motor constraints influence cultural evolution of rhythm. *Proceedings of the Royal Society B: Biological Sciences*, 287, 20202001.
Morin, O. (2016). *How Traditions Live and Die*. Oxford University Press.
Moss, J. (2007). What is imitative poetry and why is it bad? In G. R. F. Ferrari (ed.), *The Cambridge companion to Plato's Republic* (pp. 415–444). Cambridge University Press.

Murdock, B. B. Jr. (1968). Serial order effects in short-term memory. *Journal of Experimental Psychology*, *76*(4, Pt. 2), 1–15.

Nichols, S. (2002). On the genealogy of norms: A case for the role of emotion in cultural evolution. *Philosophy of Science*, *69*(2), 234–255.

Pettigrew, T. F. (1979). The ultimate attribution error: Extending Allport's cognitive analysis of prejudice. *Personality and Social Psychology Bulletin*, *5*(4), 461–476.

Plakias, A. (2018). The response model of moral disgust. *Synthese*, *195*(12), 5453–5472.

Poniewozik, J. (2013). "Vince Gilligan's *Breaking Bad* Postmortems: How Walt 'Won' Yet Lost His Soul." *Entertainment Time*. https://entertainment.time.com/2013/10/01/vince-gilligans-breaking-bad-postmortems-how-walt-won-yet-lost-his-soul/

Richerson, P., Bell, A., Demps, K., Frost, K., Hillis, V., Matthew, S., Newton, E., Narr, N., Newson, L., Ross, C., Smaldino, P., Waring, T., & Zefferman, M. (2016). Cultural group selection plays an essential role in explaining human cooperation: A sketch of the evidence. *Behavioural and Brain Sciences*, *39*, 1–68.

Richerson, P., & Boyd, R. (2005). *Not by genes alone: How culture transformed human evolution*. University of Chicago Press.

Richerson, P. J., & Henrich, J. (2012). Tribal social instincts and the cultural evolution of Institutions to solve collective action problems. *Cliodynamics: The Journal of Theoretical and Mathematical History*, *3*(1), 38–80.

Ross, L. (1977). The intuitive psychologist and his shortcomings: Distortions in the attribution process. In L. Berkowitz (Ed.), *Advances in experimental social psychology* (Vol. 10, pp. 174–221). Academic Press.

Rottman, J., DeJesus, J. M., & Gerdin, E. (2018). The social origins of disgust. In N. Strohminger & V. Kumar (Eds.), *The moral psychology of disgust* (pp. 27–52). Rowman & Littlefield.

Savage, P. E. (2019). Cultural evolution of music. *Palgrave Communications*, *5*, 16.

Schüll, N. D. (2012). *Addiction by design*. Princeton University Press.

Scott-Phillips, T., Blancke, S., & Heinz, C. (2018). Four misunderstandings about cultural attraction. *Evolutionary Anthropology*, *27*, 162–173.

Singh, M. (2025). Subjective selection, super-attractors, and the origins of the cultural manifold. *Behavioral and Brain Sciences*.

Smaldino, P. E. (2014). The cultural evolution of emergent group-level traits. *Behavioral and Brain Sciences*, *37*(3), 243–95.

Smith, M. (2022). *Engaging characters: Fiction, emotion, and the cinema* (2nd ed.). Oxford University Press.

Sperber, D. (1996). *Explaining culture: A naturalistic approach*. Blackwell.

Sperber, D. (2012). Cultural attractors. In J. Brockman (Ed.), *This will make you smarter*. Harper.

Sperber, D., & Hirschfeld, L. A. (2004). The cognitive foundations of cultural stability and diversity. *Trends in Cognitive Sciences*, *8*, 40–46.

Sterelny, K. (2012). *The evolved apprentice*. MIT Press.

Sugiyama, M. S. (2017). Oral storytelling as evidence of pedagogy in forager societies. *Frontiers in Psychology*, *8*: 471.

Tversky, A., & Kahneman, D. (1974). Judgment under uncertainty: Heuristics and biases. *Science*, *185*(4157), 1124–1131.

Vaage, M. B. (2015). *The antihero in American television*. Routledge.

Valdesolo, P., & DeSteno, D. (2007). Moral hypocrisy: Social groups and the flexibility of virtue. *Psychological Science*, *18*(8), 689–690.
Waggoner, M. (2021). Karen Stohr, *Minding the gap: Moral ideals and moral improvement*. *Journal of Moral Philosophy*, *18*(5), 546–549.
Wang, L., Zheng, J., Meng, L., Lu, Q., & Ma, Q. (2016). In-group favoritism or the black sheep effect: Perceived intentions modulate subjective responses to aggressive interactions. *Neuroscience Research*, *108*, 46–54.
Wolf, S. (1982). Moral saints. *The Journal of Philosophy*, *79*(8), 419–439.
Wylie, J., & Gantman, A. (2023). People are curious about immoral and morally ambiguous others. *Scientific Reports*, *13*, 7355.
Zagzebski, L. (2017). *Exemplarist moral theory*. Oxford University Press.
Zhu, S., & Tse, S. (2016). Possible selves, strategies and perceived likelihood among adolescents in Hong Kong: Desire and concern. *International Journal of Adolescence and Youth*, *21*(2), 135–149.

14 The Sociality Solution

Virtue, Vice, and Community

Jill B. Delston

14.1 Introduction

To paraphrase Aristotle, human beings are social animals (Aristotle, 1998). Social connection, friendship, and close relationships are a source of happiness, meaning, well-being, security, and pleasure for human beings (Haybron, 2013, p. 68; Shapiro et al., 2023). Although it is viewed as a truism, and obvious, that we are better off when we partake in significant interaction with others, measures to solve the most pressing problems blighting society often overlook this fact or contradict it. Public policy sometimes goes to great lengths to cut off human access to one another through strategies like incarceration without anticipating the inevitable problems that arise after doing so. According to the view defended here, attempting to solve problems caused by isolation through more isolation can perpetuate or exacerbate existing social ills.

In this chapter, I argue that social connectedness contributes to possible solutions for several major problems afflicting society. There is an evolutionary story to be told as to why this solution may be so effective and how it can support virtue. I argue that social connectedness prevents and resolves a great breadth of problems effectively, efficiently, and without major downsides, promoting virtue in the process. In fact, using social connection to solve one social problem may even help address other social problems in the process. For example, using social connection to address addiction could prevent or treat some of the major causes of depression. Viewing problems plaguing society as connected and having a common cause makes it possible to turn to this panacea of social connection to solve major societal problems more accurately and effectively.

The first step in the argument is to clarify the sort of social connection I advocate. I examine social connectedness both in an interpersonal and a community building sense. I employ a concept of social connection that is characterized by intimacy, vulnerability, authenticity, and honesty.

Second, I consider three issues that this model is well suited to address: healthy aging, mental health and addiction, and punishment and

DOI: 10.4324/9781003632641-19

recidivism. These sections, while only giving a short gloss on each issue, show how intimacy, vulnerability, authenticity, and honesty through social connection can contribute to addressing and preventing wrongs and harms while supporting virtue.

Third, I consider the explanatory role evolution plays in this solution. Specifically, I explore the role of sociality in the evolution of the human species as a means of explaining how social connectedness advances virtue-promoting behavior. This evolutionary explanation provides an underpinning for the argument that sociality is so effective at addressing social ills and fostering virtue. Without this evolutionary puzzle piece, sociality does not make much sense as a contributor to virtue, nor does isolation make much sense as a contributor to vice—why, after all, would sociality provide a solution to human problems? The solution appears *ad hoc* and without a foundation. With it, however, the reason sociality is necessary and effective to so many disparate concerns becomes apparent. In addition to explaining why social connection is a solution to human problems, the evolutionary story has particular import for virtue ethics due to the *eudaimonist* thesis that virtues benefit their possessor. Much of the debate in virtue ethics about the *eudaimonist* thesis focuses on the how: how do virtues benefit the possessor? This "how" sometimes seems mysterious when so many of the virtues are other-regarding and altruistic, and virtue ethicists need to explain how virtues nonetheless benefit the one expressing them. The evolutionary story, however, speaks to the why: why do virtues benefit the possessor? If virtues were selected for in a social species due to evolutionary success, then this evolutionary puzzle piece explains why they benefit the possessor.

In the last section, I consider several objections. Three are objections to the role of evolutionary psychology that address the limitations of my approach. Two are focused on the success of sociality: it is objected that some social connections are problematic, and some isolation is beneficial. The first objection catalogs the way social connection can sometimes lead to anti-social behavior and vicious behavior. However, I reply that these forms of social connection often lack the intimacy, vulnerability, authenticity, and honesty that I argue ought to characterize successful social connection. The second objection advances the idea that isolation is necessary or good. This view can often be explicit or implicit in social policy that severs social connectedness, such as incarceration, exile, and time-outs for children. However, I argue that these supposed solutions compound the problems they aim to solve. I conclude that the low-risk solutions offered by social connection far outweigh the potential downsides in using it to address societal ills.

14.2 What Is Social Connection?

I argue that preserving or increasing social connections would serve to help certain common social problems and that decreasing social connections can exacerbate them. I call my proposal to support and increase social connection the "sociality solution." To successfully defend social connection in this way, it is first necessary to define what I mean by social connection. Not all social interaction can constitute the sort of social connection that can handle major social problems because some forms of social interaction are problematic, as discussed in Section 14.7, or benign but unhelpful. For example, loneliness can occur in groups and among friends (*Social Isolation, Loneliness in Older People Pose Health Risks*, 2019). The "loneliness epidemic," which is characterized by the growing isolation in modern society, can demonstrate how loneliness can be a problem even as opportunities to connect increase. In addition, social media interaction and parasocial interaction can sometimes feel like social interaction but still lack the sorts of benefits of social connection that can characterize more meaningful relationships.[1]

The concept of social connection I consider in this chapter includes elements of intimacy, vulnerability, authenticity, and honesty. Intimacy can come in many different types and is often associated with romantic or sexual interaction. However, intimacy is not limited to physical touch or sex. Intimacy includes sharing and closeness in a reciprocal fashion in many contexts, including emotional intimacy (Sterelny, 2014)[2] and intellectual intimacy (Dover, 2022). In addition, intimacy is neither necessary nor sufficient for love. It is not necessary, because love can be felt outside of an interpersonal relationship and can have as its target someone the loving person has never met, as Sandra Langeslag has argued, such as parasocial relationships (Langeslag, 2024). It is not sufficient, because some moments of intimacy do not also include love. This closeness is often related to vulnerability, which can occur in interactions that also have trust and empathy, and which emphasizes human nature as including inherent weakness, which we can protect in one another if we share it.

Authenticity, or being true to yourself and your values in interactions, can foster social connectedness by allowing for self-expression, integrity, and openness. Authenticity, which plays a large role both in existentialism and in discussions of well-being (Anzani et al., 2019; Joshanloo & Weijers, 2024), can foster social connectedness by promoting closeness and honesty.

Honesty is an essential part of social connection because it promotes true connection, while dishonest communication can be isolating and increase loneliness instead of allaying it.[3] If authenticity is about being true to yourself, honesty describes a characteristic of being true to others. This distinction between authenticity and honesty captures two sides of the same coin, both of which I argue are integral to the understanding of sociality discussed in this chapter. In fact, authenticity without honesty or honesty without authenticity are both troubling for different reasons, discussed below. For example, being honest in dealings with another person without understanding one's true self (being authentic) can lead to feelings of hurt and betrayal. Being authentic but not honest in dealings with others poses the same risk—authentically cruel motivations may require dishonesty to achieve one's aims. Intimacy, vulnerability, authenticity, and honesty are also building blocks of trust, the development and maintenance of which are contributors to this discussion. Trust is therefore relevant to contemporary evolutionary discussions of human sociality.[4]

This characterization is meant to be value neutral. Social interaction that lacks these features is not necessarily problematic—it can be typical of the more superficial, common, or quick interaction that characterizes normal, daily life. Those sorts of interaction lacking in connection can even be valuable as offsetting loneliness and isolation (Tropp, 2025), but do not meet the higher standard of social connection that I advocate for here. Social interaction that includes these features can also be problematic, since someone can be openly malicious in ways that meet the honesty criterion. However, by narrowing social connection in this way, I hope to offer a more specific characterization that can speak to the sorts of social connection I argue ought to be fostered in public policy.

Social connection is relevant for this discussion for its two senses: interpersonal and community building. Interpersonal social connection can be found in families, friendships, acquaintances, and even in mental health settings like talk therapy. These types of social connections, when they include intimacy, vulnerability, authenticity, and honesty, can leave individuals feeling validated, safe, challenged, cared for, or loved. They can also happen outside the context of relationships and be found in moments that are nevertheless valuable in a person's life.

The NPR series, *My Unsung Hero*, follows "stories of people whose kindness left a lasting impression" ("My Unsung Hero," 2025). These transformative, validating, or even lifesaving stories often feature interactions with people who the storyteller never sees again, such as a stranger on a subway, a nurse in a hospital, or another shopper in line at the grocery store. While they can, of course, include long-term relationships, such as an employer, educator, family member, or friend, the predominance of the actions of strangers in the series exemplifies the way that social connection can occur inside or outside extended, committed relationships and can be found in momentary happenstance. By focusing on what I call social connection instead of a higher standard of love, reciprocity, and commitment, I aim to include these interactions between strangers as relevant to my thesis.

Social connection of the sort I discuss in this chapter can also be found outside of siloed interactions between individuals and be found in community contexts. In fact, fostering, supporting, and encouraging this sort of social connection can be an expressed aim of community building efforts. For example, while an individual friendship filled with social connection might play many positive roles in a person's life, it may not increase an individual's sense of belonging to a community unless it happens in a broader institutional or community context. Similarly, institutions can foster social connection that includes intimacy, vulnerability, authenticity, and honesty by giving people time, resources, and freedom to pursue or allow social connection.

In the next sections, I explore examples of the type of social connection I have defined here and the influence they can have on health and well-being across populations. It is my aim with these three case studies to argue that social connection is a preventative against vice and health problems and a boon to virtue and community building. If I can further show that this relationship between social connection, virtue, and well-being has its origin in our evolutionary history, then I will be on stronger ground for recommending social connection and resisting social isolation on an individual and public policy level. This argument has the goal of promoting social connection and minimizing social isolation to help prevent or solve major problems in applied ethics because it is appropriate or fitting to our species.

14.3 Healthy Aging

Studies on aging demonstrate the link between social connection and health and well-being. People who, in their old age, help others through caregiving for family members or volunteering in charity work tend to be healthier, more active, and live longer (Shen et al., 2017). Psychological factors like what is called a "sense of benevolence" tend to influence these caregiving activities and, in doing so, lead to healthier aging (Shen et al., 2017). In fact, studies show that having friends and volunteering can provide a stabilizing protective force in communities, preventing aging populations from moving away (Shen & Perry, 2014). These types of studies show not only the protective features of social connection like volunteering and friendship but also that people seek them out, being more likely to stay put when they have these social benefits and more likely to move when they do not.[5]

Research on longevity shows that social connection has a positive effect on health (National Institute on Aging, 2022). Loneliness and social isolation are associated with negative health outcomes including "high blood pressure, heart disease, obesity, a weakened immune system, anxiety, depression, cognitive decline, Alzheimer's disease and even death" as well as inflammation, cancer, metabolic traits, poor health behaviors and lifestyle choices, and cardiovascular disease more generally (Hu et al., 2021; Kobayashi & Steptoe, 2018; Schrempft et al., 2019; *Social Isolation, Loneliness in Older People Pose Health Risks*, 2019). For example, among older adults, social isolation and loneliness are predictive factors of decreased cognitive function (Cacioppo & Cacioppo, 2014; Donovan et al., 2017).

Empirical literatures on aging, which discuss the relationship between social connection and longevity, indicate the role of social connection on both our physical and mental health. Yet isolation often increases in old age, when common sources of meaningful interaction fall away as individuals retire and lose social connection through work, experience the death of a partner or spouse, or have health issues that make mobility and travel more difficult. Depending on its design and implementation, social policy that aims to foster, encourage, or even provide social connection would likely have a large impact on longevity, instead of social policy that allows individuals to fall through the cracks and lose community and interaction as they age.

Social policy that fosters, encourages, or facilitates social connection also has a relationship with virtue.[6] For example, the "sense of benevolence" (Shen et al., 2017) that increases both social connection and longevity is characterized through a virtue-theoretic lens in which the virtue of benevolence itself leads to social connection, which in turn fosters longevity. Similarly, the social connection that increases longevity is, to a large extent, coincident with virtuous actions, such as charity, volunteer work, and caregiving. For this reason, it is relevant to consider the ways that social policy that supports social connection also supports virtue.

14.4 Mental Health and Addiction

Social connection of the type I defined in Section 14.2 is useful not only in addressing physical health but also in addressing mental health, as shown by the mental health and addiction literature. Recreational drugs are thought to lead to a host of social ills that include the destruction of families, the loss of autonomy, inhibition of flourishing, early death, gun violence, and terrorism. For example, the relationship between endogenous opiates and bonding and the ways in which opioid addiction can threaten oxytocin levels and bonding behavior has been studied (Churchland, 2011, p. 34). This threat that psychotropic drugs pose ostensibly contributes to the explanation of some of the practices of punishment and prison sentences for the use, possession, and sale of narcotics and other recreational drugs. However, the incarceration for illegal drugs leads to its own set of social ills, independent of the direct causes of opioids and other drugs. Prison itself, discussed in Section 14.5, can indirectly contribute to the same social ills that are attributed to illegal drugs, like the breaking up of families, loss of autonomy, inhibition of human flourishing, early death, gun violence and other physical violence, as well as terrorism. In addition, the racist nature of drug laws contributes to a disproportionate impact on, for example, Black Americans by comparison to white Americans (Alexander, 2012). The high rate of recidivism, also discussed below, for those imprisoned in the American justice system only underlines these dangers. If drug addiction is a direct threat to human health and well-being, and if prison is a deterrent to drug use, these punishments may be warranted, even despite the indirect harms they cause. In this section, I challenge one understanding of the root causes of addiction as well as prison as the appropriate response. Leaving aside questions of criminalization, I argue that we have better preventative strategies, more effective deterrents, and stronger treatments at our disposal to handle the issue of drug addiction through social connection than the criminal justice system.

One understanding of addiction is that exposure to drugs is the major cause of addiction. For example, if individuals try drugs, then they will become addicted to them, and this exposure will lead to lifelong substance abuse. The "Just Say No" drug education program relies on this notion, claiming that if no children try drugs, no addiction to such drugs will be possible and children will be kept safe from the threat. Similarly, the thought that if children experiment with one illegal drug, they will fall into the wrong path leads to worries about "gateway drugs" and policies to deter even drugs deemed relatively safe, like marijuana, to prevent addiction to drugs deemed relatively unsafe, like cocaine. Once individuals are addicted to these strong, illegal drugs, the threat of social ills arises, putting whole communities in danger. According to this view, purity in society can prevent drug addiction: preventing access and exposure to illegal drugs can eliminate drug addiction entirely.

It is hard to become addicted to a drug that one can never try or never access, so preventing individuals from trying or accessing drugs has a certain logic to it. However, this understanding of the root causes of addiction, while supported by some evidence, might at worst get the story exactly backward, or at best only feature one direction of a two-way street. Substance abuse may cause social ills and psychological issues. But social ills and psychological issues may also cause substance abuse. If the former is true and the latter false, then this approach appears warranted. But if the latter is also true, then efforts to address substance abuse are sorely lacking and dangerously ignorant. In addition, because social connection can act as a deterrent to addiction and is a preventative force with few side effects, centering it in discussions of addiction is a valuable perspective.

A great deal of evidence supports the view that social ills and psychological problems contribute to drug addiction. Even when recreational drugs cause grave harm, other emotional ailments are often the root cause of the addiction, such as social isolation and poor mental health. A famous series of studies on cocaine addiction among rats that have come to be known as the "Rat Park" experiments shows that stimulating environments, rich social interaction, and intellectual challenges lead subjects to eschew even highly addictive drugs (Alexander, 2010; Alexander et al., 1978). When rats are isolated from one another in bare cages and then given access to both plain water and narcotic-laced water, they choose the narcotics repeatedly, leading to significant chemical addictions to drugs. However, when rats live in their natural social environments and are given access to social interaction on a regular basis, they avoid addiction. When these socially integrated rats are given access to both plain water and narcotic-laced water, they choose the plain water repeatedly, completely eschewing the narcotics and preventing addiction entirely. These rats tried the cocaine-laced water; they did not "just say no." But exposure to highly addictive substances was not sufficient to create addiction among the population. Instead, exposure to highly addictive substances only when combined with severe deprivation of social interaction was able to create the sort of addiction we associate with these opioids and other narcotics. Human beings are not rats, but these studies may still contribute to an understanding of social isolation and loneliness as potential contributors to addiction. Social connection, while not an antidote or treatment, remains a powerful preventative.

The journalist Johann Hari has argued that experiments like Rat Park are applicable to studies that use human beings. He argues that the opposite of addiction is not sobriety, but social connection (Hari, 2015b, 2016, 2019; Weiss, 2015). Similarly, treatment organizations like Alcoholics Anonymous (AA) use community building, mentorship, social connection, and group therapy to overcome addiction. According to this view, healthy attachments, social connections, and the ability to form and maintain them are a valuable treatment and a powerful preventative for addiction.

This evidence fits with the personal experience of many people who have avoided severe addiction and substance abuse, even after trying powerful drugs. For example, after a tooth extraction, back surgery, childbirth, or other routine medical procedure, many individuals are given the same sort of potent addictive pain medication that causes substance abuse when sold illegally without a prescription. Hari documents soldiers using heroin during the Vietnam War, yet stopping upon their return home without rehabilitation or other interventions (Hari, 2015a, 2016). According to Hari, the first taste of narcotics and opioids does not necessarily lead individuals down a path of abuse and addiction. Like the rats described above, they try the drugs, undergo sometimes extreme levels of addiction during the conditions of war, and go back to their regular lives, eschewing further opportunities for drug access when returned to their communities. Similarly, people who try Adderall, which is similar in chemical makeup to methamphetamine, do not necessarily become addicted. These substances can be addictive even when given in a legal and medical prescription, but despite their similarity to street drugs seen as highly addictive, only 1 in 10 people who try them continue toward addiction.

Given this evidence, one might argue that drug use is a necessary, though not sufficient, cause for addiction. But drug use is not necessarily required to produce the behaviors and attitudes associated with addiction. The term "dry drunk syndrome" describes a phenomenon in which an alcoholic abstains from alcohol but still exhibits the behavior and attitudes that they did while using (WebMD Editorial Contributors & DerSarkissian, 2023). Evidence of dry drunks and dry drunk syndrome, as well as post-acute withdrawal syndrome (PAWS), suggests that individuals can show a variety of addiction symptoms without the presence of the addictive substances or long after the withdrawal period (Chouinard & Chouinard, 2015; Cosci & Chouinard, 2020; Lerner & Klein, 2019). While these syndromes are not recognized by the DSM and the evidence for protracted withdrawal symptoms can be called into question, the extent to which clinicians and researchers identify these phenomena suggests that addictive substances may not be necessary or sufficient for addiction, addictive tendencies, or the negative consequences of addiction.

Treatment of addiction also offers clues as to the relationship between social connection and addiction. Treatment that includes vocational services, which allow patients to pursue productive contributions to society; family services, which promote the health and wellness of families in need and keep families together; mental health services, which include counseling and group therapy, each of which have intimate, vulnerable, authentic, honest communication as a large component of recovery; educational services, which promote the development of individual capacity; and legal services, which protect patients from a variety of different governmental interventions which can hurt long-term growth, all find some success in treating addiction. Each of these services includes creating, protecting, or promoting an individual's membership and involvement in a community with other people through social connection. That is why some experts claim that addiction might itself be a social disease in nature.

In addition to including the morally neutral characteristics of social connection, these treatments, responses to addiction, and preventative approaches also feature components of Aristotelian virtue, such as friendship, benevolence, care, prudence, temperance, and courage (Aristotle, 1999). This connection to virtue is salient also for the ways that addiction detracts from a conception of eudaimonia or flourishing and the ways that the responses to addiction contribute to it.

If involvement in one's community and the maintenance of healthy social attachments contribute to the prevention and treatment of drug addiction, then removing addicts from their communities and severing all social attachments is not likely to work. But that is precisely what imprisoning addicts for illegal use of controlled substances does. As discussed in the next section, not only is criminalization likely to be a poor fix for this issue, but also prisons can exacerbate the very problems that lead to drug abuse. Faced with shocking inadequacies in treating both mental illness and drug dependency, criminalization makes it very likely that prisons become the main "treatment centers" for mental health across the board.

As a result, far from employing the best antidotes to drug addiction, alcoholism, or substance abuse generally, most countries enact policies that promote addiction and force those least likely to be able to resist it into an environment that is statistically proven to lead to addiction.

14.5 Punishment and Recidivism

The criminal justice system punishes people in part by severing existing social connections as well as severing opportunity for social connections. Freedom and connection are so linked that the etymology bears out the relationship, as David Graeber and David Wengrow point out:

> In legal terms, at least, a slave has no family, no kin, no community; they can make no promises and forge no ongoing connections with other human beings. This is why the English word 'free' is actually derived from a root meaning 'friend.' Slaves could not have friends because they could not make commitments to others, since they were entirely under someone else's power…

Graeber and Wengrow contrast freedom with slavery, but freedom can also be contrasted with imprisonment, in which individuals' freedom is taken away as punishment for crimes committed with the same result that they lose connections with their families and communities.

Imprisonment makes sense as a preventative measure that disrupts an individual's ability to commit crime precisely because it severs them from the communities and individuals against whom they perpetrated those crimes. For example, a perpetrator of embezzlement is cut off from their resources to commit the crime again while in prison. However, incarceration does not alleviate the conditions that led to the initial crime and often makes those conditions much worse (Darwall & Darwall, 2018; Davis et al., 2022; Lippke, 2018; Shelby, 2022). Incarceration severs an individual from the kinds of social connection and meaningful relationships that make crime less likely including family and loved ones; puts individuals in connection with a variety of other criminals; makes it necessary to engage in crime to survive prison; undermines housing and employment opportunities; is a mere temporary measure; leads to a host of secondary punishments after release; and contributes to other impacts detrimental to sociality (Hoskins, 2016, 2019).[7] So far are these conditions from preventing recidivism or rehabilitating prisoners that they seem perfectly designed to increase crime.[8]

Of course, not all crime is vicious, and not all vice is criminal, since law and morality pull apart. To the extent that prison is used to reduce social ills like violence and theft, however, it is out of step with the types of policies that are likely to cause those reductions. By increasing social isolation, breaking up families, and making it difficult for prisoners to see those on the outside, social connection and community suffer. When social connection suffers, the health and well-being of prisoners suffer, too, making it difficult to get their lives back.

In addition, more social isolation awaits those imprisoned given the prevalence of solitary confinement. Though it goes by many names, including "segregation" and "protective housing," solitary confinement generally means removing all human contact from a prisoner for days to years. Solitary confinement is used as a punishment during prison or jail when a prisoner breaks rules, but also includes "protective custody" or "administrative segregation," in which a prisoner is held in solitary confinement because they are threatened by other prisoners or pose a threat to other prisoners, respectively (Merrefield, 2023).

It is difficult to track the number of people exposed to solitary confinement within the United States. Current statistics do not include those held in solitary confinement for fewer than 15 days; held in jails, military, juvenile, or immigration detention facilities; and do not include those held in solitary confinement for fewer than 22 hours a day (*Solitary Confinement in the United States: The Facts*, 2023). Even with these qualifications, tens of thousands of people live in solitary confinement just in the United States (Merrefield, 2023; *Solitary Confinement in the United States: The Facts*, 2023). The United States has one of the highest per capita imprisonment rates in the world, contributing 20% of the world's imprisoned population but only 5% of the world's population (Bieber & Johnson, 2024). Two million Americans are imprisoned, while another 5 million are jailed annually (Bieber & Johnson, 2024). That means that not only is mass incarceration a major problem in the United States (Alexander, 2012; Davis et al., 2022; Gardner & Weber, 2018) but also millions of Americans live under threat of solitary confinement.

Problems attend the criminal justice system, then, both in principle and in practice.[9] The idea that incarceration may have significant problems is a long-standing view that has gained more traction in recent years, especially through anti-carceral or non-carceral feminism (Davis, 2003; Davis et al., 2022; Emerick & Yap, 2023; Gruber, 2021; Srinivasan, 2021; Thusi, 2021). Adding to this discussion is the idea that because incarceration consists in severing social connections and isolating human beings, it not only is unlikely to alleviate the problems that lead to incarceration but is also likely to compound them.

So far, I considered three case studies of social connection: aging, addiction, and punishment. These represent a small sample of the places where social connection can lead to better health, happiness, and virtue, and social isolation can harm individuals and communities and undermine virtue. For example, social isolation can increase poverty while social connection and community building can offset it (Shelby, 2016; Werhane et al., 2020). Similarly, suburbs, racism, redlining, and other forms of social separation, segregation, and isolation contribute to environmental degradation and contribute to the climate change crisis. For reasons I discuss in Section 14.6, a sociality solution can contribute to successful child-rearing, while one that emphasizes social isolation or separation, even in punishment, can harm children.

14.6 Evolution

Why does social connection play such a large role in happiness and well-being, mental health, and physical longevity, not to mention increasing "prosocial behavior" or virtue and decreasing vice, like reductions in crime and recidivism? Why is the sociality solution effective? An evolutionary story can provide insight into the relationship between what is good for human beings, their virtues, and their origins. If human beings are characteristically social, and if sociality is central to our success (Churchland, 2011; Henrich, 2016; Wrangham, 2019), then it would make sense that they thrive in such environments, where thriving can be given a *eudaimonist* gloss that includes virtue. In the words of the editors to this volume, "...we evolved to treat virtue and desert as intrinsically good" (Wiegman & Dale, 2024).

Scholars of evolution often point to sociality as the connective tissue between virtue theory and the evolutionary perspective (Mizzoni, 2019; Narvaez, 2014; Ruse, 1990; Wiegman & Dale, 2024). For example, in his analysis of Aristotelian virtue ethics from an evolutionary perspective, Michael Ruse points out the "absolutely central recognition of human sociality" (Ruse, 1990). In "The Neurobiology and Development of Human Morality in Light of Evolution," Darcia Narvaez argues that "the *social* aspect of virtuous morality concerns how individuals and communities make meaning of social life" (Narvaez, 2014).[10]

This evolutionary approach is valuable because it offers a clear explanation for why human beings need social connection to thrive, why social connection would lead to virtue, and vice versa. In short, evolution can provide an origin story for the "sociality solution," and a useful explanation of why social connection leading to virtue can be found in so many different case studies. It can also provide a helpful bulwark against doubt that social connection really does lead to virtue and well-being in individuals and communities; knowing why they are related precludes the need to search for further explanation. It also provides a helpful recognition of the role of sociality in the evolution of our species. James Hunt argues that, "Sociality, like multicellularity, has appeared numerous times, in diverse taxa, and reached many different levels of integration" (Churchland, 2011, p. 33; Pennisi, 2009). There is nothing to gain by ignoring this evolutionary backstory to species cooperation or success.

14.7 Objections and Replies

One objection to the evolutionary perspective offered here is an is-ought problem: so what if evolution contributes to the origin of social connection leading to virtue? It is impossible to derive an "ought" (one ought to pursue social connection) from an "is" (social connection was evolutionarily advantageous for human ancestors) (Hume & Mossner, 1985, p. 521). In other words, the thesis that human need for social connection is a result of evolution does not provide the basis for any normative claims. Similarly, a Moore-type objection can be given in which the evolutionary process underpinning of my argument is ceded, but it is objected that this evolutionary background cannot justify the normative claims I have included. According to G.E. Moore, moral properties like goodness cannot be defined by or reduced to natural properties like evolution (Moore, 2012). For example, Michael Ruse states, "Whether Moore was right in thinking goodness an intuited nonnatural property, he was surely right in seeing a fallacy in the equation of 'that which has evolved' with 'that which is good'" (Ruse, 1990).

In response to this version of the naturalistic fallacy, I agree that we ought to resist the notion that an evolutionary perspective can justify the normative claims I make in this chapter. Instead, to the extent that the evolutionary perspective is useful, it is useful as an explanation, not a justification (Wiegman & Dale, 2024). This response fits with a common understanding of evolutionary psychology and sociobiology, such as that, "Sociobiologists championed evolutionary theory's explanatory power to give origin explanations of the social lives of animals, including human beings" (Smith, 2020). If the justification is ceded and the explanation is retained, then an evolutionary story also can support a traditional approach to the self in virtue ethics that virtues benefit their possessor (Annas, 2011; Hursthouse, 2001).[11] This response can also be appropriately responsive to humanity itself. For example, Patricia S. Churchland characterizes this naturalism as finding "the roots of morality in how we are, what we care about, and what matters to us—in our nature" (Churchland, 2018, p. 6). Thus, the evolutionary argument I defend is not itself a moral argument, but rather a scientific argument that underpins, supports, and explains the moral argument. In other words, evolution does not provide moral guidance on how we ought to act. However, moral theory does rely on an evolutionary argument to explain why a certain kind of virtuous sociality benefits the possessor. The stakes for virtue ethics, then, are high. If virtues do not benefit the possessor, or if they have no reason to, then an overhaul in virtue theory may be required.

A second objection resists the evolutionary story as a plausible explanation on scientific grounds because it is essentially unfalsifiable. The concern here is that there is insufficient data to justify the conclusion, in part because human beings actually did evolve this way, and thus no counterfactuals or experimental proposals are possible, and in part because it is difficult, if not impossible, to prove the evolutionary basis for certain species-level characteristics. For example, in "*Is Evolutionary Psychology Possible?*" Subrena E. Smith argues that insufficient evidence attends the notion that our current psychology is meaningfully connected to selections and adaptations over tens of thousands of years (Smith, 2020). She argues that the methodology in evolutionary psychology is not just unfounded, but impossible. In what she calls "the matching problem," Smith argues that not only must similarities be found in psychological characteristics and behavior between our present-day selves and our evolutionary ancestors, but also those similarities must have the same functions, causes, kinds, and be able to trace a consistent descendant line. The high likelihood of ontogenetic processes over a single individual lifetime, among other explanations, she argues, undermines this line of thought (Smith, 2020). For that reason, similarities are not by themselves evidence of preserved function or causal relationships. Prevalence of convergent evolution would be another example of Smith's line of argument, since the same or analogous structures would be present in multiple taxa without a common ancestor.

While this argument does generally assume the possibility and plausibility of evolutionary psychology, a field of inquiry that can certainly be undermined, it is beyond the scope of this chapter to establish the success of that discipline. In response to this objection, I bite the bullet by ceding to the opposition that the empirical evidence may be limited and that my thesis only goes as far as the empirical evidence can support it. This concession demonstrates the bounds on my approach. It is true that the empirical evidence must be limited in part because there are no controls on testing that hypothesis in the historical record (and seeking out controls would likely be unethical). But however limited it is interpreted to be, evidence does suggest that human beings evolved to be social animals that do not thrive outside of social connection and do thrive within it (Churchland, 2011; Henrich, 2016; Wrangham, 2019). Furthermore, it is not my aim to provide the empirical, evolutionary argument itself, but rather to rely on that argument to explain its role and inclusion in a virtue theory. Regardless of whether it is possible to prove that human characteristics are evolutionarily selected, that evolutionary explanation, if true, provides a significant contribution to virtue ethics and the *eudaimonist* thesis by explaining why the sociality solution is so effective and why virtues benefit their possessor. If the evolutionary story is wrong, then so much the worse for virtue ethics.

A third objection is that an evolutionary-based ethics can be used to uphold the status quo or to uphold reactionary politics. "Social Darwinism," a libertarian and highly self-interested approach to ethics, is one example of this view, as are many of the sexist claims coming out of evolutionary biology that seem to justify gendered oppression (Smith, 2020). To this objection, it must be conceded that bias can creep into the discussion. However, as I hope is clear from the discussion, that does not mean that a reactionary, conservative, self-interested, or oppressive approach to human interaction must be part of the story (Sober et al., 2003). Instead, through the emphasis on social connection, there is room for a different story that emphasizes virtue.

The last two objections I consider concern the success of sociality at achieving the aims I set out. One such objection to my view is that some types of social connection facilitate vice, inhibit the development of virtue, fracture communities, or otherwise undermine the claims I have made in this chapter about the role of social connection in identifiable societal benefits. Examples of these types of social connection include the sort of abuse and sexual assault that can sometimes thrive in communities like fraternities, schools, and cults. Since cultures of abuse can take hold in such communities, and since those communities are presumably founded on the sorts of social connection that allow them to continue, it seems that social connection can foster both virtue and vice.

In response, I argue that these sorts of communities are characterized by social interaction that lacks the sort of features I outline in the definition of social connection, including intimacy, authenticity, and honesty. The attitudes toward sex, in particular, as they play a role in rape culture, are antithetical to intimacy, emotional connection, vulnerability, and honesty. These communities foster interaction without intimacy, physical vulnerability without emotional vulnerability, or authenticity without honesty. For example, the crime and vice of rape and rape culture incorporate elements of dishonesty, especially toward the victims of rape and sexual assault. If rape culture expresses an individual's true, authentic self and sexual expression, or if it becomes authentic to them through the social contagion present in their community, then the individual would have to hide or suppress this characteristic as a means toward achieving the goal of rape or sexual assault. Similarly, if the individual is suppressing their preferences or identity to conform to the rape culture present in their community, their social interactions would be lacking in both honesty and authenticity.

The last objection I consider is the role of isolation in better achieving the virtues of community and social benefits. If social isolation is a net benefit in pursuit of virtue or other social gains, then my argument would be overstating and overemphasizing the role of social connection in these same gains. Examples of using social isolation to treat vice and other socially undesirable behavior are myriad, from solitary confinement in prisons, to prisons themselves, to time outs in child-rearing. I have already discussed some of these in Section 14.4 concerning prison and other forms of punishment.

Time outs for children, though plausible as an example of temporary social isolation, are a controversial component of child-rearing (Holden et al., 2022; James, 2020). One feature of note about time outs for child-rearing is that it is generally agreed upon by its defendants to be effective or valuable only in the context of a larger intimacy and deep connection between parent and child.[12] In fact, advocates of time outs for children are unlikely to view their methodology as an example of social isolation. Effective use of time outs generally happens in larger contexts of interaction and is intended to be combined with (1) warnings to the child; (2) expressed reasons for a time out; and (3) a duration and endpoint (Everett et al., 2010). Thus, defenders of time outs would likely agree that it is the social connection that is valuable as well as what makes time outs effective, which is precisely why taking it away is deemed a negative consequence for behavior.

A major danger of time outs or social isolation is that they make the unwanted behavior more likely. In other words, if a child is acting out of a need for social connection, then withholding social connection is unlikely to be an effective tool in achieving the desired result.

In addition, research defending the value of time outs is generally coincident with comparisons to and benefits over other alternatives to behavior modification that are much worse by all accounts, such as corporal punishment and other physical assaults on the child (Everett et al., 2010). The success of time outs by comparison to spanking, for example, would be questionable. In fact, when time outs are compared to alternatives that emphasize social connection, such as "time ins," they can fare less favorably according to parents in terms of ease of use and successful outcomes (Holden et al., 2022). Whereas a time out involves a child sitting alone for a few minutes as a punishment for their behavior, a time in involves a caregiver working with the child to listen to, empathize with, or otherwise validate the child's emotions and give the child tools for emotional regulation that hopefully will help behavior in the future. A successful moral upbringing that teaches children right from wrong, imparts to them knowledge of manners and mores, and does so in a way without harming them is a larger discussion (Delston, 2024). However, I hope to have responded to the objection to my view that social isolation is a necessary component of child-rearing or is likely to lead to virtue.

While "good" social isolation to pursue virtue or community is a plausible objection, it is not successful enough empirically to warrant throwing out social connection as the primary and most valuable means toward community and virtue. In addition, isolation's role as an objection is only possible because adherents value social connection so supremely.

If the suggested remedies based on social connection and leading to virtue are tried and do not lower the incidence of the social ills or vice, then my thesis, which relies on the empirical evidence supporting the relationship between social connection and virtue, also fails. For this reason, the thesis is distinct from views that offer in-principle reasons against the practices in question. For example, some argue that prison is in violation of human dignity (Brooks, 2018; Davis, 2003; Davis et al., 2022; Derek Jeffreys, 2018; Emerick & Yap, 2023). According to this view, the success of an anti-carceral position in reducing crime is distinct from the arguments against incarceration. The view offered in this chapter is much more limited and is focused on the relationship between social connection and virtue.

14.8 Conclusion

Human beings need social connection, and we need to be enmeshed in a community. These features of our social nature contribute to our mental and physical health, our happiness and well-being, and our virtue. These features are also likely to have an evolutionary origin as a social species. However, we often overlook, undermine, or minimize these features of the self in public policy and individual action, to our consistent detriment, using social isolation, separation, and segregation when social connection is called for. In this chapter, I argued that promoting social connection, especially of the type that includes intimacy and vulnerability as well as authenticity and honesty, is a low-risk, high-reward investment. We ignore our need for social connection at our peril.[13]

Notes

1 In *All About Love,* bell hooks describes love as, "a combination of care, commitment, trust, knowledge, responsibility, and respect" (hooks, 2018, p. 8). She adheres to M. Scott Peck's definition in *The Road Less Traveled:* "the will to extend one's self for the purpose of nurturing one's own or another's spiritual growth" (Peck, 2003, p. 81). While these accounts are instructive, love has to be a higher standard than social connection because it (hopefully) characterizes the most important relationships in our lives, while social connection can be a broader and more inclusive concept. One can have a social connection to someone without love, as when making a new friend, or love without social connection, as when a loved one is far away or out of touch. However, hooks' definition of love is both relevant and helpful to an investigation of social connection.

2 Kim Sterelny discusses Krebs-Dawkins signals, which are signals aimed not at sharing information but instead at altering the actions of the sender or recipient (Dawkins & Krebs, 1978). These signals raise the specter of dishonest communication as these signals can deceive receivers into actions that have selfish value for senders. He considers the motivational value of these signals and their emotional content: "So joint action in emotionally amplified situations reinforces mutual bonds more powerfully than collective action in calmer emotional waters" (Sterelny, 2014, p. 110).

3 For discussions of the benefits and harms of conversation, see Ross (2022, 2025).

4 Thank you to Charlie Kurth for this point.

5 This point is not to overlook the significant work and sometimes burden of caregiving. Thank you to Charlie Kurth for this point. For more on this issue and the problem of care exploitation, see McKittrick-Sweitzer (2021a, 2021b).
6 Not all caregiving would be classified as altruistic from an evolutionary perspective, since care for one's offspring increases one's fitness. For this reason, a distinction between psychological altruism and evolutionary altruism is useful (Ruse, 1990; Sober et al., 2003; Stich, 2007).
7 Prison does provide opportunities to create new social connections. I thank Jason Gardner for this point. The opportunity to create new social connection may be hindered by the severing of all previous contacts and the grief that may cause, however.
8 This argument will not reach the retributivist, who is not motivated by prevention or reduction in crime but instead by desert. I accept this limitation of the argument, with the caveat that retributivists may be separately interested in ways to reduce crime.
9 For an account of the relationship between retributive intuitions and evolutionary processes, see Wiegman (2017, 2019). For a discussion of punishment and evolutionary processes, see Churchland (2011, pp. 81–86). An evolutionary underpinning of retribution and revenge as intrinsically valuable or even punishment as valuable need not justify prison.
10 Inclusion in society would be insufficient to establish my thesis, since one could live in society without achieving social connection. I thank Howard Curzer for this point. However, its support for the sociality solution is still relevant to my discussion.
11 In the words of Bernard Williams, "the important point is that evolutionary biology is not at all directly concerned with the well-being of the individual, but with fitness, which is the likelihood of that individual's leaving offspring" (Williams, 1985). Benefits to the individual, then, would be relevant insofar as they support fitness.
12 This feature places time outs in a different category than traditional punishments like the criminal justice system and prison. While the latter can include rehabilitative aims, they need not. To the extent that the two are similar, however, my response to each will work for the other.
13 I owe a debt of gratitude to many people for feedback on this chapter. Thank you to Isaac Wiegman for incredibly useful comments at all stages of the process. Thank you also to Michael Dale for valuable input. Thanks also go to Jason Gardner, Charlie Kurth, and Howard Curzer for detailed and helpful suggestions and recommendations. I would also like to thank the NextGen Precision Health Initiative for their generous support of my research.

References

Alexander, B. K. (2010). *The myth of drug-induced addiction*. http://www.parl.gc.ca/Content/SEN/Committee/371/ille/presentation/alexender-e.htm

Alexander, B. K., Coambs, R. B., & Hadaway, P. F. (1978). The effect of housing and gender on morphine self-administration in rats. *Psychopharmacology*, *58*(2), 175–179.

Alexander, M. (2012). *The new Jim Crow: Mass incarceration in the age of colorblindness*. The New Press.

Annas, J. (2011). *Intelligent virtue* (1st ed.). Oxford University Press.

Anzani, A., Morris, E. R., & Galupo, M. P. (2019). From absence of microaggressions to seeing authentic gender: Transgender clients' experiences with

microaffirmations in therapy. *Journal of LGBT Issues in Counseling*, *13*(4), 258–275. https://doi.org/10.1080/15538605.2019.1662359
Aristotle (1998). *Politics*. Hackett Publishing Company.
Aristotle (1999). *Nicomachean ethics* (2nd ed.). Hackett Publishing Co.
Bieber, C., & Johnson, J. (2024, May 9). Mass Incarceration in America: Stats and Facts. *Forbes Advisor*. https://www.forbes.com/advisor/legal/criminal-defense/mass-incarceration-in-america-stats-and-facts/
Brooks, T. (2018). Restorative justice and punitive restoration. In M. Gardner & M. Weber (Eds.), *The ethics of policing and imprisonment*. Springer International Publishing. https://link-springer-com.ezproxy.umsl.edu/chapter/10.1007/978-3-319-97770-6_8
Cacioppo, J. T., & Cacioppo, S. (2014). Older adults reporting social isolation or loneliness show poorer cognitive function 4 years later. *Evidence-Based Nursing*, *17*(2), 59–60. https://doi.org/10.1136/eb-2013-101379
Chouinard, G., & Chouinard, V.-A. (2015). New classification of selective serotonin reuptake inhibitor withdrawal. *Psychotherapy and Psychosomatics*, *84*(2), 63–71. https://doi.org/10.1159/000371865
Churchland, P. S. (2011). *Braintrust: What neuroscience tells us about morality*. Princeton University Press. https://doi.org/10.2307/j.ctt1wx93nm
Cosci, F., & Chouinard, G. (2020). Acute and persistent withdrawal syndromes following discontinuation of psychotropic medications. *Psychotherapy and Psychosomatics*, *89*(5), 283–306. https://doi.org/10.1159/000506868
Darwall, S., & Darwall, W. (2018). Criminal process as mutual accountability: Mass incarceration, carcerality, and abolition. In: *The ethics of policing and imprisonment* (pp. 187–212). Palgrave Macmillan. https://doi.org/10.1007/978-3-319-97770-6_11
Davis, A. Y. (2003). *Are prisons obsolete?* Seven Stories Press.
Davis, A. Y., Dent, G., Meiners, E. R., & Richie, B. E. (2022). *Abolition. Feminism. Now*. Haymarket Books.
Dawkins, R., & Krebs, J. R. (1978). Animal signals: Information or manipulation? In J. R. Krebs & N. B. Davies (Eds.), *Behavioural ecology: An evolutionary approach* (pp. 282–309). Blackwell Scientific.
Donovan, N. J., Wu, Q., Rentz, D. M., Sperling, R. A., Marshall, G. A., & Glymour, M. M. (2017). Loneliness, depression and cognitive function in older U.S. adults. *International Journal of Geriatric Psychiatry*, *32*(5), 564–573. https://doi.org/10.1002/gps.4495
Dover, D. (2022). The conversational self. *Mind*, *131*(521), 193–230. https://doi.org/10.1093/mind/fzab069
Emerick, B., & Yap, A. (2023). *Not giving up on people: A feminist case for prison abolition*. Rowman & Littlefield Publishers.
Everett, G. E., Hupp, S. D. A., & Olmi, D. J. (2010). Time-out with parents: A descriptive analysis of 30 years of research. *Education and Treatment of Children*, *33*(2), 235–259.
Gardner, M., & Weber, M. (Eds.). (2018). *The ethics of policing and imprisonment*. Springer International Publishing. https://doi.org/10.1007/978-3-319-97770-6
Gruber, A. (2021). *The feminist war on crime: The unexpected role of women's liberation in mass incarceration*. University of California Press. https://www.ucpress.edu/books/the-feminist-war-on-crime/paper
Hari, J. (2015a, January 20). *The likely cause of addiction has been discovered, and it is not what you think*. HuffPost. https://www.huffpost.com/entry/the-real-cause-of-addicti_b_6506936

Hari, J. (2015b, June). *Everything you think you know about addiction is wrong* [TEDTalk]. https://www.ted.com/talks/johann_hari_everything_you_think_you_know_about_addiction_is_wrong
Hari, J. (2016). *Chasing the scream: The opposite of addiction is connection* (reprint ed.). Bloomsbury USA.
Hari, J. (2019). *Lost connections: Uncovering the real causes of depression – and the unexpected solutions*. Bloomsbury Circus.
Haybron, D. M. (2013). *Happiness: A very short introduction* (1st ed.). Oxford University Press.
Henrich, J. (2016). *The secret of our success: How culture is driving human evolution, domesticating our species, and making us smarter*. Princeton University Press.
Holden, G. W., Gower, T., Wee, S. E., Gaspar, R., & Ashraf, R. (2022). Is it time for "time-in"?: A pilot text for the child-rearing technique. *Pediatric Reports*, *14*(2), 244–253. https://doi.org/10.3390/pediatric14020032
Hooks, B. (2018). *All about love: New visions*. William Morrow Paperbacks.
Hoskins, Z. (2016). Collateral restrictions. In: *The new philosophy of criminal law*. Rowman & Littlefield International. http://proxy.library.umkc.edu/login?url=http://search.ebscohost.com/login.aspx?direct=true&scope=site&db=e900xww&AN=1130798
Hoskins, Z. (2019). *Beyond punishment?: A normative account of the collateral legal consequences of conviction*. Oxford University Press.
Hu, J., Fitzgerald, S. M., Owen, A. J., Ryan, J., Joyce, J., Chowdhury, E., Reid, C. M., Britt, C., Woods, R. L., McNeil, J. J., & Freak-Poli, R. (2021). Social isolation, social support, loneliness and cardiovascular disease risk factors: A cross-sectional study among older adults. *International Journal of Geriatric Psychiatry*, *36*(11), 1795–1809. https://doi.org/10.1002/gps.5601
Hursthouse, R. (2001). The virtues benefit their possessor. In R. Hursthouse (Ed.), *On virtue ethics*. Oxford University Press. https://doi.org/10.1093/0199247994.003.0009
James, C. (2020, February 6). *Child development: The time-out controversy: Effective or harmful?* Blogs. https://medicine.iu.edu/blogs/pediatrics/child-development-the-time-out-controversy-effective-or-harmful
Jeffreys, D. (2018). *America's jails: The search for human dignity in an age of mass incarceration*. NYU Press. https://search.ebscohost.com/login.aspx?direct=true&AuthType=ip,sso&db=e900xww&AN=1612001&site=eds-live&scope=site&custid=096-820
Joshanloo, M., & Weijers, D. (2024). Ideal personhood through the ages: Tracing the genealogy of the modern concepts of wellbeing. *Frontiers in Psychology*, *15*, 1494506. https://doi.org/10.3389/fpsyg.2024.1494506
Kobayashi, L. C., & Steptoe, A. (2018). Social isolation, loneliness, and health behaviors at older ages: Longitudinal cohort study. *Annals of Behavioral Medicine: A Publication of the Society of Behavioral Medicine*, *52*(7), 582–593. https://doi.org/10.1093/abm/kax033
Langeslag, S. J. E. (2024). Refuting six misconceptions about romantic love. *Behavioral Sciences*, *14*(5). https://doi.org/10.3390/bs14050383
Lerner, A., & Klein, M. (2019). Dependence, withdrawal and rebound of CNS drugs: An update and regulatory considerations for new drugs development. *Brain Communications*, *1*(1), fcz025. https://doi.org/10.1093/braincomms/fcz025
Lippke, R. L. (2018). The case against jails. In: *The ethics of policing and imprisonment* (pp. 109–128). Palgrave Macmillan. https://doi.org/10.1007/978-3-319-97770-6_7

McKittrick-Sweitzer, L. (2021a). *Care exploitation: Recognizing and preventing a pervasive injustice*. The Ohio State University. https://etd.ohiolink.edu/acprod/odb_etd/etd/r/1501/10?clear=10&p10_accession_num=osu1618235524273382
McKittrick-Sweitzer, L. (2021b, May 18). The care exploitation of essential workers. *Blog of the APA*. https://blog.apaonline.org/2021/05/18/the-care-exploitation-of-essential-workers/
Merrefield, C. (2023, June 6). Solitary confinement: A primer on its use and consequences in the U.S. *The Journalist's Resource*. https://journalistsresource.org/home/solitary-confinement-research-primer/
Mizzoni, J. (2019). Evolution and (Aristotelian) virtue ethics. *Human Affairs*, *29*(2), 199–206. https://doi.org/10.1515/humaff-2019-0016
Moore, G. E. (2012). *Principia Ethica*. Digireads.com.
My Unsung Hero. (2025, July 21). [Broadcast]. In *NPR*. https://www.npr.org/series/1134955065/my-unsung-hero-from-hidden-brain
Narvaez, D. (2014). *Neurobiology and the development of human morality: Evolution, culture, and wisdom (Norton Series on Interpersonal Neurobiology)*. W. W. Norton & Company.
National Institute on Aging (2022). *What do we know about healthy aging?* National Institutes of Health. https://www.nia.nih.gov/health/healthy-aging/what-do-we-know-about-healthy-aging
Peck, M. S. (2003). *The road less traveled: A new psychology of love, traditional values and spiritual growth*. Touchstone.
Pennisi, E. (2009). On the origin of cooperation. *Science*, *325*(5945), 1196–1199. https://doi.org/10.1126/science.325_1196
Ross, S. (2022, September 13). *How do good conversations work? Philosophy has something to say*. Psyche Ideas. https://psyche.co/ideas/how-do-good-conversations-work-philosophy-has-something-to-say
Ross, S. (2025). On conversation, its modes, pleasures, and perils. *Contemporary Aesthetics*, *23*. https://contempaesthetics.org/2025/09/14/on-conversation-its-modes-pleasures-and-perils/
Ruse, M. (1990). Evolutionary ethics and the search for predecessors: Kant, Hume, and all the way back to Aristotle? *Social Philosophy & Policy*, *8*(1), 59–85. https://doi.org/10.1017/S0265052500003745
Schrempft, S., Jackowska, M., Hamer, M., & Steptoe, A. (2019). Associations between social isolation, loneliness, and objective physical activity in older men and women. *BMC Public Health*, *19*(1), 74. https://doi.org/10.1186/s12889-019-6424-y
Shapiro, A., Lim, M., & Intagliata, C. (2023, January 15). What's the #1 thing to change to be happier? A happiness researcher weighs in. *NPR*. https://www.npr.org/2023/01/15/1149063567/whats-the-1-thing-to-change-to-be-happier-a-top-happiness-researcher-weighs-in
Shelby, T. (2016). *Dark ghettos: Injustice, dissent, and reform*. An Imprint of Harvard University Press.
Shelby, T. (2022). *The idea of prison abolition*. Princeton University Press.
Shen, H.-W., Delston, J. B., & Wang, Y. (2017). Does a sense of benevolence influence volunteering and caregiving among older people? *Social Work Research*, *41*(3), 155–166. https://doi.org/10.1093/swr/svx012
Shen, H.-W., & Perry, T. E. (2014). Giving back and staying put: Volunteering as a stabilizing force in relocation. *Journal of Housing for the Elderly*, *28*(3), 310–328. https://doi.org/10.1080/02763893.2014.930368
Smith, S. E. (2020). Is evolutionary psychology possible? *Biological Theory*, *15*(1), 39–49. https://doi.org/10.1007/s13752-019-00336-4

Sober, P. E., Wilson, P. D. S., & Wilson, D. S. (2003). *Unto others: The evolution and psychology of unselfish behavior*. Harvard University Press.

Social isolation, loneliness in older people pose health risks. (2019, April 23). National Institute on Aging. https://www.nia.nih.gov/news/social-isolation-loneliness-older-people-pose-health-risks

Solitary Confinement in the United States: The Facts. (2023, June). Solitary Watch. https://solitarywatch.org/facts/faq/

Srinivasan, A. (2021). *The right to sex: Feminism in the twenty-first century*. Farrar, Straus and Giroux.

Sterelny, K. (2014). *The evolved apprentice: How evolution made humans unique*. MIT Press.

Stich, S. (2007). Evolution, altruism and cognitive architecture: A critique of Sober and Wilson's argument for psychological altruism. *Biology & Philosophy*, *22*(2), 267–281. https://doi.org/10.1007/s10539-006-9030-1

Thusi, I. (2021). Feminist scripts for punishment. *Harvard Law Review*, *134*(7), 2449–2484.

Tropp, L. R. (2025, May 21). Making eye contact and small talk with strangers is more than just being polite – the social benefits of psychological generosity. *The Conversation*. http://theconversation.com/making-eye-contact-and-small-talk-with-strangers-is-more-than-just-being-polite-the-social-benefits-of-psychological-generosity-252477

WebMD Editorial Contributors, & DerSarkissian, C. (2023, July 8). What to know about dry drunk syndrome. *WebMD*. https://www.webmd.com/mental-health/addiction/what-to-know-dry-drunk-syndrome

Weiss, R. (2015, September 30). The opposite of addiction is connection: New addiction research brings surprising discoveries. *Psychology Today*. https://www.psychologytoday.com/us/blog/love-and-sex-in-the-digital-age/201509/the-opposite-addiction-is-connection

Werhane, P. H., Newton, L., & Wolfe, R. (2020). *Alleviating poverty through profitable partnerships: Globalization, markets, and economic well-being* (2nd ed.). Routledge. https://doi.org/10.4324/9780429329043

Wiegman, I. (2017). The evolution of retribution: Intuitions undermined. *Pacific Philosophical Quarterly*, *98*(2), 490–510.

Wiegman, I. (2019). Payback without bookkeeping: The origins of revenge and retaliation. *Philosophical Psychology*, *32*(7), 1100–1128.

Wiegman, I., & Dale, M. T. (2024). The co-evolution of virtue and desert: Debunking intuitions about intrinsic value. *Synthese*, *204*(4), 1–18. https://doi.org/10.1007/s11229-024-04751-4

Williams, B. A. O. (1985). *Ethics and the limits of philosophy*. Harvard University Press.

Wrangham, R. (2019). *The goodness paradox: The strange relationship between virtue and violence in human evolution*. Vintage.

15 The Tribalism Trap

Overcoming Political Polarization in an Age of Outrage

Joshua May

Unlike snow leopards, orangutans, and numerous other mammals, humans are deeply social creatures who evolved in tight-knit groups. For hundreds of thousands of years, our ancestors formed tribes that competed against one another. The group that sticks together survives together—and, importantly, produces offspring. It's no wonder we naturally value traits like loyalty, compassion, and trust that tend to be reserved more for members of one's inner circle (Greene, 2013; Haidt, 2012). As natural selection instilled preferences for group allegiance and social status, humans also developed moral and religious systems that further reinforced these tendencies through strong cultural identities (Henrich, 2015; Kumar & Campbell, 2022), which have also been the basis of war and other tribal conflicts.

These groupish tendencies aren't fixed or "hard-wired" and they can be virtuous in certain contexts. Loyalty can bind group members together to yield monumental feats of cooperation (Hammond & Axelrod, 2006). Us-vs-Them thinking can be useful in the "environment of evolutionary adaptedness," when aggressive groups competed for scarce resources in the Pleistocene. As humans have become more prosperous and peaceful, their moral circles have expanded to include universal human rights and even some consideration for non-human animals (Buchanan & Powell, 2018; Kumar & Campbell, 2022).

In modern societies, however, tribal tendencies can become vices and incite culture wars. Certain cultural and political groups have become more tribalistic as personalized media has fragmented voters into echo chambers and social media algorithms amplify contempt and distrust of opposing coalitions. There is, it seems, an evolutionary mismatch between human psychology and modern ecology. The problem is similar to how obesity and Type 2 diabetes soar in some wealthy nations: while powerful desires for salt, sugar, and fat help humans flourish in the wild, they become maladaptive when addictive calorie-dense foods are affordable and plentiful (Bateson et al., 2004). Likewise, the *tribalism trap* is the temptation to let

DOI: 10.4324/9781003632641-20

our modern milieu inflame tribal instincts into excess, transforming loyalty and moral conviction into incivility and arrogance.

Is such tribalism inevitable? Prominent theories imply that it's either an ingrained feature of human nature or unavoidable in modern societies connected by divisive algorithms and ideologies. But this chapter argues otherwise. We start with research on political polarization and put it in the context of eroding virtues, which suggests a way forward that doesn't rely on banning the internet or going back to a pre-industrial age. Compassion, respect, and intellectual humility can (and should) be developed in the modern world with a combination of altering one's media environment and improving one's patterns of thought to promote virtues of civility and humility. Just as we can exercise temperance in the glistening aisles of the grocery store, it's possible to be humble and civil while consuming news from the internet and staying connected with others on social media.

15.1 Polarization Problems

Let's start by clarifying tribalism, with its inherently negative connotations. While loyalty is a virtue, *tribalism* involves excessive in-group favoritism, out-group hostility, and a tendency to prioritize group identity over impartial judgment. Some tribes are defined by their location on a map or genetic makeup, but our focus will be on groups defined by moral or political ideologies, which form an intellectual tribe. Rallying with one's political teammates and dunking on opponents feels just as good as scoring points on the basketball court, but politics isn't a game, even if our modern world presents it as one. Such tribalism involves not only moral vices, such as callousness and incivility, but also intellectual vices, such as arrogance and dogmatism (Cassam, 2019). We'll assume these are problematic and ask how we can avoid them in modern societies amid algorithms and norms that make tribalism so appetizing.

Drawing moral conclusions from evolutionary mismatch is difficult. Some ethicists have argued ambitiously that the foundations of human morality are now completely outdated and should be jettisoned in favor of more utilitarian systems that simply calculate which actions and policies will promote the greater good (Greene, 2013; Singer, 2005). Yet such sweeping arguments, which ask us to abandon fundamental moral commitments to individual rights, face formidable challenges (see, e.g., Kumar & May, 2019; May & Baker, forthcoming). A much less controversial conclusion is that tribal tendencies in modern societies have led to dogmatism, callousness, and other traits widely considered vices by our own lights.

We'll focus on how tribalism arises in contemporary forms of political polarization rising in various countries, especially in the United States.

Social scientists have studied polarization extensively in recent decades (e.g., Klein, 2020; Kleinfeld, 2023; Talisse, 2019), and they've come to identify two main forms. *Ideological polarization* involves more extreme political views, which have become more common among politicians and the most politically engaged citizens. Among the general electorate, we find more *affective polarization*: animosity toward people with opposing political identities, often accompanied by negative stereotypes and prejudice. Research suggests that many supporters of America's two major political parties hold fairly similar views about public policy, yet increasingly despise the opposing political party and its members (Mason, 2018; see Figure 15.1). In recent decades, partisans who strongly identify with the major parties are particularly inclined to view the other side as dishonest, immoral, unintelligent, and close-minded (Pew Research Center, 2022).

The two forms of polarization are connected in many people, particularly the most politically engaged. If you dislike and distrust people on the other side, you'll come to find their worldview even more wrongheaded, their information sources even less reliable. Negative feelings can then lead to adopting more extreme viewpoints that lie even further from the people whose moral views you despise. (Of course, this is the generic "you." You, dear reader, might be intellectually humble, charitable, civil, compassionate toward political opponents, and more. If so, read the second person as referring to other readers who, like myself, aren't so virtuous.)

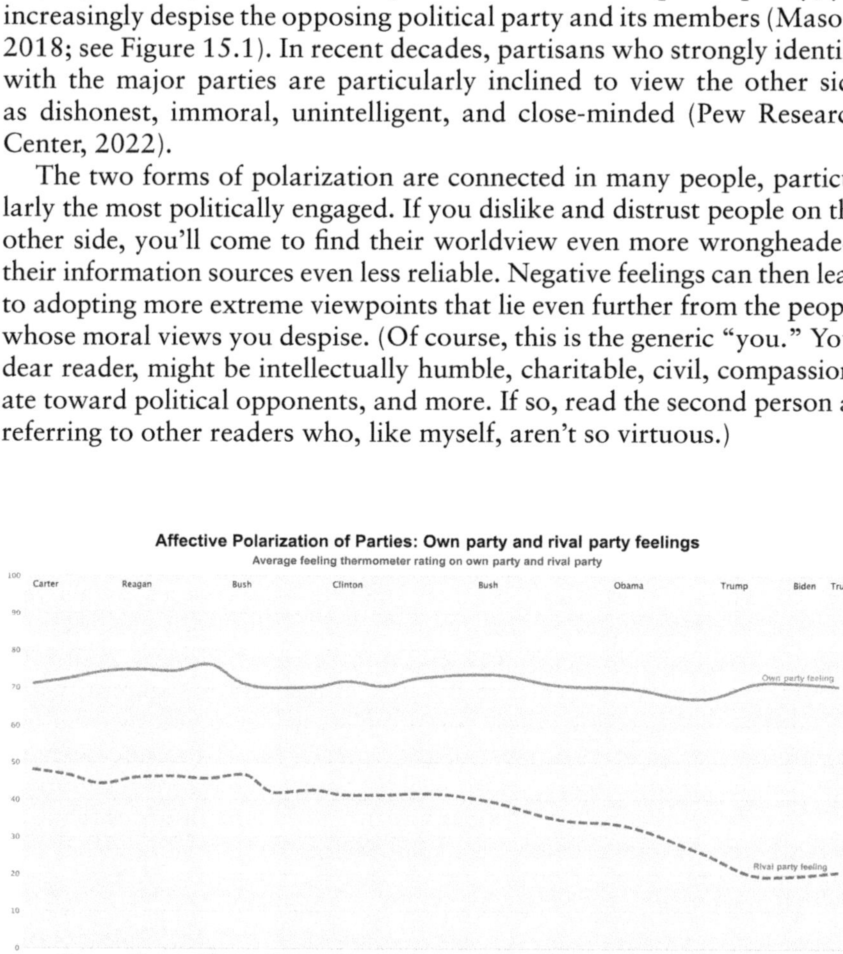

Figure 15.1 Average feeling thermometer ratings of own party and rival party, 1980–2024

Source: American National Election Studies (electionstudies.org)

Some of this is *rational*, at least in a narrow sense of the term. Humans are not cold robots designed to dispassionately reason alone. We are deeply social animals who evolved to reason with others who display mutual trust and respect—attitudes that evolved long ago to help humans cooperate and learn from one another (Kumar & Campbell, 2022). Moreover, as many philosophers have come to recognize in social epistemology, no one person can be an expert on every topic. We all must rely on others as sources of information. And it can be quite reasonable to rely on information from those you trust, which may well be those in your echo chamber, not the apparent fools outside of it (Levy, 2022; Rini, 2017). If you espouse the liberal worldview propounded by Democrats, it's understandable to dismiss a climate skeptic, just as it makes little sense for a staunch conservative to defer to a Marxist economist. However, even if in some sense rational or understandable, it's not virtuous to let oneself become trapped in an echo chamber that distorts one's view of opponents. Intellectual humility requires the skill of recognizing and avoiding conditions ripe for dogmatism and incivility, especially in the face of widespread disagreement.

The problem of polarization isn't just a matter of isolated feelings and ideas; these have consequences for personal relationships, elections, public policy, and democracy. Polarization has led people to lose friends and family over politics (Jurkowitz & Mitchell, 2020; Savage, 2021). During an interview with *The New York Times*, one Republican voter mentioned that her support of Donald Trump has led to losing "two friendships of over 20 years," which she described as "heartbreaking" (America in Focus 2025). It's easier than ever for people to consume partisan media and interact with like-minded others on the internet, forming echo chambers in which contrary opinions and sources are actively discredited (Nguyen, 2020). Increasingly, Americans don't even want to live near people with opposing views and so move to more politically homogeneous neighborhoods (Kaysen & Singer, 2024). Animosity and misunderstanding result in less compromise, less cooperation, and less functional democratic systems. In the worst cases, tribalism fuels extremism, political violence, profound distrust, and ruthless retribution, which undermine the shared narratives and faith in institutions required to hold democracies together (e.g., Buchanan, 2020; Haidt, 2022; Sullivan, 2017; Sunstein, 2017).

Some forms of political polarization are certainly healthy. We do need citizens to have moral convictions in politics and to hash out their differences. Abolitionists and civil rights leaders were passionate and once considered both polarizing and extreme (Sunstein, 2017, p. 86). The problem of tribalism, however, isn't the passionate expression of unconventional views. It's when politics is invariably viewed as a battle between good and evil that requires contempt for the enemy and profound distrust toward political opponents—not just the worst politicians but everyone within the coalition, including fellow citizens who throw their vote away to the "dark side." Such attitudes weren't prominent features of the civil rights movement. Indeed, fervent tribalism of this sort may be restricted to certain segments of society today, like the most politically engaged. Even so, our question is whether such polarization can be curbed where it exists in problematic forms.

Consider some examples of tribalism in action. Certain issues are so politicized that being a good member of the Left or the Right requires a particular stance—what Dan Williams (2022) calls *identity-defining beliefs*. A number of possibilities from contemporary politics come to mind, including views on: immigration; affirmative action and DEI; defunding the police in the wake of the Black Lives Matter protests; mass shootings and gun control; mask mandates and school closures during the COVID-19 pandemic; and the lab-leak theory about its origins. Politically engaged readers will readily recognize the party lines on these issues. But let's dwell on climate change, which has become a highly politicized issue in many parts of the world.

Whether humans are significantly contributing to global warming is a largely scientific question, even if what to do about it is a moral one. Yet there are partisan views on the science that have impeded productive approaches to the issue. Republicans have largely disregarded scientific consensus and denied that there is an urgent threat, while Democrats have tended to catastrophize, arguably exaggerating the problem and sometimes misrepresenting the science (Heath, 2025). There is a palpable sense that to be a good liberal or progressive is to view climate change as an existential threat, while good conservatives see it as largely overblown.

Identity-defining beliefs can be revealed by imagining a group member breaking ranks. Ezra Klein (2020) considers what would happen if Sean Hannity, a prominent conservative on Fox News, decided that climate change was a serious threat to humanity and publicized this change in view:

> He would lose friendships, viewers, and money. He could ultimately lose his job. And along the way he would cause himself immense personal pain as he systematically alienated his closest political and professional allies. The world would have to update its understanding of who Sean Hannity is and what he believes, and so, too, would Sean Hannity. Changing your identity is a psychologically and socially brutal process.
>
> (Klein, 2020, pp. 95–96)

This is a human, not a conservative, phenomenon. I once attended a sustainability workshop with dozens of other academics, where one of the most exuberant and inspiring presentations was about how we urgently need to stop burning fossil fuels and "electrify everything!" During the break, while everyone was chatting in the courtyard, one audience member expressed concern that the talk was so one-sided. He was curious about the perspective of those who think we don't need to move completely and immediately away from fossil fuels. Everyone around him shot back with anger, ridiculing the opposing view as not worth any airtime and appalling to even bring up. Although I'm personally in favor of moving away from fossil fuels, this smacked of tribalism on the Left.

The indignant defense of one's identity-defining beliefs is perfectly natural. Once a belief becomes central to one's identity, reasoning kicks in to defend it, regardless of one's political leanings (Clark et al., 2019; Ditto et al., 2025; Williams, 2023c). And identity groups even reward rationalizations that support the party line (Williams, 2023b). These are arguably forms of motivated reasoning, in which one's evaluation of information and arguments is shaped by one's goals and desires, such as maintaining status within a valued group (May, 2018a; Williams, 2022).

Motivated reasoning isn't necessarily toxic. Even scientists are motivated to defend their pet theories or produce findings that garner social status (May, 2021). Sometimes biases lead to questionable research practices, but other times moral values help scientists dig more deeply into questions that others ignore (Anderson, 2004). Under the right conditions, biased reasoning can lead to knowledge when multiple viewpoints are put into respectful dialog to yield a marketplace of ideas (Mercier & Sperber, 2017). Tribalism, on the other hand, yields dogmatism and echo chambers, along with incivility and contempt that shut down productive dialog and cooperation.

Contrast the coordinated effort to repair the ozone layer in the Earth's stratosphere. Scientists discovered in the 1970s that chemicals used in aerosols and refrigeration (CFCs) were depleting the ozone layer that protects the planet from ultraviolet radiation. By 1987, an international treaty—the Montreal Protocol—phased out these damaging substances. Now the ozone layer has already recovered substantially, with a full recovery expected within this century (Chu, 2025). Healthy disagreement heals the ozone layer; tribalism produces climate change culture wars.

To be fair, the comparison isn't perfect. Banning a chemical used in a relatively small number of products is much easier (and less contentious) than completely reshaping how energy is used and created. Even so, political tribalism hasn't only cropped up over fossil fuel bans or taxes on carbon emissions but also in the development of more sustainable sources of power, such as wind turbines and solar panels. Humanity's response to climate change could have been closer to the ozone crisis. Despite disagreements about the Montreal Protocol, we came together to discern the facts and collaborated to achieve a common goal. That was back in the 1980s, one might say, before so much political polarization. So what changed?

15.2 Potential Causes

Solutions to the problem of polarization require understanding its causes. Most analyses focus on *structural features* of modern society, such as new information technology and shifting political norms. Some are more credible or more powerful causes than others, so let's briefly examine prominent contenders.

A natural thought is that *economic downturns* trigger exclusionary tribal instincts. Allen Buchanan and Rachel Powell (2018) have argued that humans are naturally inclusive and trustworthy but that the perception of scarcity and disease makes us more tribalistic. Tribalism is an "adaptively plastic" trait, similar to how water fleas can grow armor or spines when threatened by predators. Although Buchanan and Powell don't purport to explain political polarization, perhaps events like the 2008 Global Financial Crisis and the COVID-19 pandemic of 2020 have triggered more tribalism. Changes in the economy are probably a factor (isn't it always?), but this doesn't quite capture how polarization began to increase well before this century. It also doesn't explain why other major economic crises of the previous century didn't result in present levels of polarization within societies.

Some theorists point instead to divisive *ideologies* that foment tribalism and degrade political discourse. White nationalism, critical race theory, and cancel culture have been charged with pitting racial, ethnic, and moral tribes against one another in a war of good versus evil (Buchanan, 2020; Lukianoff & Schlott, 2023; Mounk, 2023). These ideas, propounded by influential people and institutions, find their way into individual minds and breed distrust and disdain toward out-groups. Although this explanation might better fit the timeline of increased political polarization, divisive ideologies could be more a consequence of polarization than a cause. In the absence of growing contempt for political opponents, such ideologies would struggle to gain acceptance. At the very least, there are other prominent factors to consider.

One common theory is that the internet and social media platforms have facilitated the spread of *misinformation* about politically charged issues, from vaccine mandates to climate change. Perhaps believing more outlandish claims results in more extreme political views and animosity toward those who disagree. There is some evidence that misinformation is a problem within some circles (Sunstein, 2017), but much of this has been overblown and supported by flawed research (Williams, 2023a). Some people do certainly believe outright lies, speculative claims, and dubious conspiracy theories, but these aren't common enough to explain the increase in problems of polarization. Most people aren't that gullible (Mercier, 2020), and many simply report outlandish beliefs to signal group identity or bad-mouth opponents (Funkhouser, 2022; Hannon, 2021). When many Trump supporters said that his inauguration crowd was larger than Barack Obama's, it's more likely that these loyalists were signaling political allegiance than accepting obvious falsehoods (or hallucinating).

What's more plausible is that *modern media* has radically changed the way people consume and discuss political information. Humans evolved in small groups of people whom they know personally and interact with face-to-face. Email, text, and video calls can wonderfully connect distant family and friends, but they aren't the best medium for communicating subtle thoughts about emotionally charged topics (even with the addition of emojis). Social media platforms, in particular, tend to prop up and perpetuate echo chambers with algorithms that supercharge hostility and serve up content that tantalizes one's biases (Haidt, 2022; Klein, 2020; Sunstein, 2017). With social media being free for users, the business model called for products that keep users on the platform so they can be shown the advertisements that pay the bills. Even just introducing "like" buttons and algorithmic news feeds incentivized the creation of inflammatory posts and comments that go viral or otherwise spur engagement, particularly about divisive issues. In the past decade, consumption of news from social media has increased around the world. In 2013, only 23% of Americans reported using social media platforms as a source of news; now it's 54% (Graves & Jenkins, 2025). While more than social media is to blame for political dysfunction and irrationality, it has arguably contributed to affective polarization.

Prior changes to *political institutions* also play a role. Back in the 1980s, for example, American political parties became more ideologically homogeneous, politicians spent less time socializing with members of opposing political parties, and the death of the Fairness Doctrine meant that news outlets didn't need to present both sides of controversial issues. As Klein (2020) has put it, these kinds of changes to political institutions and norms have led people to develop strong "mega-identities" tied to political parties and surrounding ideologies (see also Buchanan, 2020). As broad ideologies become a strong part of one's identity, human nature breeds mistrust of non-members and defense of the group against intellectual attacks. Combined with increasingly isolated media echo chambers, it's no wonder citizens become more polarized, both in their views and their feelings toward opponents.

15.3 Potential Solutions

No one factor is the cause of political polarization. Rather, there are many contributing factors, just as with any complex social phenomenon—from the decline in teen mental health to the rise in drug overdoses. Many of these contributing factors are features of political and media systems, and so it seems "the problem is structural," as Jonathan Haidt (2022) puts it. Many theorists have accordingly called for policies that prevent partisan media, homogeneous political parties, and divisive ideologies from turning natural tendencies into tribalism. Allen Buchanan represents this view when declaring with emphasis: "*an effective response to tribalism must be institutional*" (2020, p. 248; see also Sauer, 2019). Klein similarly urges us to "zoom out from the individuals to better see interlocking systems that surround them" (Klein, 2020, p. XVI).

However, political systems are hard to change, as are the products of big tech companies—especially when nations are polarized and mired in gridlock. Reforms and regulations have been few and minimal, and the business models of media platforms are well entrenched. Newspapers and other legacy media companies have already collapsed, and the ones still standing are losing subscribers to more personalized media provided by YouTube, TikTok, and other platforms apparently designed to keep users engaged and enraged. It seems there's no going back to the days of bipartisan media and ideologically diverse political parties.

It would certainly help for our social environments to change. Politicians, policy makers, and tech companies should absolutely consider relevant reforms. Yet we shouldn't focus solely on these sources of political polarization. Individuals can and should engage in the project of self-improvement to combat tribalism in their own lives. Indeed, as with all forms of social change, individual and structural factors are deeply linked (Brownstein et al., 2025). Political systems won't change unless voters demand cooperation over animosity; divisive ideologies won't lose ground unless people recognize and recoil from them; social media companies won't develop new business models or healthier platforms if consumers keep using current versions of these products. Conversely, improving political systems and tech products won't cure tribalism if individuals remain too lazy to see past groupthink or overcome the appetite for political junk food (which Klein does recognize, if not emphasize).

Individuals aren't completely blameless because they aren't powerless. We have some responsibility to cultivate civility, open-mindedness, and similar virtues, even when it's difficult (Cassam, 2019; Vallor, 2016). Citizens can support and vote for politicians who unite rather than divide. They can get more involved in local politics (Klein, 2020), rather than national politics, which increasingly feels like a drama-filled reality show—*The Real Politicians of Washington DC*, as Tim Urban (2023) describes America's series.

There are limits to these particular strategies. As the political system incentivizes dividers, there are few, if any, uniters to support. National debates increasingly influence local politics (think immigration, tariffs, transgender athletes), and many people have now befriended or moved next to mostly co-partisans. It's hard to distance yourself from the drama of national politics if most of your friends and neighbors are fellow liberals or conservatives (or libertarians or transhumanists or what have you).

Perhaps we should talk less about politics and increase non-political interactions with members of other political tribes, as Robert Talisse (2019) recommends. Conservatives should play soccer with progressives; democratic socialists should go hunting with populists. As you befriend those with other viewpoints, you can learn to like them, thereby reducing your animosity and promoting a culture of collaboration and cooperation across party lines. However, it's hard to see how this can happen if we talk about politics less. You may like Joey who is a Trump supporter, but you can still find his political stance idiotic and appalling while admiring his jump shot. Indeed, the development of cancel culture and constraints on free speech have already made it more difficult to talk politics with others (Corvino, 2024; Lukianoff & Schlott, 2023). Trust and respect require mutual understanding.

And there is still the concern that algorithms and personalized news sources will reinforce echo chambers amid these friendships. Haidt (2022) maintains that social media has been a major cause of polarization, particularly in America, by incentivizing conflict and empowering trolls and provocateurs. Of course, completely ditching social media is easier said than done. If you want to see photos of your child at camp or at soccer practice, you probably need an Instagram account. If you want to keep up with your hiking group or old friends back home, you might need to scroll through Facebook.

Some people get news from sources other than social media, such as news apps, TV programs, podcasts, and even radio stations. We could follow the counsel of philosophers who urge individuals to diversify their media diets (Worsnip, 2018). Perhaps if we consume partisan media from all sides, we can escape our echo chamber. However, as C. Thi Nguyen (2020) points out, mere exposure to alternative viewpoints doesn't guarantee any change in one's views or feelings toward the other side. Many liberals who watch Fox News are liable to develop *more* animosity toward the Right and adopt views even further to the Left. Echo chambers are propped up more by distrust of opponents, and disdain for them, than by ignorance of their views. As Nguyen puts it, other relevant voices are not merely "excluded through omission" but "actively discredited" simply because they represent opposing viewpoints.

Nugyen argues that escaping an echo chamber may require a complete *social-epistemic reboot*. Unlike Descartes during his meditations, you needn't distrust your senses or the testimony of others. You must, however, throw out all your previous beliefs about who is and isn't a trustworthy source and start anew, unbiased by ideologies. Any such reboot is, as Nugyen acknowledges, extremely difficult if not literally impossible. Although a handful of former neo-Nazis and conspiracy theorists have done it, wholesale reboots are especially difficult to pull off in a world of fragmented and personalized media, algorithms designed to promote extremity, and political mega-identities.

So, can we escape tribalism in the modern world? Each of the prominent solutions to political polarization has limitations, but it's unfair to only consider each in isolation. A multi-pronged approach would recommend making the changes together: minimize your social media presence, diversify your news sources, support politicians who aim to unite while distancing yourself from the drama of politics, *and* befriend people who have different viewpoints (while talking to them occasionally about politics). There are many environmental and psychological barriers, to be sure. The trick is not a wholesale reboot, but an Aristotelian approach to modifying one's environment, patterns of thought, and interactions with others. In short: ecological control, humility, and civility.

15.4 Ecological Control

Let's start with feasible changes to one's media environment. You could, for example, diversify your media diet while being intentional about the amount of political media you consume, such as switching off notifications for apps that tend to yield unproductive exchanges or other engagement. That doesn't require abandoning all beliefs about who is trustworthy, but it's the start of a healthier relationship to politics.

A useful model can be found in other attempts to navigate the mismatch between our hominid brains and modern technology. As Cal Newport conceives it, *digital minimalism* is "a philosophy of technology use in which you focus your online time on a small number of carefully selected and optimized activities that strongly support things you value" (Newport, 2019, p. 28). Applied to politics, this would mean choosing to minimize one's exposure to news and social media that foment tribalism. As with digital minimalism, the goal is not necessarily to abstain from political media entirely but to only allow in what's psychologically and morally healthy for you—what promotes virtues, not vices.

Maintaining some digital presence doesn't necessitate tribalism. Take the case of Megan Phelps-Roper, once a devoted member of the Westboro Baptist Church in Kansas—a group notorious for its hateful denunciations of homosexuality and for picketing the funerals of fallen soldiers regarded as fighting for a morally corrupt nation. Phelps-Roper (2019) describes her dramatic move to leave this fiercely tribal church as resulting from respectful and humble conversations with critics on the social media site Twitter, where individuals responded to her inflammatory posts not with outrage, but with patient and thoughtful engagement. They asked, politely, how she interprets the Bible and whether passages were inconsistent with Westboro's actions and worldview. Eventually, her online conversations didn't just change her mind about the church; she became less vitriolic and more understanding of intellectual opponents.

Phelps-Roper is proof that breaking out of a powerful echo chamber is possible without a wholesale reboot. Of course, Twitter (now X) is usually a hotbed of political tribalism, so she was lucky to encounter alternative viewpoints from people who challenged her with genuine compassion and respect. Better to shape one's environment to ensure these results more reliably, which typically requires fewer online interactions. As sociologist Sherry Turkle (2015) argues, face-to-face conversation is where we learn to listen and empathize, and where we can feel most heard and understood. Indeed, Phelps-Roper's transition didn't become fully complete until she met some of her critics in person and they became friends—one of whom she eventually fell in love with and married.

Altering one's media environment to help promote one's values amounts to exercising agency to indirectly shape one's habits. The digital minimalist is like the healthnut who keeps junk food out of the house. Sure, she could try to use willpower in the moment to overcome the temptation to eat candy and chips, but that's not very effective. Instead, she can indirectly minimize the cravings by removing cues from her immediate environment. Exerting such "ecological control" (Clark, 2007) is similar to the niche construction found throughout the animal kingdom. Humans, with our relatively small teeth and weak jaws, have developed external scaffolding for digestion—pots, pans, and fires—to essentially pre-digest our food (Sterelny, 2010). The same tactic works for changes in attitude or belief. Long ago, Blaise Pascal recommended that doubters cultivate a belief in God by going to church, performing the rituals, and surrounding oneself with theists. Shaping one's beliefs and values invariably involves such *social* niche construction.

Ecological control is powerful, but so are the habits that underlie tribalism. Once one has already built a strong moral and political identity, it's easy to take it offline into real-life interactions with others. Dogmatism, callousness, and incivility can remain for a conservative who deletes his social media accounts, plays soccer with liberals, and reads *The New York Times*. It's all too easy to dismiss and seethe while reading or listening to alternative viewpoints. We need the right motivations, feelings, and thought patterns—key elements of virtue, as emphasized by ancient thinkers from Aristotle to Confucius (Hursthouse & Pettigrove, 2023; Vallor, 2016). Shaping one's environment can help improve thoughts and feelings, but it's not necessarily sufficient.

Part of the problem is that social environments involve other people, many of whom are leaning into tribalism. It's easy to heed Pascal's advice to join and emulate Christians; harder to avoid Us-versus-Them thinking when it pervades the culture. Improving one's intellectual environment is thus something of a collective action problem, where other people's decisions shape your own incentives. So we have to approach social niche construction with strategies for productively interacting with people viewed as outsiders, especially when their tribalism is activated. Phelps-Roper arguably came to Twitter with genuine intellectual curiosity, as did her interlocutors. Part of the solution is to cultivate habits of humility and civility, which encourage others to reciprocate.

15.5 Humility

A core component of intellectual humility is recognizing one's own limitations (Whitcomb et al., 2017). When it comes to the tribalism of ethics and politics, a considerable limitation is "my-side bias"—the tendency to defend the views of one's own tribe while discrediting the opposition (Mercier & Sperber, 2017). My-side bias, arguably a form of motivated reasoning, is what makes it possible for a conservative journalist to dig deeply into claims about a Democratic president's purported scandal while Left-leaning journalists hardly report it (or vice versa). Journalists, voters, and other inquirers might aspire to objectivity, and the best of us can get somewhat close, but it's psychologically impossible to completely and permanently remove the glasses that filter one's perception of reality.

Fortunately, intellectual humility requires no such feat. Virtuous reasoners simply seek to recognize their biases, even if they can't eliminate them completely. Of course, recognition isn't enough. Just as the courageous go beyond recognizing their fear to overcoming it, the intellectually humble try to minimize their biases and cultivate an openness to being mistaken. But how?

A useful distinction between two mindsets is drawn by Julia Galef (2021). Our default is the *soldier mindset*, which seeks to fortify one's viewpoint and defend it against attack. This defensive posture is useful in some contexts, such as the courtroom, when one is duty bound to defend a client. However, in politics we're aiming to seek the truth—to determine whether climate change is an existential threat, whether a presidential candidate sexually assaulted a staffer, whether school closures during a pandemic do more harm than good, and so on. The soldier mindset is great for coming up with arguments, but defensiveness can too easily lead to dubious rationalizations and blind one to inconvenient truths. Especially amid political polarization, we should be wary of aiming primarily to defend the tribe.

There is an alternative, however. The *scout mindset* prioritizes seeking the truth and following the evidence where it leads. Rather than engage in wishful thinking and other forms of biased reasoning, scouts fully recognize that their preconceived notions could be wrong. Rather than adopting the soldier's defensive posture and feeling pained at the thought of being wrong, scouts welcome correction, even revel in it. A conservative scout would rather play devil's advocate than sit around with her friends talking about how awful the Democrats are. A liberal scout doesn't reluctantly read conservative magazines or listen to libertarian podcasts, out of a sense of duty, but rather does so out of genuine curiosity. That doesn't mean one must enjoy or seek out patently lazy reasoning or non-sequiturs. The scout mindset demands only that one read and listen charitably, searching for the best versions of alternative viewpoints on the controversy at hand.

Of course, most of us think we're already adopting the scout mindset; it's only those fools on the other side who engage in wishful thinking and posturing to protect their precious identity. That, of course, falls short of intellectual humility by failing to recognize one's own limitations. True scouts appreciate what Kathryn Schulz (2011) says about being wrong: *it feels just like being right*.

There are dangers to being too humble. It's not virtuous to give credence to seriously outlandish views or crackpot theories (Fantl, 2018; Levy, 2023). As the saying goes, it's good to keep an open mind, but not so open that your brain falls out. Nor is it virtuous to become apathetic toward injustice and public policy (Hannon & Kidd, 2024), or to give up on loyalty and group allegiances (Faria et al., 2024).

However, intellectual humility doesn't require a lack of conviction, changing your mind, taking seriously every position in logical space, or giving up your values and identities. It just requires appreciating one's limitations enough to be open to persuasion on controversial issues, just as a president preparing to drop a bomb should be open to hearing the rationale for alternative courses of action (Rini, 2018). We should similarly be open to considering opposing points of view about contentious issues in ethics and politics, if only to better understand, to reduce animosity, and to work with others toward shared goals (Táíwò, 2020). Our tribal tendencies are only objectionable when they veer into tribalism.

Let's consider a specific example of someone who exhibits intellectual humility without sacrificing her identity or passion for politics. Mónica Guzmán (2022) was shocked to find that her parents, both Mexican immigrants, voted for Trump in 2016 (and then again in 2020). On election night in 2016, when even Republicans seemed surprised that Trump won, Guzmán felt America's democracy was broken. She called her parents and explained her fears and dismay. Her mother listened and then explained her own perspective. In Mexico, she and many others firmly believed their votes didn't count, that the system was rigged. Guzmán's mother explained why she felt democracy was working: people voted and this man, this underdog, won. *La democracia jaló, Mónica. Yo voté, y jaló.* "Democracy worked, Mónica. I voted, and it worked" (Guzmán, 2022, p. 60). Guzmán, whose deep love for her mother prompted genuine curiosity and humility, writes: "I never thought of it that way." That bit of insight—and others like it—have helped her avoid political tribalism. Despite being a leading expert in crossing political divides, Guzmán still counts herself as a progressive, though she has plenty of conservative friends.

Rather than requiring a change in belief or identity, intellectual humility can often change your interlocutor's mind. The people who engaged politely, respectfully, and sincerely with Phelps-Roper could only do so if they thought they might learn something from her about why she believes what she does. By engaging open-mindedly, they didn't change their minds about the hateful teachings of the Westboro Baptist Church. Instead, they encouraged Phelps-Roper to also let down her defenses so they could explore the issue together. Humility requires only a willingness to acknowledge one's shortcomings and minimize their influence. Maybe someone changes their mind, maybe not. Either way, tribalism is mitigated, if not fully extinguished.

15.6 Civility

Close relationships, like that of a mother and daughter, often bring with them humility and respectful disagreement. This is the truth in recommendations, like Talisse's, to build civic friendships with people in one's community who don't share the same politics. If we're to learn about the origins and nuances of someone's viewpoints, however, we must be able to talk to them about these differences. Yet talking politics activates relevant identities and the soldier mindset to protect them. So another crucial step to escaping echo chambers involves changing the ways we interact with others—family, friends, and strangers. A scout mindset is a fine starting place for sifting through the news, but it can be ineffective in conversation with others, coming off as cold and distant if executed poorly.

Civility, or something close to it, is a good candidate for the missing piece. As ancient Chinese philosophers and cultures around the world have found, civility is a social glue that binds us together while its absence divides (Olberding, 2019). It provides the conditions for active listening and mutual understanding. After all, you're not going to listen or open up to someone who acts as though she despises or distrusts you (May, 2023; Kumar & May, forthcoming). Controlled studies have confirmed that respectful dialog is necessary for reducing polarization about contemporary moral issues. In a set of landmark field studies using trained political canvassers who talk with hundreds of voters on their doorsteps, researchers have found that the key is to share a personal story and approach the topic in a non-judgmental, respectful manner (Broockman & Kalla, 2016; Kalla & Broockman, 2023). These approaches have been found to slightly but durably reduce prejudice against immigrants and transgender people. More to the point, even if voters don't change their minds about policy issues, they open up and listen to the other side.

A shining example of this approach can be found in Daryl Davis, a Black jazz musician who regularly maintains mutual respect with the seemingly unreachable: members of the KKK. Davis simply treats these Klan members like people and talks to them about their lives, their families, and race, too. The Klan members come to like Davis and hear him out; they often change their minds about racial separation and superiority, too. Famously, Davis convinced Roger Kelly, among others, to leave the Klan (Davis, 2017). In 1996, Kelly told CNN, "We don't agree on everything, but at least [Daryl] respects me to sit down and listen to me, and I respect him to sit down and listen to him." Indeed, Davis and Kelly became friends, a process Davis has repeated many times with other Klan members. Part of his strategy is to keep in mind their common humanity. Every person, he says, wants five things: to be loved, to be respected, to be heard, to be treated fairly, and for their family to be safe and happy (TED Interview 2024). That may go beyond mere civility, but the point is that one's comportment, choice of words, and demeanor can transform a tense exchange into a productive dialog, perhaps even a friendship and a conversion.

Of course, members of marginalized groups aren't responsible for befriending or convincing bigots. Some people really don't deserve one's trust or respect (Zamalin, 2022). But tribalism exaggerates the size of that group, and Davis's example demonstrates the power of respect to defuse affective polarization. Some ideologies and their proponents do warrant anger, as well (Cherry, 2021; Srinivasan, 2018). Yet civility and respectful dialog preclude only contempt, not anger (Brooks, 2019; May, 2023). As with any friendship, Davis occasionally expresses anger, irritation, and frustration toward his racist friends.

Of course, people of opposing moral tribes can communicate respect toward others without really having any. Civility can be a thin veneer while contempt lies underneath. Nevertheless, civility provides the conditions for genuine respect and goodwill to develop. As Colin Marshall (2026) puts it, a good starting point is to simply *make your warmth obvious*. Warmth communicates that you're not a threat, which lowers defenses and promotes a scout mindset. In contrast, consider how some pro-life activists display images of aborted fetuses in public. While shock and disgust might garner attention, the tactic often backfires with supporters of abortion and those on the fence (May, 2018b). Not only is this a dubious approach to persuasion, but it also makes vivid the importance of carefully packaging one's message to promote goodwill and ultimately respect, where appropriate.

Attempts to persuade while talking politics can be disrespectful, even when done warmly. Political persuasion can amount to communicating that someone's most fundamental values, beliefs, and identity need changing (Tsai, 2014; Marshall, 2025). Aggressive atheists display this kind of disrespect when communicating to Christians, Muslims, and other theists that their beliefs are delusions and that their cultural identities are pointless.

Yet political dialog can be aimed at persuasion if one is likewise open to being persuaded oneself, which exhibits *mutual* respect. That means Daryl Davis has to be open to members of the KKK making some valid points, at least about their own experiences and concerns. And Phelps-Roper's interlocutors had to take her hateful views seriously. Only then are all parties exhibiting genuine respect (Rini, 2018), not merely paying lip service to it. Rather than make one vulnerable to toxic ideas, being open to persuasion promotes mutual understanding and depolarizes interlocutors even if no one fully changes their mind.

15.7 Conclusion

Human nature makes tribalism particularly tempting in the modern world. Much like devouring fried foods and pastries, it's pleasurable to affirm one's political beliefs or ridicule opponents for their seeming ignorance or irrationality. However, just as we can and should eat more wholesome foods despite the availability of junk, we can and should consume less partisan media and develop healthier approaches to discussing controversial moral issues. The analogy isn't perfect, as food choices might not be as socially determined as one's moral and political beliefs (though see May & Kumar, 2023). Nevertheless, the comparison illustrates how individual decisions can overcome and reshape one's conditions.

Virtue cultivation provides a useful framework for integrating relevant individual and structural changes. Traits and habits are, after all, shaped not only by one's own intentions but also by one's environment. Here again the analogy to temperance is fitting. Eating healthier requires the right environment (no junk food in the house, access to fresh produce) in addition to the right patterns of thought and behavior (acquire cooking skills, adopt a dietary philosophy, recognize and defuse rationalizations). Similarly, combating tribalism requires a combination of strategies to change one's environment, thoughts, and behavior—through, e.g., digital minimalism, humility, and civility. Importantly, these habits are infectious: examples like Daryl Davis illustrate how treating others with respect encourages reciprocation, leading to a virtuous shift in the social dynamic.

Overcoming the excesses of tribalism ultimately requires other virtues as well. When our tribe crosses a line, for example, we need the *courage* to get up in arms about it, rather than concoct justifications. Doing so risks social sanctions—often the most feared punishment for communal creatures like ourselves. Yet, as Aristotle would emphasize, we're also rational animals who seek the truth and genuine reasons for our beliefs. While leaning on our communities for safety, meaning, and knowledge, we can also resist dogmatism, arrogance, and incivility.

References

America in Focus (2025). The Economy Still Isn't What These 11 Latino Trump Voters Had in Mind. *The New York Times*. https://www.nytimes.com/interactive/2025/07/28/opinion/focus-group-latino-trump-voters.html?smid=nytcore-ios-share&referringSource=articleShare

Anderson, E. (2004). Uses of value judgments in science: A general argument, with lessons from a case study of feminist research on divorce. *Hypatia*, *19*(1), 1–24.

Bateson, P., Barker, D., Clutton-Brock, T., Deb, D., D'Udine, B., Foley, R. A., Gluckman, P., Godfrey, K., Kirkwood, T., Mirazón Lahr, M., McNamara, J., Metcalfe, N. B., Monaghan, P., Spencer, H. G., & Sultan, S. E. (2004). Developmental plasticity and human health. *Nature*, *430*(6998), 419–421.

Broockman, D., & Kalla, J. (2016). Durably reducing transphobia: A field experiment on door-to-door canvassing. *Science*, *352*(6282), 220–224.

Brooks, A. C. (2019). *Love your enemies: How decent people can save America from the culture of contempt*. Broadside Books.

Brownstein, M., Kelly, D., & Madva, A. (2025). *Somebody should do something: How anyone can help create social change*. MIT Press.

Buchanan, A. (2020). *Our moral fate: Evolution and the escape from tribalism.* MIT Press.
Buchanan, A., & Powell, R. (2018). *The evolution of moral progress: A biocultural theory.* Oxford University Press.
Cassam, Q. (2019). *Vices of the mind: From the intellectual to the political.* Oxford University Press.
Cherry, M. (2021). *The case for rage: Why anger is essential to anti-racist struggle.* Oxford University Press.
Chu, J. (2025). Study: The ozone hole is healing, thanks to global reduction of CFCs. *MIT News.* https://news.mit.edu/2025/study-healing-ozone-hole-global-reduction-cfcs-0305
Clark, A. (2007). Soft selves and ecological control. In D. Spurrett, D. Ross, H. Kincaid, & L. Stephens (Eds.), *Distributed cognition and the will.* MIT Press.
Clark, C. J., Liu, B. S., Winegard, B. M., & Ditto, P. H. (2019). Tribalism is human nature. *Current Directions in Psychological Science*, *28*(6), 587–592.
Corvino, J. (2024). *What debating gay marriage opponents taught me about the limits of free speech.* Detroit Free Press. https://www.freep.com/story/opinion/contributors/2024/01/17/limits-of-free-speech-debating-gay-marriage/72242814007/
Davis, D. (2017). Why I, as a black man, attend KKK rallies. *TEDx Talks.* https://www.youtube.com/watch?v=ORp3q1Oaezw
Ditto, P. H., Celniker, J. B., Siddiqi, S. S., Güngör, M., & Relihan, D. P. (2025). Partisan bias in political judgment. *Annual Review of Psychology*, *76*, 717–740.
Fantl, J. (2018). *The limitations of the open mind.* Oxford University Press.
Faria, F. N., Anomaly, J., & Willy, C. (2024). Enlightened tribalism. *Controversial Ideas*, *4*(2), 9. doi:10.35995/jci04020009
Funkhouser, E. (2022). A tribal mind: Beliefs that signal group identity or commitment. *Mind & Language*, *37*(3), 444–464.
Galef, J. (2021). *The scout mindset: Why some people see things clearly and others don't.* Portfolio.
Graves, L., & Jenkins, J. (2025). Digital News Report: United States. *Reuters Institute for the Study of Journalism.* https://reutersinstitute.politics.ox.ac.uk/digital-news-report/2025/united-states
Greene, J. (2013). *Moral tribes: Emotion, reason, and the gap between us and them.* Penguin.
Guzmán, M. (2022). *I never thought of it that way: How to have fearlessly curious conversations in dangerously divided times.* BenBella Books.
Haidt, J. (2012). *The righteous mind: Why good people are divided by politics and religion.* Vintage.
Haidt, J. (2022). Why the past 10 years of american life have been uniquely stupid. *The Atlantic.* https://www.theatlantic.com/magazine/archive/2022/05/social-media-democracy-trust-babel/629369/
Hammond, R. A., & Axelrod, R. (2006). The evolution of ethnocentrism. *Journal of Conflict Resolution*, *50*(6), 926–936.
Hannon, M. (2021). Disagreement or badmouthing? The role of expressive discourse in politics. In E. Edenberg & M. Hannon (Eds.), *Political epistemology* (pp. 297–318). Oxford University Press.
Hannon, M., & Kidd, I. J. (2024). Is intellectual humility compatible with political conviction? *Journal of Ethics & Social Philosophy*, *27*, 211.
Heath, J. (2025). Highbrow climate misinformation. *In Due Course.* https://josephheath.substack.com/p/highbrow-climate-misinformation

Henrich, J. (2015). *The secret of our success*. Princeton University Press.
Hursthouse, R., & Pettigrove, G. (2023). Virtue ethics. In E. N. Zalta & U. Nodelman (Eds.), *The Stanford encyclopedia of philosophy*. https://plato.stanford.edu/archives/fall2023/entries/ethics-virtue/
Jurkowitz, M., & Mitchell, A. (2020). A sore subject: Almost half of Americans have stopped talking politics with someone. *Pew Research Center*. https://www.pewresearch.org/journalism/2020/02/05/a-sore-subject-almost-half-of-americans-have-stopped-talking-politics-with-someone/
Kalla, J. L., & Broockman, D. E. (2023). Which narrative strategies durably reduce prejudice? Evidence from field and survey experiments supporting the efficacy of perspective-getting. *American Journal of Political Science*, *67*(1), 185–204.
Kaysen, R., & Singer, E. (2024). Millions of Movers Reveal American Polarization in Action. *The New York Times*. https://www.nytimes.com/interactive/2024/10/30/upshot/voters-moving-polarization.html
Klein, E. (2020). *Why we're polarized*. Avid Reader Press/Simon & Schuster.
Kleinfeld, R. (2023). Polarization, Democracy, and Political Violence in the United States: What the Research Says. *Carnegie Endowment for International Peace*. https://carnegieendowment.org/research/2023/09/polarization-democracy-and-political-violence-in-the-united-states-what-the-research-says?lang=en
Kumar, V., & Campbell, R. (2022). *A better ape: The evolution of the moral mind and how it made us human*. Oxford University Press.
Kumar, V., & May, J. (forthcoming). moral reasoning and moral progress. In D. Copp & C. Rosati (Eds.), *Oxford handbook of meta-ethics*. Oxford University Press.
Kumar, V., & May, J. (2019). How to debunk moral beliefs. In J. Suikkanen & A. Kauppinen (Eds.), *Methodology and moral philosophy*. Routledge.
Levy, N. (2022). *Bad beliefs: Why they happen to good people*. Oxford University Press.
Levy, N. (2023). Too humble for words. *Philosophical Studies*, *180*(10), 3141–3160.
Lukianoff, G., & Schlott, R. (2023). *The canceling of the American mind: Cancel culture undermines trust, destroys institutions, and threatens us all—but there is a solution*. Simon & Schuster.
Marshall, C. (2026). *Just hear me out: How to change minds without lies, bullshit, or moral compromise*. University of Chicago Press.
Marshall, C. (2025). Political persuasion is prima facie disrespectful. *Journal of Moral Philosophy*, 22, 584–617.
Mason, L. (2018). *Uncivil agreement: How politics became our identity*. University of Chicago Press.
May, J. (2018a). *Regard for reason in the moral mind*. Oxford University Press.
May, J. (2018b). The limits of appealing to disgust. In N. Strohminger & V. Kumar (Eds.), *The moral psychology of disgust*. Rowman & Littlefield.
May, J. (2021). Bias in science: Natural and social. *Synthese*, *199*, 3345–3366.
May, J. (2023). Can Relationship Science Teach Us to Talk Politics Again? *The Prindle Post*. https://www.prindleinstitute.org/2023/08/can-relationship-science-teach-us-to-talk-politics-again/
May, J., & Baker, K. (forthcoming). Rights aren't rationalizations. In S. Summers & A. Sajó (Eds.), *Human rights restated: An investigation of mind and rights*. Cambridge University Press.
May, J., & Kumar, V. (2023). Harnessing moral psychology to reduce meat consumption. *Journal of the American Philosophical Association*, *9*(2), 367–387.

Mercier, H. (2020). *Not born yesterday: The science of who we trust and what we believe*. Princeton University Press.
Mercier, H., & Sperber, D. (2017). *The enigma of reason*. Harvard University Press.
Mounk, Y. (2023). *The identity trap: A story of ideas and power in our time*. Penguin Press.
Newport, C. (2019). *Digital minimalism: Choosing a focused life in a noisy world*. Portfolio.
Nguyen, C. T. (2020). Echo chambers and epistemic bubbles. *Episteme*, *17*(2), 141–161.
Olberding, A. (2019). *The wrong of rudeness: Learning modern civility from ancient Chinese philosophy*. Oxford University Press.
Pew Research Center (2022). As Partisan Hostility Grows, Signs of Frustration with the Two-Party System. https://www.pewresearch.org/politics/2022/08/09/as-partisan-hostility-grows-signs-of-frustration-with-the-two-party-system/
Phelps-Roper, M. (2019). *Unfollow: A memoir of loving and leaving the Westboro Baptist Church*. Farrar, Straus and Giroux.
Rini, R. (2017). Fake news and partisan epistemology. *Kennedy Institute of Ethics Journal*, *27*, 43–64.
Rini, R. (2018). Abortion, Ultrasound, and Moral Persuasion. *Philosophers' Imprint*, 18, 1–20. http://hdl.handle.net/2027/spo.3521354.0018.006
Sauer, H. (2019). Butchering benevolence moral progress beyond the expanding circle. *Ethical Theory and Moral Practice*, *22*(1), 153–167.
Savage, M. (2021). Family estrangement: Why adults are cutting off their parents. *BBC*. https://www.bbc.com/worklife/article/20211201-family-estrangement-why-adults-are-cutting-off-their-parents
Schulz, K. (2011). On Being Wrong. *TED*. https://youtu.be/QleRgTBMX88
Singer, P. (2005). Ethics and intuitions. *The Journal of Ethics*, *9*, 331–352.
Srinivasan, A. (2018). The aptness of anger. *Journal of Political Philosophy*, *26*(2), 123–144.
Sterelny, K. (2010). Minds: extended or scaffolded? *Phenomenology and the Cognitive Sciences*, *9*(4), 465–481.
Sullivan, A. (2017). America Wasn't Built for Humans. *New York Magazine*. https://nymag.com/intelligencer/2017/09/can-democracy-survive-tribalism.html
Sunstein, C. R. (2017). *#Republic: Divided democracy in the age of social media*. Princeton University Press.
Táíwò, O. (2020). Identity Politics and Elite Capture. *Boston Review*. https://www.bostonreview.net/articles/olufemi-o-taiwo-identity-politics-and-elite-capture/
Talisse, R. B. (2019). *Overdoing democracy: Why we must put politics in its place*. Oxford University Press.
TED Interview (2024). How to fight hatred with curiosity with Daryl Davis. https://www.ted.com/pages/how-to-fight-hatred-with-curiosity-with-daryl-davis-transcript
Tsai, G. (2014). Rational persuasion as paternalism. *Philosophy & Public Affairs*, *42*(1), 78–112.
Turkle, S. (2015). *Reclaiming conversation: The power of talk in a digital age*. Penguin.
Urban, T. (2023). *What's our problem? A self-help book for societies*. Wait But Why.
Vallor, S. (2016). *Technology and the virtues: A philosophical guide to a future worth wanting*. Oxford University Press.

Whitcomb, D., Battaly, H., Baehr, J., & Howard-Snyder, D. (2017). Intellectual humility. *Philosophy and Phenomenological Research*, *94*(3), 509–539.

Williams, D. (2022). Identity-defining beliefs on social media. *Philosophical Topics*, *50*(2), 41–64.

Williams, D. (2023a). The fake news about fake news. *The Boston Review*. https://www.bostonreview.net/articles/the-fake-news-about-fake-news/

Williams, D. (2023b). The marketplace of rationalizations. *Economics & Philosophy*, *39*(1), 99–123.

Williams, D. (2023c). The case for partisan motivated reasoning. *Synthese*, *202*(3), 89.

Worsnip, A. (2018). The obligation to diversify one's sources: Against epistemic partisanship in the consumption of news media. In J. Saunders & C. Fox (Eds.), *Media ethics, free speech, and the requirements of democracy* (pp. 240–264). Routledge.

Zamalin, A. (2022). *Against civility: The hidden racism in our obsession with civility*. Penguin Random House.

16 Borg Virtue Ethics

Howard J. Curzer and Joel D. Velasco

16.1 Introduction

Biologists and philosophers writing about human evolution tend to be backward-looking. They describe past evolution and use it to give explanations of how we came to be the way we are now. The evolution of humans has not stopped, but genetic changes are slow and relatively minor. Certainly nothing that ethicists need to pay attention to. However, major evolutionary shifts are in the process of happening right now, merely in a very different way. This chapter will be forward-looking, focused on the very near future. We shall argue that two new species are evolving out of humanity right now.

This raises natural questions for virtue ethics. Are the members of these new species the sort of beings that can have virtues and vices? Do they have the same virtues and vices as humans? If not, how do the virtue lists differ?

Some caveats and cautions will be necessary. First, we are going to rely on Aristotle for inspiration and direction at several points. But we do not appeal to the authority of Aristotle, let alone accept everything Aristotle says. Second, this chapter is exploratory. We are going to push a highly controversial line as far as we can. Third, we cannot argue thoroughly for all, or even most, of the components of our view. Our view will have to stand or fall on the appeal of the picture we sketch. If we end up with a plausible picture, then further work along this line might be useful. But even if this picture turns out to be implausible, we will have probed the structure of virtue ethics more generally, and by examining transhuman ethics, we will understand virtue theory and the *human* virtues better.

16.2 Human Virtue

A quick and dirty version of a *Contemporary Aristotelian Virtue Ethics* (VE) account of virtue for humans is this. *Flourishing* consists in (a) the

DOI: 10.4324/9781003632641-21

excellent exercise of all essential human abilities, plus (b) crucial, intrinsically valuable goods of fortune, (c) throughout a complete life.

a Excellent exercise of all essential human abilities is defined in terms of virtues. Each *virtue* is a disposition to respond excellently to situations of a certain sort – a certain sphere of human life. The response to a situation includes not just action, but also perception of the situation, passions prompted by the situation, planning and choice of action, etc. Courage is a disposition to respond excellently to situations involving risk; temperance is a disposition to respond excellently to situations involving sensual pleasure; justice is a disposition to respond excellently to situations involving the distribution of scarce goods; and so on. There are good and bad ways of dealing with every situation, so together, all of the virtues cover all of life.
b Some maintain that, beyond the instrumental goods necessary for the acquisition and exercise of virtue, no further goods of fortune are necessary for flourishing. Following Aristotle, we disagree. He maintains that some goods of fortune are components of, and not just means to, flourishing. He lists good birth, good children, and good looks as examples of intrinsic goods of fortune (Aristotle, *Nicomachean Ethics* 1099b2–6). Some people might reject these examples, but we believe that members of a severely disparaged class, parents of sociopathic children, or terribly disfigured people are unlikely to flourish.
c A life is *complete through time* if it lasts for a long stretch of life (decades), though not necessarily from birth to death, and it constitutes a coherent narrative. That is, it comes to some sort of closure. It has a beginning, middle, and end. A life is *complete at a point in time* if, over a short period of time (weeks), the essential human abilities are excellently exercised throughout all of the spheres of human life – situations involving risk, situations involving sensual pleasure, situations involving distribution of scarce resources, etc. A complete life is both complete through time and complete at each point in time.

Thus, flourishing is virtuous activity together with crucial intrinsically valuable goods of fortune throughout a complete life.

Let us briefly guard against some potential misconstruals about which actions, passions, etc. are virtuous. (1) Aristotle is a commonsense philosopher, but he allows that common sense sometimes goes wrong. Like Aristotle, we deny that virtuous actions and passions are just the ones conventionally considered to be good. (2) Virtues are not defined in terms of objectively morally good actions, either. In virtue ethics, morally good acts, if they are defined at all, must be defined (perhaps in some complex way) in terms of the virtues, and not the other way around.

(3) Aristotelian virtues prescribe actions and passions that lie in a mean, but one cannot use Aristotle's golden mean to determine which actions are virtuous. Instead, to identify virtuous actions, look to what virtuous people characteristically do.

Virtues are character traits that benefit their possessors. However, (4) it is not the case that every virtuous act forwards flourishing, or (5) that virtuous agents are generally motivated by their own flourishing when they perform virtuous acts, or (6) that the possession of virtues guarantees a flourishing life.

Aristotle's thesis is more moderate. Virtues are the way to bet, even though they sometimes induce their possessors to act in counter to their own interests. Agents are more likely to flourish if they are courageous, temperate, and just than if they are cowardly, intemperate, and unjust. Although there are no guarantees, *the virtues are character traits most conducive to flourishing in each sphere of human life.*

Virtues forward flourishing in different ways. Each character virtue is a disposition to excellently exercise some of the abilities humans share with other animals, meshed with and modified by practical reasoning. Intellectual virtues are dispositions to exercise the parts of reason excellently. Virtuous activity is pleasant and meaningful. It also makes for a longer life, avoids regrets, and provides certain crucial goods of fortune. Most importantly, it is intrinsically valuable. Virtuous activity is the core of any flourishing life.

It will be noticed that virtues are defined in terms of flourishing, and flourishing is defined in terms of virtues, and both rest upon an account of the essential human abilities (i.e., human nature) which we shall provide below. VE maintains that the circle is hermeneutic rather than vicious. That is, we begin with hazy minimal conceptions of both flourishing and virtue, and as each informs the other, we come to amplify and clarify both conceptions.

16.3 Human Nature

It is clear that an account of human flourishing depends on an account of human nature. What counts as good for us depends on what kinds of creatures we are. Since flourishing and the virtues are metaphysically intertwined, the virtues depend on human nature as well.

It might be hoped that the virtue ethicist could rely on a well-accepted account of human nature and then go from there. But there is no standard, well-accepted account. And in fact, one prominent view popular among philosophers of biology is that there is no such thing as human nature at all. One way to reach this conclusion is to have strong requirements on what could count as human nature. Start with the idea that human nature consists in our essential traits – the traits that are necessary and sufficient for being human. Reflection on the nature of evolution tells us that there are not likely to be any such traits and so, on this conception, no such thing as human nature (Hull, 1986). Other authors see an important role for human nature and so weaken the conception. For example, Machery (2008) argues that human nature is just the set of properties that humans tend to possess as a result of the evolution of their species, while Ramsey (2023) develops a trait cluster approach. In our view, "human nature" refers to different things in different contexts, and there is no reason to expect a single, best account for all purposes. Kronfeldner (2018) defends a pluralistic view like this. She argues that there are different concepts of human nature that satisfy different purposes, including different classificatory, descriptive, and explanatory purposes in different sciences.

In order to see what kind of account of human nature will play the right kind of role in our project, we'll briefly examine Aristotle's picture and then see what use it is put to in an account of flourishing.

One depiction of Aristotle's view of human nature (in Figure 16.1) is a multi-level pyramid. Level #1 is plant nature. The abilities to take in nourishment, grow, and reproduce are shared by all plants, although different plants do these things in different ways. Level #1 plus level #2 is animal nature. Animals are plants plus more. In addition to eating, growing, and reproducing, all animals can desire, move, and sense. Sensation includes both gaining information about the world and experiencing sensual pleasure and pain. Levels #1, #2, and #3 constitute human nature. Humans are animals plus more. We are rational animals.[1] Aristotle divides reason – the distinctive ability of human beings – into three components: theoretical, practical, and productive reason (Aristotle *Metaphysics* 1064a1–18).

Each of these levels is transformed by the level above it. The ways in which plants and animals take in nourishment are very different from each other *because* animals can desire, move, and sense. Similarly, the ways in which humans and other animals desire are very different from each other

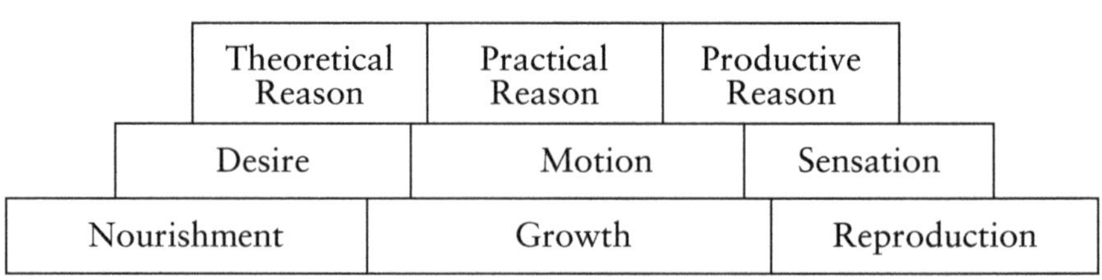

Figure 16.1 Human nature

because humans can reason. Productive reason produces artifacts that improve motion, sensation, etc. Practical reason formulates plans to achieve one's goals. Theoretical reason generates understanding of the way the world works. All three sorts of reasoning distinguish human desires, motions, etc. from the ways in which animals desire, move, etc.

16.4 Two New Species

Several trends, attributable largely to the way in which people use the internet, are the basis for the emergence of two *trans-human species*. First, relationships networks among individuals are getting wider and wider, incorporating more and more people. Social media, news media, social networking, blogs, dating apps, texting, massively multiplayer online games, etc. have recently fueled a dramatic increase in communication among people. Each person interacts socially, economically, politically, etc. with a huge number of others. In this sense, societies are now much more integrated than they used to be.

Second, various aspects of ordinary reasoning and decision-making are fundamentally changing due to instantly available information. Our daily scheduling and planning are controlled by electronic calendars and reminders. As we travel, we are directed by a GPS without any understanding of where we are or where we are going. We follow detailed "how-to" videos rather than figuring anything out for ourselves. And we don't memorize facts or remember anything that we have done – we will just look it up again next time. So we don't just happen to use the web to communicate with others; we *need* to use it.

Once we decide to do something, we use the web to aid us. But the web also tells us, either directly or indirectly, what to do. People are under increasing social pressure. Their choices are almost completely limited to socially endorsed, or at least socially acceptable, options by the shared *worldview* of society. The term "worldview" suggests a set of interconnected beliefs, but we intend the term to also include passions and moods, perceptions and patterns of thinking, goals, and values. As we are using the term, a worldview includes not only a picture of the way the world is but also a prescription of how to react to it, and a detailed stipulation of what is valuable within every sphere of human life. Worldviews are not the products of a few individuals; instead, they emerge out of the millions of (primarily electronic) communications made by individuals to each other.

Using the above VE account of human nature and flourishing as a starting point, we shall sketch accounts of the nature and flourishing of two new species. Using the VE method of deriving human virtue sketched above, we shall derive accounts of their virtues below.

16.5 Cyborg Nature

The primary way in which members of society communicate with other members is through the web. But people do not just *communicate* electronically. Deliberately or incidentally, people create and then identify with a persona – an *e-self*. Increasingly, a person consists of a human self plus an e-self. People live on the web as well as in houses! These person/e-person combinations are the new normal. They are familiar; we do not gape at them in amazement. But we should. They consist of human beings enhanced by artificial parts.[2] They are *cyborgs*,[3] a trans-human species.

Just as animals are plants plus more, and humans are animals plus more, so cyborgs are humans plus more (as depicted in Figure 16.2). The additional abilities are provided by the cyborg's artificial parts. These are abilities for interacting with other e-persons, and with the web as a whole. The existence of the e-self does not simply add new abilities to our nature. It transforms them.

We claimed that cyborgs are a new species. To many readers, this will seem absurd. Yet it is the right term to use here. Let us explain why.

The natural place to look for definitions of "species" is in biology or perhaps philosophy of biology. Historically, the most prominent definition of a species is the Biological Species Concept: species are groups of actually or potentially interbreeding natural populations that are reproductively isolated from other such groups (Mayr, 1942). This definition is "forward looking" – if populations are reproductively isolated now, we can expect that over time, they will change independently and become increasingly distinct over time. However, many contemporary systematists think that biological classification should reflect evolutionary history and the biological species concept can obscure this. Instead, they propose some version of a Phylogenetic Species Concept where species are defined by shared evolutionary history (Baum & Donoghue, 1995; Velasco, 2008). These views are "backward looking." While these two definitions are the most

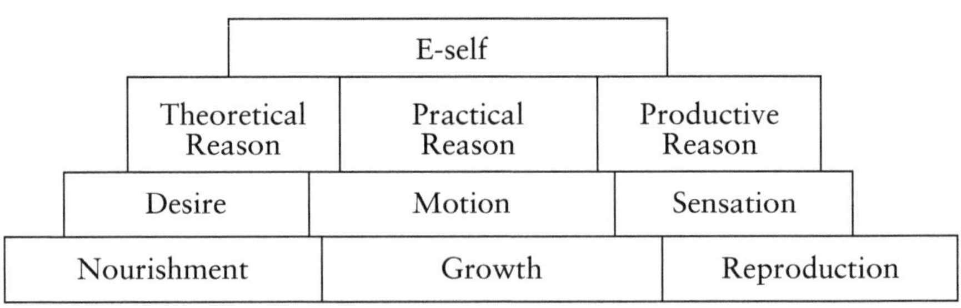

Figure 16.2 Cyborg nature

prominent, there are many more defended in the literature. An increasingly popular view is that we shouldn't choose at all – instead, there is really a shared, underlying agreement on what species are: lineages through time (de Queiroz, 2007).

Cyborgs, as they exist now at least, are parts of populations that share recent genetic history with other humans, and they are not reproductively isolated from other humans. Thus, they will share descendants in the future. In evolutionary terms, we are part of the same lineage. So any of the three views noted above would classify cyborgs as humans. We agree – cyborgs should be classified as *Homo sapiens* for purposes of biological taxonomy. We don't mean to be defending a new view of taxonomic species.

Our purposes are different. In our account of virtue ethics, the virtues and flourishing are definitionally intertwined and dependent upon human nature. We immediately think of *human* flourishing because we are human. But any rational creature will have virtues which are definitionally tied up with their flourishing. It is easy to imagine creatures which meet these definitions of *Homo sapiens* yet are so different from us that their flourishing would be quite distinct from ours. Imagine technology advances to the point where humans can simply download their minds into other artificially constructed bodies at will with memories and personalities intact. Individual minds could live an arbitrarily long time; individual bodies would have many new abilities. Society would undoubtedly be very different. Still, these are rational creatures and there is still such a thing as flourishing and living well – but it is different than human flourishing because their nature is different. Thus, their virtues would be different as well.

Weren't other changes in the way we think and make decisions (writing, calculator) also transformative enough to form cyborgs? An extreme version of the extended mind hypothesis (Clark & Chalmers, 1998) implies that people have been cyborgs since the first hominid put a mark on a rock to remind herself of something. But we prefer to maintain distinctions among (a) mere tools, (b) artificial parts that merely replace, or moderately enhance something that a purely organic mind could already do, and (c) artificial parts that enable a purely organic mind to do something radically different than it could already do. It is these radical enhancements that turn humans into cyborgs. The web is such an artificial part.

Previous changes certainly increased our ability to keep track of things and to solve problems, but the web eliminates the need to do such things. These differences in degree are so great that they become differences in kind. They fundamentally change the kind of creatures we are. Thirty years ago, we were still human. However, we contend that the recent changes – in particular, contemporary internet communication and the changes in our basic practical reasoning that it has led to – are so significant that they change *who we are* in a fundamental way. Indeed, they have *already* changed many of us.

We contend that cyborgs have a different nature and therefore distinct virtues. For Aristotle, the nature of the thing is its "eidos" – what makes it the kind of thing that it is. This is commonly translated as "species." If cyborg natures are distinct from human nature, then it is perfectly appropriate to call them a different species, despite how it sounds to our ears.

16.6 Cyborg Virtue

Each cyborg virtue is a disposition to exercise the set of essential human abilities in combination with, and transformed by, the essential cyborg abilities excellently within a certain sphere of cyborg life. Various thinkers, such as Vallor (2016), have tried to work out the virtues of humans in the face of new technologies such as the web. Vallor argues that in order to flourish in our current technologically saturated environment, we need to cultivate a new kind of moral character; one that expresses what she calls the "technomoral virtues." As an example, the virtue of justice becomes technomoral justice which is two interrelated character traits. First, "a reliable disposition to seek a fair and equitable distribution of the benefits and risks of emerging technologies." And second, "a characteristic concern for how emerging technologies impact the basic rights, dignity, or welfare of individuals and groups" (p. 128). In detail, this appears to be a new virtue defined in relation to our technological environment. But it, like all the others, is deeply connected to the traditional virtues. As she puts it, "These will not be radically new traits of character, for they must remain consistent with the basic moral psychology of our species. Rather the technomoral virtues are new alignments of our existing moral capacities, adapted to a rapidly changing environment that increasingly calls for collective moral wisdom on a global scale" (p. 10).

Our approach is rather different. We maintain that since the abilities of the e-self transform the lower-level abilities, the virtues of a cyborg are quite different from the virtues of a human. We shall offer a few examples. While open-mindedness, self-reliance, and generosity are human virtues, we argue that their cyborg correlates are rather different character traits.

Through the web, cyborgs encounter vast quantities of knowledge, much of it emanating from experts, and abundant falsehoods, most of them emanating from systematic campaigns of deception. Virtuous cyborgs neither ignore nor embrace new claims. They cope by developing information filters, primarily within their e-selves. They don't try to assess all novel claims on their own by considering each new claim on its merits. No cyborg possesses the necessary skills, fund of knowledge, or spare time to evaluate everything seen on the web. The attempt would undermine rather than enhance understanding. Cyborgs would be taken in by clever conspiracy theories, scammed by devious swindlers, and overwhelmed with countless claims to check (Buzzell & Rini, 2023). Traditional open-mindedness and self-reliance are vices for cyborgs. In their place, the cyborg virtues include the virtues of *e-collaborative-open-mindedness* and *e-crowdsourcery*. Virtuous cyborgs deal with novel claims by (a) holding themselves closed to novelty outside of their area of expertise, (b) relying on the judgment of trusted experts within their e-community to evaluate such novelty,[4] and (c) evaluating novel claims that fall within their own area of expertise. Each new thing is evaluated by experts within their e-community, while the rest of the cyborgs rely upon the judgment of these experts. Thus, as a collective, cyborgs are open-minded and self-reliant, but each virtuous individual is closed-minded about almost everything and makes their own decisions about almost nothing.

Generosity requires empathy, but empathy is dangerous to cyborgs. The web is filled with swindlers and troublemakers. Unlike the self, the e-self must be constantly alert for scams and hoaxes, and also ego-defended against offense and provocation. Sadly, suspicion and indifference to criticism are both necessary and incompatible with empathy. Furthermore, the web is filled with tales of terrible suffering and injustice. Giving attention to even a fraction of these would spread empathy quite thin. Just as the dangers of information overload and fake news have made information filters necessary, so the dangers of trickery, trolling, and tragedy have made empathy filters necessary. Cyborgs should focus their empathy and energy on a small, verifiable, manageable set of evils while ignoring the plight of everyone else. This is *e-generosity*. Different cyborgs will focus on different evils; many of those in need will receive help from someone. Thus, as a collective, cyborgs are generous, though individual virtuous cyborgs help only those in their focus.

16.7 Cyborg Hives

The second transhuman species is even stranger and more controversial than the first. As humans change, so human societies are changing as well. Past human societies are derivative entities; they are best understood as collections of humans. Perhaps some of these collections have an ontological status all their own. They are, so to speak, more than merely the sum of their parts. But even so, their existence is grounded in and understood in terms of their parts.

We maintain that as the 21st century unfolds, taking individual people as basic entities while societies are merely derivative is increasingly misleading. Individuals are becoming mere parts of societies, just as organs are parts of people. A heart can't exist or be understood on its own; it works and makes sense only as a part of a circulatory system which, in turn, works and makes sense only as a part of a human being. Similarly, individual cyborgs also can only properly be understood as parts of their societies.

Let's take a breath. The term "society" is used in many ways. By "society" we mean a collection of people together with their shared institutions, practices, and comprehensive worldview. Some states and societies have the same members, but many do not. Bifurcated news media, social media, political and religious affiliations, ways of speaking and dress, marriage and child raising practices, etc. suggest that Western society is currently fissioning into two societies: Liberal and Conservative. There are smaller societies, too. Some of these are properly contained within one of these two large societies; others intersect with both.

A few past thinkers have considered societies to be decision-making entities.[5] But how to take this claim has always been something of an interpretive puzzle. Was it meant metaphorically, or mystically, or somehow literally? Regardless, now things have changed. While society used to be merely a collection of people with a bit of organizational structure, the organizational structure is becoming more and more immense, integrated, and influential. Admittedly, this is a difference of degree, but when differences of degree become great enough, they become differences in kind. It is now becoming increasingly plausible to consider contemporary societies to be organisms.[6] What sort of organisms? Aristotle lumps humans together with bees and ants as "political animals" (Aristotle, *Politics* 1253a7–9). We shall argue that Aristotle was prescient. His claim was a stretch when he made it, but humans have become ever more similar to these eusocial insects. Contemporary societies are human/cyborg societies. As the percentage of people who lack e-selves declines, these are well on the way to becoming simply cyborg societies, and *cyborg societies are cyborg hives.* These super-organisms of eusocial animals constitute a second trans-human species.

16.8 Are Contemporary Societies (Almost) Cyborg Hives?

The social insects (ants, bees, wasps, and termites) live in complex social structures called "colonies." While the individual insects are living organisms, understanding their behavior requires seeing them as part of the colony. As humans become cyborgs, their social structures increasingly resemble such colonies. As evidence that contemporary societies are hives, we shall list 12 features that contemporary societies share with insect colonies.[7]

1 Unlike packs, colonies are *leaderless*. Although ants are guided by pheromones emitted by the queen, there is no sense in which she, or any other individual, is a decision-maker. Similarly, what unites a contemporary society is not the sort of hierarchical decision-making structures that unite countries. Societies do not have rulers, bureaucrats, constitutions, legal systems, etc. Like other colonies, cyborg societies are leaderless.
2 Unlike migrating flocks, colonies *persist* for long periods of time. Similarly, unlike crowds, societies persist for long periods of time.
3 Unlike herds, colonies appear to be rational. That is, there seems to be a *hive mind* that responds to situations in complex, rational ways. It seems to grasp situations, classify them, construct plans in response to them, and finally implement its plans. Similarly, contemporary societies engage in collective decision-making, applying sets of beliefs contained within worldviews to perceived situations, resulting in actions. They, too, seem to have hive minds.
4 Each colony consists of numerous members divided into *multiple castes*, performing different tasks. There is little or no significant switching from one job to another. People have considerably more social mobility than ants, but considerably less than the "American dream" suggests. Certainly, children do not always grow up to work in the family business or follow in the footsteps of their parents when choosing trades or professions. But children seldom leave their parents' caste (economic and social status). People born into the working class tend to stay in the working class, although they may opt for different working-class jobs than their parents, and they may switch from one working-class job to another during their lives. Similarly, people born into the professional class stay in the professional class. And so on (Urhan et al., 2012, p. 2).
5 Colony members are self-interested; they strive to protect their own lives when they perceive themselves to be threatened. They flee or fight. On the other hand, they routinely *sacrifice themselves* for the good of the colony. They do not act reluctantly in these cases. They do not seem to see themselves as making sacrifices.[8] People seem considerably more self-interested than ants. But like ants, they often act with seeming obliviousness against their self-interest because of social pressures or incentives. For example, people buy products and services that they do not need or even want. They support governmental policies and social practices that hurt them in many ways,

even to the point of shortening their lives. Heidegger highlights the fact that people are in denial about their own deaths (Heidegger, 1962, pp. 252–260). We go further by suggesting that people deceive themselves about many harmful practices and choices. Consumerism, self-destructive support, and certain self-deceptions are fanned by the web and best explained in terms of their benefit for the hive, for they harm the agent.

6 Colony members are in nearly constant *communication* with each other. Their communication enables and motivates near-universal *cooperation*. Similarly, although people's communication is mostly chit-chat, the web does enable and motivate social cooperation on a vast scale.

7 Colony members share the *same overall goals*. Their interests are bound up with the flourishing of the colony. Each ant reacts in much the same way to situations. Similarly, people identify with their society. Their interests align to some degree with the interests of society, and they react in similar ways to situations.

Because we are concerned with character virtue and intellectual virtue, and virtue is concerned with choice, we shall focus on several sorts of decision-making within colonies.

8 The colony generally operates on *autopilot*. Colony members do routine things almost all of the time. They do not even see themselves as making decisions. At first glance, the idea that society is a hive seems absurd because, unlike ants, people seem to make lots of non-routine choices. However, we suggest that this is an illusion; the vast majority of our choices are routinized. People do not even consider them to be decisions. For example, people wear clothes to work and sit on chairs in seminar rooms automatically rather than by choice.

9 Colony members routinely make *trivial decisions*. An ant may "choose" whether to turn right or left when it leaves the anthill. Similarly, people choose such things as the color of their clothes and where in the room to sit. But, of course, these are trivial choices. Most of our decision-making is like the ants' choices to zig left or right.

10 Members do sometimes make non-trivial choices, but these are severely *constrained*. An ant may choose among different, standard tasks that are useful to the colony. But it will not choose non-standard tasks. Ants seldom start revolutions, found new colonies, pioneer practices, etc. People have many more choices than ants. However, even the big decisions are, in an important sense, not full-blown choices. Individual people do choose careers, mates, houses, etc. But as the existentialists emphasize, society severely limits the range of live options of action, belief, perception, passion, etc. available to its members by inculcating

its worldview. Some options are impossible within the contemporary world (e.g., finding a job as court jester or starship pilot) because they can exist only when embedded in a social structure that does not exist; others do make sense, but are off the table because of a lack of certain goods (high school dropouts becoming professors); yet others are live, but disparaged options (working as a prostitute). People choose from a very limited menu of options. People seldom act without thought, but thought is constrained within a very narrow range by their society's worldview.

11 A few individual ants do choose proscribed options. They deviate, either because they try and fail to follow the routine or for some other reason. But these non-conformists' choices are uncommon. Among people, some non-conformists are losers who cannot do what society expects of them. Others are misled, deluded about the world, or even mentally ill. Some are iconoclasts; the sort of people Nietzsche describes as lions. Others are driven in different directions by creative impulses; the sort of people Nietzsche describes as children (Nietzsche, 1954, pp. 25–27). And of course, a few are people who see through society's worldview to the truth; the sort of people Plato describes as philosophers (Plato *Republic*, 520a5), and contemporary activists might call "woke."

12 On rare occasions, crises demand that the colony, itself, make decisions (e.g., where to move the colony when it must relocate). But these extraordinary choices are not made by a single individual. Nor are they made by multiple individuals who arrive at compromises or take votes. Rather, the colony members communicate with each other by dances and pheromones, and a *communal choice emerges* out of information-sharing among individuals. Similarly, crises demand that society, itself, make decisions. But even extraordinary decision-making by society is communal. It is not made by a single individual or a committee. Instead, a decision emerges from zillions of interactions among members. For example, a few years ago, Liberal society decided to endorse gay marriage. For a long time, only a few individuals accepted gay marriage. Then a few more. And so on. Eventually, a tipping point was reached. The legitimacy of gay marriage became accepted as an obvious truth by Liberal society.

Weren't other changes in the way we communicate with people (e.g., printing press, telephone) and the way we travel (trains, planes, and automobiles) also transformative enough to form a hive mind?

Previous changes certainly increased the number of people reached, and the speed with which they were reached, but that was not enough.

The web made it possible to reach almost everyone instantaneously, and that wide, fast reach took society to the level of eusocial animal.

16.9 Are Cyborg Societies a New Species?

We have argued that cyborg societies are hives. A hive is a kind of colony – a complex, structured society. In a very weak sense, it is very clearly an individual thing – a particular entity. But what kind of individual is it?

Organisms are very special kinds of things – highly functionally organized and physiologically integrated biological entities. Insect colonies are also functionally organized and physiologically integrated, leading many to suggest that they are a kind of superorganism (Wilson & Sober, 1989; Hölldobler & Wilson, 2009). In their influential work *The Major Transitions in Evolution*, Maynard Smith and Szathmáry (1995) treated the origin of eusociality and the formation of these insect colonies as a *major transition* – a new fundamental level of biological organization. Whether or not they should count as superorganisms or even as organisms in their own right (as in Martens, 2010; Okasha, 2011), they are cohesive, biological individuals that form lineages and are some kind of unit of natural selection (Hamilton et al., 2009; Haber, 2013).

Each of these authors approaches the metaphysics of insect colonies from the point of view of evolutionary theory. When we focus on the transmission of genetic information through time, the division of reproductive labor in the social insects looms large. Unlike the social insects and their sterile castes, cyborgs (at least as they currently exist) still maintain their own reproductive independence. This is certainly important. But with a sufficiently broad understanding of evolution, we can see that the true novelty and importance of human societies and cyborg hives lie in their distinct form of information transmission through language. Maynard Smith and Szathmáry (1995) recognized language as a major transition, and any sufficiently broad understanding of evolution that looks at persistence and change through time will recognize the importance of cultural transmission. Here, human societies are clearly different than any other biological entity on the planet.

We wish to push this one step further. We aren't just claiming that human societies with language and cultural transmission are a new kind of thing. We are claiming that cyborg hives are yet another new thing distinct from previous human societies. Another major transition.

Clearly, cyborg hives are individual entities and thus are a distinct kind of thing with their own natures. In this very weak sense, a cyborg hive can have virtues just as a hammer or a hamster can be disposed to exercise its abilities well. But can it have virtues in a stronger sense? Can it flourish?

We argue that it can. We will present an account of a cyborg hive's flourishing, its nature, and a few of its virtues.

16.10 Flourishing for Societies

It may be useful to contrast societies with states.[9] Aristotle thinks that people have natural abilities which enable them to live in states, and natural tendencies to develop states. In this sense, Aristotle considers states to be natural. But Aristotle primarily thinks of states as artificial. They are tools for accomplishing certain goals constructed by the deliberate choices of individuals. Founders write constitutions. Politicians make decisions based on reasons.

According to Aristotle, the goal of a state is to make flourishing lives available to its citizens. Aristotle acknowledges that different states are built around different conceptions of a flourishing life. But Aristotle thinks that most of these conceptions are mistaken. Some states aim to enable their citizens to lead lives focused on wealth or sensual pleasure, but such lives are not flourishing lives. Aristotle allows only two versions of flourishing: contemplative lives oriented around theoretical reasoning, and practical lives oriented around practical reasoning plus the animal abilities. This sort of life consists of courageous acts, temperate acts, just acts, generous acts, etc. Just moving through life doing the right thing almost all of the time. Modern thinkers agree that the state is an artifact, and they joust with Aristotle, and with each other, about what counts as flourishing, and which conditions of flourishing (if any) the state should provide.

Societies are like states in some respects. For example, they include individuals organized into social structures. Cyborg societies make decisions based on reasons. Like states, cyborg societies have goals. But unlike states, these goals are not imposed on cyborg societies by founders, and the choices are made by hive minds rather than committees. Cyborg societies are not artifacts: they develop by nature. They are best understood as organisms. So understood, the goal of a society is not the flourishing of its members. Like other organisms, a cyborg society's goal is its own flourishing.

We call the flourishing of any sort of society a *golden age*. "Golden age" is, of course, a vague term, and any historical examples of societies experiencing golden ages can be contested. We take a society to be experiencing a golden age when it is growing – rapidly increasing and improving – along many different parameters: number of people, health and wealth of the society's members, amount of territory, security and stability, degree of influence on surrounding societies, scientific knowledge, technological knowledge, fine arts and beauty, creativity of all sorts, etc.

Some ways of growing involve conquest, coercion, colonization, exploitation, resource extraction, and other repugnant activities. But these are not the only ways, or even the best ways, for societies to grow. Western society has insinuated itself throughout much of the world through economic, technological, and cultural ties, for example. In most of the world, people can eat at McDonald's, carry iPhones, and watch Hollywood movies. Along with these things come a certain worldview and certain practices, institutions, and rituals.

For members of the intelligentsia (such as our presumed readers), it is tempting to picture golden ages solely in terms of advances in the sciences and the arts. Indeed, these are very important parameters, but not absolutely necessary ones. To experience a golden age, a society must advance along many parameters, but not every parameter. It is possible for a society to advance along enough parameters to experience a golden age even though it does not advance along the parameters of the sciences and the arts. For example, Mongol society from Genghis Khan through Kublai Khan experienced a golden age without significant advances in science or art.

Growth along any parameter might come about by luck, but in order for a society to grow along many parameters all at once, the society must be doing many things right. In VE terms, the society must be excellently exercising its essential abilities.

16.11 The Nature of Contemporary Societies

We have established that societies are currently (or at least currently becoming) cyborg hives. We have sketched an account of flourishing for societies and observed that this consists in the excellent exercise of the society's essential abilities – i.e., the hive's nature. What are the essential abilities of a society? Of a cyborg hive?

Aristotle does not offer a canonical way to divide states into parts. At one point, he lists six essential components of the state: farmers, craftsmen, soldiers and weapon-makers, wealth-makers, priests, legislator/judges (Aristotle, *Politics* 1328b4–14). Perhaps he thinks that these parts correspond to the essential functions of the state.

Like states, there is no canonical way to divide societies into parts, no canonical list of society's essential abilities. Plato suggests in the *Republic* that a society is just a person writ large with the same nature and virtues (Plato, *Republic* 441c-d). Is contemporary society similarly a cyborg writ large? We get to "Yes" only by stretching definitions considerably. But using human nature as a basis for cyborg nature, and cyborg nature as a basis for the nature of contemporary society does provide continuity of exposition.

1 Nourishment: Societies do not live on sunlight like plants, or by consuming other organisms like animals. They live on political, economic, and military power.

2 Reproduction: Reproduction for humans does not consist in making children, but rather in raising children. Child-raising is good for humans. It generally makes people's lives better in multiple ways. However, societies make, but do not raise their offspring. New societies (roughly individuated by the possession of new worldviews) emerge from existing ones in several ways. Societies reproduce by budding, fission, or fusion. But these are not essential abilities of societies. They are not even good for the parent societies.
3 Growth: Societies grow along various parameters. For example, they grow in population through births above the replacement rate and by assimilating members of other societies. They grow in knowledge as discoveries in various fields enter their worldviews. They do not grow in wealth when the average standard of living of their members rises, but rather when the wealth available for social projects increases. And so on.
4 Motion: Theoretically, societies can migrate, but it is not an essential ability. Moreover, migration is not currently feasible for the huge societies we have in mind.
5 Desire: Societies desire many of the same goods of fortune as humans (e.g., wealth, status). There are also many differences among objects of desire. A particularly important difference is that the exercise of human abilities is pleasant and desirable for humans, but societies don't feel or desire pleasure.
6 Intra-hive Communication: One distinctive feature of cyborg societies is the ability of their members to communicate with each other rapidly. This enables collective consciousness to form and to create and communicate worldviews. It also enables individuals to respond to changes in the circumstances of their societies.
7 Sensation: Societies do not experience sensual pleasure. But societies do gain information about the world via senses as people communicate what they see, hear, etc. to other members of society. Cyborg societies sense things much more quickly than human societies. Cyborgs can communicate their sensations immediately to everyone by uploading pictures, videos, music, etc. to social media and news media.
8 Theoretical Reason: Societies have many thoughts (shared beliefs) about the world. Again, cyborg societies are considerably quicker at sharing thoughts than human societies, and they are arguably more thoughtful as well. Beliefs can become part of a cyborg society's worldview with the speed of a viral meme. And while previously most ideas were shared with a few people and went no further, in cyborg societies ideas are uploaded and so considered by vast numbers of people all at once. Indeed, the main activity of the hive minds of cyborg societies is the creation, maintenance, and modification of their worldviews.

9 Practical Reason: Societies do almost no deliberating and choosing. Deliberation, such as it is, occurs only in extraordinary situations, and then only minimally. Society's practical reasoning consists in drawing straightforward, even trivial conclusions from the worldview about how to proceed. Decisions generally emerge without any thought process. Formerly acceptable practices become unacceptable, for example. Society may settle on a line of reasoning to support its decisions, but this is after-the-fact reasoning, and usually mere rationalizing.
10 Productive Reason: Parallel with practical reasoning, societies do almost no productive reasoning.
11 Mobilization: Worldviews marshal a huge number of members of each society to undertake multiple joint projects, which enhance the flourishing of that society. The worldview makes certain actions the obvious things to do.
12 Motivation: Worldviews also inspire members to engage in these projects eagerly rather than grudgingly.

16.12 Collective Virtue

Armed with this account of the essential abilities, the nature of society, we shall turn to a sketch of the virtues of society. Does it even make sense to talk of the virtues of a collective? A summative approach to the issue of collective virtues maintains that choices are made, and virtues are possessed only by individuals. A collective may be said to have virtues, but this is just a shorthand way of saying that each (or most) of its members have those virtues. The virtues of the collective not only reduce to but also match the virtues of each (or most) of the individuals who are its members. To say that a collective is courageous is just a shorthand way of saying that its members are courageous.

However, societies do not automatically possess the virtues of their members, nor do the members automatically inherit the virtues of society. We must be particularly on guard to avoid the fallacies of composition and division.

Taking a step away from the summative approach, one might maintain that the virtues of a collective match the virtues of its members *qua* members, although these people may not possess these virtues *qua* individuals. For example, when thinking and acting as a corporate executive, Alphonso is ruthless just like the corporation itself, but in the rest of his life, he is a sweetheart. To say that a collective is courageous on this marginally non-summative approach is just a shorthand way of saying that (at least some of) its members are courageous insofar as they are members (Fricker, 2010, pp. 246–247; Byerly and Byerly, 2016, pp. 39–40).

A somewhat stronger non-summative approach maintains that the virtues of a collective do not necessarily match the virtues of its members in any sense. The collective may possess virtues that none of its members possess. Its decisions may be compromises that no one likes. Its virtues may consist of components distributed among the members. Begonia acts courageously, despite feeling excessive fear; Chadwick has courageous feelings, but misperceives situations; Dahlia perceives situations of risk correctly, but reacts wrongly to them; and so on. None of these is courageous. Yet when properly combined, Begonia, Chadwick, and Dahlia constitute a courageous collective.

However, even this stronger non-summative approach is insufficiently radical. It still makes the virtues of a collective rest upon the virtues of individuals in some way or another and ignores the fact that members of society have collective-dependent traits. Worse, it assumes that the traits which count as virtues for individuals and societies are the same. But this is not the case. A hive is a different kind of thing than a cyborg. It has a different nature. A flourishing hive does not flourish in the same way that a cyborg flourishes. Thus, we should not expect the virtues of a hive to be the same as the virtues of a cyborg.

16.13 Virtues of Society

Providing even quick-and-dirty accounts of a few of the virtues of society is going to be tricky. One must avoid the temptation to conflate the character traits of society with the traits of a roughly coextensive state, or the traits of the members of society. *The virtues of society are dispositions conducive to cultivating a golden age by creating, maintaining, and changing society's worldview, institutions, practices, rituals, and other social structures.*

Many of the familiar human virtues are also virtues of societies (e.g., courage, good-temper). However, since societies have different essential abilities than humans, the list of their virtues diverges from the list of human virtues. We offer a few highly speculative examples of differences between hive and human virtues.

1 Goods of fortune, such as wealth and influence, are *nourishment* for societies. Societies use such things to flourish. Nourishment is power, and a flourishing society needs power. The more powerful, the better. Too little food and too much food are both bad for people, but there is no such thing as too much nourishment for societies. Moderation with respect to nourishment is not a virtue of societies. Instead, the trait governing goods of fortune that is conducive to the cultivation of a

golden age is a limitless desire and tendency to pursue various goods of fortune. In other words, *greed* is a virtue for societies.

2 Knowledge is instrumentally as well as intrinsically valuable to humans. Thus, the epistemic virtues are virtues for humans. For societies, things are different. Societies use their faculties of *reason* primarily in fashioning worldviews. Worldviews are vehicles for mobilizing, motivating, and managing hive members. But the truth about the world is often uninspiring (depressing, boring, etc.), complicated, and paralyzing. Thus, true worldviews are often poor mobilizers, motivators, and managers. What makes a worldview good (good for society, conducive to bring about a golden age) is not truth. Thus, the epistemic virtues are not virtues of societies. Instead, the character traits conducive to the formation of good worldviews – inspiring, intelligible, incentivizing pictures of the world and stories about the culture – are a suite of dispositions to use rhetorical and raconteurial skills to fashion worldviews. One might call these the *virtues of inspirational mythmaking*.

3 Relatedly, because the *growth* of societies is sometimes accomplished by absorbing other societies and individuals, a good society is an effective assimilator. Assimilation is accomplished by persuasion. Truth is not particularly important for persuasion, so honesty is not a virtue of society. Instead, *persuasiveness* is a virtue of societies.

4 Coping with rapid changes requires quick adaptation to new situations, so *flexibility* is a virtue for societies. But flexibility for a society is the ability to make quick changes to worldviews and other social structures. This requires hive members who are able to pivot from one claim to an incompatible claim. Societies with flexibility can induce their members to accept claims such as "Oceania had always been at war with Eurasia" (Orwell, 1949, p. 34). A good memory is an obstacle to flexibility; a good forgettery is desirable. Critical thinking is an obstacle to flexibility; a disposition to make snap judgments is desirable. And so on.

5 Appropriate ambition is a virtue for humans. It lies in a mean. Humans should set demanding, but attainable goals for themselves. Neither aiming at easily achievable goals nor aiming at nearly impossible goals is conducive to human flourishing. However, aiming high pays for societies. Members of society need an accurate sense of their own limitations, but a belief in the limitless possibilities of society stimulates them to ever greater efforts. Thus, *limitless ambition* is a virtue for societies.

6 For humans (and states), justice is valuable in itself, and also because it enables people (and states) to cooperate, or at least avoid conflict with each other. But cooperation among societies is not an issue. What about fostering justice among members of society? Are just social structures

conducive to the cultivation of golden ages? That is, if a society's worldview encourages its members to treat each other more justly, and if its social institutions, practices, rituals, etc. become more just, will the society move closer to a golden age? Perhaps. Consider one of many aspects of this question. In order to flourish, it is crucial for societies to ensure that creative people are highly motivated, equipped with the right tools, and free from distracting burdens. Perhaps this can be accomplished justly. Give each member of society sufficient opportunities and goods of fortune and the creatives among the members of society will be able and eager to create. On the other hand, historically, societies have enabled the creatives in a different way. A large number of society's members have been exploited so as to provide the creatives with what they need. If this is the only way, or the best way to foster creativity, and similar aspects of the question are settled similarly, then justice is not a virtue of society.

16.14 Conclusion

Virtue ethics gives an account of the virtues and of living well. These are definitionally intertwined, and both depend on human nature. We have argued that humans are in the process of undergoing a radical evolutionary shift – our very natures are changing. We are becoming cyborgs. Cyborg nature is human nature plus more. Cyborgs are in near-constant communication with each other, and their reasoning is transformed by being part of a hive. We therefore sketched an introductory account of cyborg flourishing and cyborg virtues. Cyborg virtues are modified human virtues. Roughly speaking, human virtues and cyborg virtues aim at the same goals. Cyborgs accomplish those goals collectively, facilitated by the internet, rather than accomplishing them as isolated individuals. But both humans and cyborgs flourish by being (different versions of) open-minded, self-reliant, generous, etc.

Next, we argued that the way that cyborgs interact with the internet and with each other is becoming increasingly analogous to the way that the social insects form colonies. These cyborg hives are becoming rational creatures in their own right with a different list of virtues based upon their own, unique way of flourishing. Human virtues and the virtues of society are rather different and aim at different goals because society's nature is not augmented human nature, or even augmented cyborg nature. Virtuous societies are greedy rather than temperate. They generate inspiration and simplicity rather than truth. Persuasiveness and flexibility are virtues of society; honesty is not. Virtuous societies are boundlessly ambitious rather than moderately ambitious, and (arguably) exploitive rather than just.

Admittedly, much of what we say about contemporary persons and societies is not quite right – yet. No one knows what the future holds, but some trends seem clear. It is only a matter of time before virtue ethics for *humans* is a thing of the past.

Notes

1 Aristotle, *De Anima* 413a23–25, b11–24, 414a28–b20. Imagination's place is contested. It may be subsumed within other faculties (Aristotle, *De Anima* 433b28–434a7).
2 The e-self is a part of the cyborg individual, though not a physical part. We want to leave open whether smartphones (or other internet-accessing devices) should be considered parts of the cyborg as well.
3 Numerous authors, including Haraway (1985), Warwick (2002), Clark (2003), and many others, have used the term "cyborg" in related contexts before. Sometimes there is a requirement of new abilities due to technology that is integrated into the body. Other times, restored function via artificial parts such as prosthetic limbs suffices. Perhaps even tool use suffices for being a cyborg. Depending on the view, humans have all been cyborgs for thousands of years; alternatively, many (but not all of us) are cyborgs now, or finally, maybe none of us are now, but in the future we likely will be. We want to draw attention to what we see as a change in our nature due to a change in our essential abilities. Since this change is driven by technologies, "cyborg" is a natural name for this new type of creature. But we do not want to require that our artificial parts be physically integrated into our bodies, nor do we intend to use the term "cyborg" in reference to any other particular author's views or terminology.
4 If no credentialed experts are available, then cyborgs rely on the next best thing – the individuals within the group best placed to evaluate the novelty in question.
5 For example, Rousseau's general will, Confucius's will of heaven, and Hegel's world spirit are collectives which engage in rational decision-making.
6 This view that society is an organism is taken seriously by some within the scientific community. Some social psychologists and biologists take this view to be an illuminating way to assemble, organize, and explain various bits of data (e.g., Kesebi, 2012; Stearns, 2007; Wilson et al., 2008).
7 The term "hive" sometimes specifically refers to the structure that bees live in. "Colony" is the more general term, but it is vague. We shall use "hive" in order to capture the connotations of the sci-fi term "hive mind."
8 This is an important reason for taking colonies to be superorganisms.
9 To be clear, we are not suggesting that societies grow out of states. They are different sorts of entities.

References

Aristotle (2016). *Metaphysics* (C. D. C. Reeve, Trans.). Hackett.
Aristotle (2014). *Nicomachean Ethics* (C. D. C. Reeve, Trans.). Hackett.
Aristotle (2017). *De Anima* (C. D. C. Reeve, Trans.). Hackett.
Aristotle (2017). *Politics* (C. D. C. Reeve, Trans.). Hackett.
Baum, D. A., & Donoghue, M. J. (1995). Choosing among alternative 'phylogenetic' species concepts. *Systematic Botany*, *20*(4), 560–573.

Buzzell, A., & Rini, R. (2023). Doing your own research and other impossible acts of epistemic superheroism. *Philosophical Psychology*, *36*, 906–930. https://doi.org/10.1080/09515089.2022.2138019

Byerly, R., & Byerly, M. (2016). Collective virtue. *Journal of Value Inquiry*, *50*, 33–50.

Clark, A. (2003). *Natural-born cyborgs: Minds, technologies, and the future of human intelligence*. Oxford University Press.

Clark, A., & Chalmers, D. J. (1998). The extended mind. *Analysis*, *58*, 7–19.

de Queiroz, K. (2007). Species concepts and species delimitation. *Systematic Biology*, *56*, 879–866.

Fricker, M. (2010). Can there be institutional virtues? In T. S. Gendler & J. Hawthorne (Eds.), *Oxford Studies in epistemology* (vol. 3, 235–252). Oxford University Press.

Haber, M. (2013). Colonies are individuals: Revisiting the superorganism revival. In F. Bouchard & P. Huneman (Eds.), *From groups to individuals: Evolution and emerging individuality* (195–218). MIT Press.

Hamilton, A., Smith, N. R., & Haber, M. H. (2009). Social insects and the individuality thesis: Cohesion and the colony as a selectable individual. In J. Gadau & J. Fewell (Eds.), *Organization of insect societies* (572–589). Harvard University Press.

Haraway, D. J. (1985). A cyborg manifesto: Science, technology, and socialist-feminism in the late twentieth century. *Socialist Review*, *80*, 65–108.

Heidegger, M. (1962). *Being and Time* (J. Macquarrie & E. Robinson, Trans.). Harper and Row.

Hölldobler, B., & Wilson, E. O. (2009). *The superorganism*. W. W. Norton & Company, Inc.

Hull, D. (1986). On human nature, *PSA: Proceedings of the Biennial Meeting of the Philosophy of Science Association*, *1986*(2), 3–13.

Kesebi, S. (2012). The superorganism account of human sociality: How and when human groups are like beehives. *Personality and Social Psychology Review*, *16*, 233–261.

Kronfeldner, M. (2018). *What's left of human nature? A post-essentialist, pluralist, and interactive account of a contested concept*. MIT Press.

Machery, E. (2008). A plea for human nature. *Philosophical Psychology*, *21*(3), 321–329.

Martens, J. (2010). Organisms in evolution. *History and Philosophy of the Life Sciences*, *32*(2–3), 373–400.

Maynard Smith, J., & Szathmáry, E. (1995). *The major transitions in evolution*. Oxford University Press.

Mayr, E. (1942). *Systematics and the origin of species from the viewpoint of a zoologist*. Columbia University Press.

Nietzsche, F. (1954). *Thus spoke Zarathustra* (W. Kaufmann, Trans.). Penguin.

Okasha, S. (2011). Biological ontology and hierarchical organization: A defense of rank freedom. In B. Calcott & K. Sterelny (Eds.), *The major transitions in evolution revisited* (53–64) (Vienna Series in Theoretical Biology). MIT Press.

Orwell, G. (1949). *Nineteen eighty four*. Penguin.

Plato. (2004). *Republic* (C. D. C. Reeve, Trans.). Hackett.

Ramsey, G. (2023). *Human nature*. Cambridge University Press.

Stearns, S. (2007). Are we stalled part way through a major evolutionary transition from individual to group? *Evolution*, *61*, 2275–2280.

Urhan, S., Currier, E., Elliott, D., Wechsler, L., Wilson, D., & Colbert, D. (2012). "Pursuing the American Dream: Economic Mobility Across Generations." Pew

Charitable Trusts. http://www.pewtrusts.org/~/media/legacy/uploadedfiles/pcs_assets/2012/pursuingamericandreampdf.pdf
Vallor, S. (2016). *Technology and the virtues: A philosophical guide to a future worth wanting*. Oxford University Press.
Velasco, J. D. (2008). Species concepts should not conflict with evolutionary history, but often do. *Studies in the History and Philosophy of Biological and Biomedical Sciences*, *39*, 407–414.
Warwick, K. (2002). *I, cyborg*. University of Illinois Press.
Wilson, D. S., & Sober, E. (1989). Reviving the superorganism. *Journal of Theoretical Biology*, *136*, 337–356.
Wilson, D. S., van Vugt, M., & O'Gorman, R. (2008). Multilevel selection and major evolutionary transitions: Implications for psychological science. *Current Directions in Psychological Science*, *17*, 6–9.

Index

Note: Page references in *italics* refer to the figure and "n" refer to the endnotes.

For Product Safety Concerns and Information please contact our EU
representative GPSR@taylorandfrancis.com
Taylor & Francis Verlag GmbH, Kaufingerstraße 24, 80331 München, Germany

www.ingramcontent.com/pod-product-compliance
Ingram Content Group UK Ltd.
Pitfield, Milton Keynes, MK11 3LW, UK
UKHW022312250926
475702UK00046B/707

* 9 7 8 1 0 4 1 0 5 8 3 9 7 *